EVOLUTION AFTER DARWIN

THE UNIVERSITY OF CHICAGO CENTENNIAL

VOLUME II

THE EVOLUTION OF MAN

SOL TAX, EDITOR

EVOLUTION AFTER DARWIN

THE UNIVERSITY OF CHICAGO CENTENNIAL

VOLUME II

THE EVOLUTION OF MAN

MAN, CULTURE AND SOCIETY

THE
UNIVERSITY OF
CHICAGO PRESS

EVOLUTION AFTER DARWIN

THE UNIVERSITY OF CHICAGO CENTENNIAL

VOLUME I

THE EVOLUTION OF LIFE

EDITED BY SOL TAX

VOLUME II

THE EVOLUTION OF MAN

EDITED BY SOL TAX

VOLUME III

ISSUES IN EVOLUTION

EDITED BY SOL TAX AND CHARLES CALLENDER

Library of Congress Catalog Number: 60-10575

The University of Chicago Press, Chicago 37
Cambridge University Press, London, N.W. 1, England
The University of Toronto Press, Toronto 5, Canada

Printed in the U.S.A.

PREFACE

On November 24, 1859, Charles Darwin at last saw in print the manuscript over which he had labored for almost a quarter of a century, the book whose ponderous title has become the familiar *Origin of Species*. The world had been waiting, and in a single day the first edition of 1,500 was sold out. One hundred years later, the day was celebrated as marking one of those events that influence the career of man by changing his perspective of himself and his place in the universe.

In December of 1955 the University of Chicago began planning its celebration of the centenary in the most appropriate manner—bringing to bear, on the subject of evolution, current knowledge from a variety of relevant fields, thus advancing once more our understanding of the world and man.

About fifty scientists were selected during 1956, and their themes were agreed upon; during 1957 and 1958, they developed the papers that are published here. As these were completed, they were exchanged among the authors. Armed with new information and insights, all but five of the authors met at the University on November 22, 1959, to prepare for panel discussions of the issues in evolution which were to be held for the public during the five-day Celebration, beginning on the Centennial of the publication date of *Origin of Species*. The discussions were based on the papers that had been distributed in advance, but the papers themselves were not delivered at the Celebration.

The present volume, *The Evolution of Man,* and its companion volume, *The Evolution of Life,* contain most of the collected Darwin Centennial papers. A small group of papers, concerned with the relationship between science and spiritual values, appears in a third volume, *Issues in Evolution,* which also contains the record of the Celebration itself, including the discussions, and an index to all three volumes. Collectively the work is called *Evolution after Darwin*.

Alfred L. Kroeber's essay which opens the present volume serves as an introduction to the human phase of evolution. The papers that follow, through that of Julian H. Steward, trace the emergence of *Homo sapiens* and the gradual development of civilization. With Professor Magoun's essay on the early views of brain function, the volume turns its attention to the nature of behavior and mental activity, beginning with Professor von Muralt's paper on the evolution of nervous conduction and extending through four essays on comparative learning from individual orientations. The volume then returns to its consideration of man's unique psychological development and his control, through society and culture, of himself and his destiny.

For the selection of scientists to prepare these papers of Volume II, Ilza Veith, S. L. Washburn, and the late Robert Redfield share with the Editor (on behalf of the Darwin Centennial Celebration Committee) major responsibility. The authors themselves are, of course, fully responsible for their respective contributions.

SOL TAX

CHICAGO
February 1960

CONTENTS

A. L. KROEBER

EVOLUTION, HISTORY, AND CULTURE

There is a sort of huge disproportion between Darwin's specific contribution to science—the origination and substantiation of the principle of natural selection—and the overwhelming effect which the establishment of this purely biological principle came to have on total science. There was evidently a particular historic concatenation in the world's thought which enabled Darwin's discovery to trigger off consequences so great. This concatenation and its effects is my subject of inquiry.

In the first place, just because Darwin's contribution was so essentially biological, we tend to forget how slow its complete absorption into total biology was, while the world around was being shaken by the revolution which the biological innovation induced.

It was indeed a revolution, against which orthodox religion promptly mobilized its full strength, the dissenting factions perhaps even more bitterly than the long-established and prosperous. The final offensive effort in the long retreat came in the Scopes trial at Dayton in Tennessee sixty-six years after publication of the *Origin of Species.* There was of course no valid reason why the biological sphere should not always be autonomously ruled by biological decisions. Yet the minority of age of biological science had lasted so long that the guardianship and authority of dogma over it had come to be accepted as almost perpetual and inherent.

Then there was the important fringing corona of rationalism, whose interest had largely shifted from God to man. Insofar as this element of opinion was also anticlerical, it tended to find comfort in the alarm with which organized religion viewed "Darwinism." On the other hand, insofar as the intellectual element posited its values in specifically human qualities like spirituality or the categorical imperative,

A. L. KROEBER is Professor of Anthropology at the University of California and has conducted extensive ethnological exploration in California since 1900. Founder and past president of the American Anthropological Association, Professor Kroeber is also past president of the American Folk Lore Society and the Linguistic Society of America. He was the Huxley Memorial Medalist in 1945 and was awarded the Viking Medal in 1946. Among his innumerable publications are the books: *Configurations of Culture Growth* (1944), *Anthropology* (1948), and *The Nature of Culture* (1952).

the evolutionary bridging between body and soul was quickly sensed as disconcerting. After all, Darwin in person took this inevitable step within twelve years with *The Descent of Man,* as remorseless a demonstration as it was a restrained one.

From outside, it probably looked as if all must be elation in the biological realm so suddenly freed from its immemorial dependent or colonial condition. But actually the sledding was rough, in spite of ardent patriot progagandists like Thomas Huxley and Haeckel. Sexual selection aroused doubts early; the non-functionality of organs still in the emergent stage was a troublesome problem; use and disuse as a contributing causal factor, which Darwin himself as well as Spencer continued to accept, was a rotten and treacherous prop. Above all, there was little systematic knowledge of heredity and variation, the two grand factors on which selection was to operate. It was obvious enough that heredity and variation existed, but only as averages or trends of masses of phenomena, evident as wholes, but unanalyzed and unmeasured.

This was still the situation when Mendel was rediscovered around the turn of the century. Overwhelmingly, biologists had been accepting evolution because there was nothing else for them to do; but they had not proved it to their own satisfaction. The developing science of genetics for a while contributed perhaps more new difficulties than new aids; as Huxley says,[1] it took time to reconcile the new fundamental principle of genetic discontinuity with the continuity of phenotypic evolution. The result is that the achievement of successful Neo-Darwinism barely antedates the battle of Dayton, after which reaction could no longer be mobilized.

As regards the grand over-all effect of Darwin's life and work on the world's thinking, it seems to me that he did more than anyone else to establish a historical approach as valid in science, and that the significance of this effect may be even greater, and perhaps will be more enduring, than what he accomplished for biology as such. This is what I had in mind when in opening I referred to what Darwin "triggered." He may have been mainly unconscious of this larger effect, though I doubt if he was wholly unconscious of it. At any rate, I now come to my central task, which is to consider how and why this triggering occurred. That in turn involves consideration of Darwin's place in general or total science. Under "general science" I would include any organization of knowledge and understanding based solely on naturalistic principles, irrespective of subject matter.

[1] "The Emergence of Darwinism," pp. 1–21 in *Evolution after Darwin. Vol. I, The Evolution of Life* (The University of Chicago Press, 1960).

The Historical Approach to Science

Can human history, in the ordinary sense of the word as history is written, taught, and studied, be included in total science? I think it must be so included. It is recorded and sifted knowledge, always organized with reference to time and space, usually organized also with regard to significance, and as far as possible concerned with cause and effect, though admittedly only subjectively and fumblingly. The significances stressed may be predominantly moral, or again patriotic; they can also be enhanced by the manner of writing, which fact involves an aesthetic element. There are intellectual problems contained in the materials of historiography, ranging from considerations of human nature to those of progress and destiny; and the range of consideration can be very broad or intensively narrow. It is evident that the manners of doing historiography are varied and plastic, and its findings equally so. There is in it little of refined or special technique; and this condition is borne out by the fact that fairly competent history was being written both in China and on the Mediterranean between two and three thousand years ago. In the time since then, most or all other branches of science have been developed, refined, and enriched enormously, even though in jerks and starts and by no means continuously. In other words, the study of human history still operates by essentially primitive methods: it is our archives, libraries, decipherment of strange writings, and control of foreign languages, our footnote and reference techniques that have progressed in bulk and organization since Thucydides, rather than the interpretations made by historians. Historiographic research, almost alone, remains without systematic and "theoretic" results. Some would say that it is knowledge but not science because it remains on a concrete level and does not abstract.

On the other hand, historiography obviously deals successfully with the description of events in the flow of time and operates with an ease and success that it took all the remaining branches of science a very long period of apprenticeship even to approximate. Darwin's is a household name not because housewives and householders are deeply and clearly concerned about natural selection, but because Darwin is the symbol and was in large measure the agent of natural science finally achieving the historic approach of evolution, of being willing and able to operate in biology without reservation or constraint in the dimension of time. In retrospect, it seems very difficult to realize that this achievement could have been mainly delayed until only a century ago. As a historian of sorts, I am impelled to reflect upon and analyze this anomaly of the intellectual history of our civilization, and in fact of the human species and its attribute of culture.

I am driven to begin with a hoary pair of opposites, the static and dynamic. It is a platitude, though a significant one, that static situations are generally easier to analyze and understand than dynamic ones. By definition, they exclude change; they do not involve the dimension of time; there are no events. In physics, statics preceded dynamics; in mathematics, geometry developed before algebra, and in teaching beginners, this order is still largely followed.

However, historical inquiry, though lacking any serious systematization of principles or generalizations, and therefore generally accounted as being on the most rudimentary and retarded level of science, is obviously overwhelmingly dynamic: its subject matter is events, that is, changes!

This anomaly is evidently the result of its subject matter, which is the interactions of the most complicated and variable of units known to occur in nature—human beings—and these viewed preferably not quiescent and at rest (when historiography also slumbers) but in full and multiple interaction. The bigger the interpersonal and mass shocks and the greater their effects, the more significant and interesting does history become.

Now, not only are human beings inherently the most variable of natural units, but they have further developed a faculty of increasing their variance with originality, creatively and exponentially. The product of this exponentially varying faculty is what we have come to call human culture or civilization. It has the added scientific disadvantage of being extremely difficult to isolate from context and environment for manipulation into experiment and test. It is no wonder that having chosen such a tough, subtle, and refractory subject matter, historiography has remained literally primitive in its scientific method: as regards general theory, the more easily productive fields of research had been preempted by physicists, chemists, and biologists.

As a matter of fact, historians every so often leave off their prevalent narrative flow in order to hold a moment or brief period steady while they review the state of the institutions, economies, arts, manners, attitudes, and values, at the moment selected, in the country or area they are dealing with. They call such a treatment "topical." The examination is placed in time, but it is not diachronic internally. What further distinguishes such a topical discursus (or independent essay), is that the endlessly varying events of narrative history are now left out of consideration in favor of the pervasive regularities of human behavior in the period and area in question. In other words, description of institutions in the widest sense of that word, or, more generally, of "culture" or form of civilization, replaces the usual narrative of particular and unforeseeable doings and happenings. Essentially such topical treatments by historians are sociological or anthropological in their nature. In dealing with what is customary, predictable within its limits, and therefore regular, such topical history must certainly be included within the confines of natural science even by those who would deny such inclusion to merely narrative historiography.

In such a "topical" description of a culture, whether by a historian or an anthropologist, time does not enter because the "moment" chosen, even though it cover a century or two, is short enough for changes in it to remain unimportant relative to stable conditions persisting through it. If due consideration is given to the interrelation of the several aspects of the culture and society, one to another, the description will rise from the merely enumerative level to being functionally integrated; in other words, an over-all pattern is now being distinguished.

In short, there exists a kind of history, or spontaneous outgrowth of historiography, which by renouncing narration of the endless particularity of the primary data of human history, and confining its span to a virtual moment, attains to some of the generalization of natural science. This gain is achieved by foregoing a dynamic approach for a static one. The topic or moment of culture dealt with is viewed in equilibrium.

Pre-Darwinian Social Thought on Progress

This examination of historiography has been made because the concept of evolution which Darwin set off with such astonishing success is, in the larger sense, a historical concept: a process operating with change through time, and mainly irreversible and non-repetitive.

For the actual history of science, it must be significant that, in the half to three-quarter century preceding 1859, the notion of evolution was more widely held as applicable to social man than to animals and plants. This was certainly expressly true from Condorcet on. This may be so in part because the time span of man is so much briefer than that of most of nature. And yet man's evolution was early conceived as going on through the whole of his existence. In any event, the actual priority of a widespread conviction of human evolution, before much belief in the evolution of life was held, is undubitable.

This socioculturally oriented evolution of man's condition and achievement was viewed as progressive, and included the comparative value judgment of change for the better, of improvement. In fact it was probably based primarily on this value judgment, whereas Darwin basically argued only that species "originated," that is, life forms could change instead of being immutably fixed.

The time of Condorcet and his fellows and followers was too early for them to be properly describable by the modern term "social scientists;" but they certainly were not natural scientists. The idea of progress may have originated with the discoveries and extensions of knowledge in the age of explorations that began in the fifteenth century. It was certainly connected with the Quarrel of Ancients and Moderns that arose in the seventeenth century as a result of the political and cultural florescence of France under Louis XIV, and from there spread through Europe, with the balance gradually inclining more and more in favor of the moderns, until in the eighteenth century the verdict crystallized as "Enlightenment," newly achieved. Condorcet was followed by Comte with his three stages, and in England by Spencer, who is reckoned as a pre-Darwinian evolutionist, though his naturalistic knowledge was second-hand and his system largely speculative even though sober.

It is presumably in this stream of social and philosophic thought that the idea of the permanent effect of use and disuse originated and was then injected into biological thinking as "inheritance of acquired characters"—a doctrine that Spencer invoked, Darwin refused to reject, and that died hard—Freud being the last great name to cling to it. In recent biology it is often called Lamarckism, though Lamarck himself used it to explain secondary evolutionary modifications or deviations from the immanent pattern or eternal configuration of the organic realm.[2] Transmission of acquired characters of course actually does take place on the sociocultural level, and on a large scale. The phrase is a somewhat crude but largely sound description of the normal process of culture change and history. Culture does alter by use and disuse, much of it does accumulate, all culture is always acquired by learning. The mistake lay in the transfer and application of the idea to the genetically dominated realm of the organic.

Preconditions of Evolutionary Biology

As we look back to 1859, the wonder today is that Darwin's triggering could have been historically postponed as long as it was. There is not only his own hesitation of a score of years. It would seem that the geological evidence alone, still more as it was re-enforced by the paleontological, would have sufficed to force a break-through decades earlier. Religious dogma was an influence, but no longer the decisive

[2] G. G. Simpson, "The History of Life," pp. 117–80 in *Evolution after Darwin. Vol. I, The Evolution of Life* (The University of Chicago Press, 1960).

one in nineteenth-century European science. It may be that it just comes harder for human beings, including scientists, to face a changing cosmos than a fixed one. Hellenic religion prescribed next to no dogma with which an evolutionary or diachronic way of thinking would have conflicted; yet Greek science also cheerfully confined itself to a fixed world.

As a matter of fact a review of the total history of science reveals very little real concern with a historical approach until after 1750, except of course for human historiography. Such a concern is therefore a late and rare phenomenon in the history of science.

However, to substantiate this finding it is necessary to distinguish two aspects within the dynamic approach: a microdynamic, which seeks for regularity within process, and a macrodynamic, which alone is wide-open and genuinely historic in interest.

The phenomena of night and day, of the moon, of the year are characteristically short-range, dynamic, and repetitive. To transpose slightly, it might be said *Plus ça va vite, plus ça dure longtemps:* things spin and reel, but their order and result stay the same. The situation is dynamic indeed, but it is also dynamically fixed. Even the Copernican overturn, Kepler's laws, Galileo and Newton did not alter this basic outlook. The first break was speculative, when LaPlace and then Kant tried to imagine how the solar system had come about and devised a nebular hypothesis in explanation of its history. An empirically historical science of astronomy had to await more powerful telescopes, photography, and the spectroscope—in short, a vast increase of knowledge due to technological progress.

The earth sciences of geology and paleontology confront obvious phenomena that seem to us, who have acquired historical orientation, almost to demand historical interpretation—the superposition of strata, for instance, and comparison of similar fossil-bearing strata in different parts of the earth. Yet as long as information on these matters was scant, spotty, irregular, there was little sure sequence of conditions to be got out of the data. So guesses as to the primacy of this or that favored set of processes—neptunic, volcanic, diluvial, catastrophic—simple "origin" hypotheses—were advanced and argued. It was only as systematized and coherent knowledge—contextual knowledge—grew, that arbitrary lunges at seizing a basic process barehanded were crowded out by the growing mass of information which almost enforced a system of interpretation by its own mass. The geological frame of factual knowledge and process appears to have been substantially ready for macrodynamic and evolutionary understanding some decades before Darwin—and an understanding not too different from the present-day one. I can only guess tentatively what prevented the consummation. Perhaps no amount of piling on of technical evidence alone could suffice. Some great public event was needed that both touched dogma and released dramatic human affects. This quality perhaps was supplied by Darwin's dealing with organisms which by implication included mankind and so stimulated human interest more than any merely geologic process could have. If this suggestion is inadequate, those whose knowledge of the history of geology is fuller can amend it.

Within biological science, Linnaeus was a necessary precondition to Darwin; especially so, Linnaeus' system as modified into the "natural" system of classification of life forms, based on investigation into total structure. In fundamental theory, Cuvier was the antithesis of Darwin; but his "types," converted into genetic phyla, still hold largely today, underneath the accumulated mass of re-enforcing knowledge.

In celestial, terrestrial, and organic science alike, the first beginnings toward use of what became a historic approach occurred about the mid-eighteenth century, and achieved acknowledgment about mid-nineteenth, centering around Darwin, who made the first conscious and directed break-through.[3] It had by then become clear that all natural phenomena had a potential history if they could be placed in designable space and time.

An Exception: Philology [4]

A lone exception to the rule of pre-Darwinian absence of macrodynamic approach appears in philology. In the late 1780's Sir William Jones recognized the common origin and diverging descent of the Indo-European languages—an insight which grouped species of idioms into genera and genera into a family, resulting in a genuine phylogeny, perhaps the first in any field of knowledge. Strictly speaking, Jones' finding was preceded by some recognitions of similarities within Indo-European subdivisions and within some non-Indo-European groups of languages. But these tended to be construed rather as variations or "corruptions" of one extant language, whereas the large gap in time and space between Sanskrit and the earliest recorded

[3] The latter half of the eighteenth century also saw the foundation of scientific chemistry, which however developed into a fundamental and ahistoric science in increasingly close association with physics. The causes for the lateness of chemistry can therefore not well be the same as for the historic sciences, but they may be allied. I hazard that the nature of the phenomena of chemistry is such, as compared to those amenable to the approach by physics, as to require a larger organized corpus of knowledge before sound interpretation can be effected.

[4] The four paragraphs in this section were not in the paper as originally prepared for the Darwin Centennial, but were added subsequent to the five panel sessions.

languages of Europe shifted construal of the Indo-European situation into one of descent of daughter languages from a reconstructed common mother tongue.

From the point of view of general science, several significances characterize this recognition of Indo-European descent. First, the recognition was derived empirically, forced by the evidence, not speculative. Second, it was not concerned with social thought about progress. Third, it originated in the humanities, and dealt with language which is a segment of culture, although a clearly defined concept of culture was not elaborated until about two generations later. Fourth, the principle of inferring common descent from parallel similarity of form and structure proved applicable to other groups of languages, because specific patterns of linguistic change were found to possess significant regularity even though their specific causation remained unknown.

It would seem now that this priority of the genetic scheme of Indo-European descent was in some degree the consequence of the particular nature of language. Comparison has not yet provided equally consistent demonstration of lines of descent or regularity of change for the remainder of culture. Nor has any discovery been made which explains causally the mechanism of linguistic change in a manner analogous to the light which the science of genetics has thrown on hereditary conservation and change in biological evolution. The early discovery of linguistic phylogeny thus remains an unexplained anomaly or sport in the history of science. But though puzzling, it is indubitable and remarkable, and its precise raison d'être will no doubt be discerned before long.

I can only suggest a favoring condition: the existence of a well established and thorough organization of relevant knowledge, in the shape of accurate grammars and lexica of Greek, Latin, and the principal mediaeval and modern vernaculars of Europe. This accurately classified corpus of data would necessarily add much weight to the significance of similarities discernible on comparison with an older and remote Asian language such as Sanskrit. Jones had available a well-worked-out philological taxonomy, as Darwin had available the post-Linnaean cumulative taxonomy of plants and animals.

Darwin's Influence on Anthropology

When the revolution so long planned and still deferred by Darwin finally came, most of the sciences immediately underlying biology were affected powerfully and rapidly. But the overlying sciences dealing with man were much less stimulated or influenced by him. Historiography and economics were long almost unaffected, as was concrete sociology. Theoretical sociology must be excepted—but then it had had its own pre-Darwinian progress theory. Psychology was still so unextricated from philosophy as to be too immature for much influencing. Philosophy itself was surprisingly little affected. It did increasingly drift away from its traditional problems into becoming the commentator and appraiser of natural science; but to date it has recognized very little in science beyond what can be conformed straitly to the model of physics.

The one science of man whose course reacted sharply to the impact of Darwin was anthropology, which even in its formative days had always contained a definite "natural history" interest—as it still insists on retaining an avowedly biological beach-head—and which was perhaps the least encumbered by normative and ameliorative aims.

In the 1840's it looked as if the several strands of anthropology might co-ordinate themselves as "ethnology." Societies were formed under that name in Paris, London, and New York. In Germany, since 1843, Klemm was assiduously describing assembled ethnographic data under the title of *Cultur-geschichte* and *Culturwissenschaft,* and he was certainly dealing with cultural phenomena though he seems to have been unable to formulate a clear-cut concept of culture. In England, this was the period of Prichard and Latham and their natural history approach. Data were growing, though theory was still weak. Morgan's first book, the descriptive *League of the Iroquois* (1851) belongs in this stream, and is distinctive chiefly in being based mainly on direct inquiries among the people whose culture is depicted.

Such was the sort of hesitant, unspectacular drift which characterized what was to become anthropology, for some fifteen years before the *Origin of Species* was published. The effect of this event was tremendous: a crop of founding fathers of a science of anthropology sprang up within two years, and within the dozen years to *The Descent of Man* many of the basic books of "Classical Evolutionary Anthropology"—as it later came to be called—had appeared. Here is the list:

1861: Bachofen, *The Matriarchate* (*Mutterrecht*)
1861: Maine, *Ancient Law*
1864: Fustel de Coulanges, *The Ancient City*
1865: McLennan, *Primitive Marriage*
1865: Tylor, *Researches into the Early History of Mankind*
1870: Lubbock, *Origin of Civilization*
1871: Morgan, *Systems of Consanguinity*
1871: Tylor, *Primitive Culture*

The principal additions that might be made to this early cluster would be Morgan's *Ancient Society* in 1877 and Tylor's *Anthropology* of 1881.

What had these books, half of them by jurists and only one (Lubbock's) by a banker-naturalist,[5] to do with Darwin's natural selection as a mechanism of biological change? Nothing whatever, directly. But their authors were evidently stimulated by the idea of evolution to which Darwin had given a foundation. If it was now revealed that there had been great changes in the forms and structures of animals and plants, there were likely to have been great alterations also in man's status and condition, in his institutions, customs, and mode of life. What Darwin had done for life in general—giving it a history—they were proposing to do for human culture.

Limitations of Early Anthropology

There were two reasons why these first anthropologists could not achieve a solid success similar to that of Darwin—why, in fact, the course of the evolution of culture still remains much less certainly ascertained than that of life. The first of these reasons is that Darwin inherited a highly accurate, solid, and comprehensive classification of animals and plants which had developed by systematic cooperation among biologists since Linnaeus, more than a century before. As against this, the would-be anthropologists had a helter-skelter miscellany of travelers' tales and missionaries' accounts, from which obtruded some picturesque features: the couvade, matriliny, divine kings sacrificed, houses of snow—somewhat like elephants' trunks and armadillos' armor in pre-Linnaean natural history. An added difficulty became clear only gradually: the family tree which outlines the history of life is throughout a one way affair: once two life forms have diverged a very little, they cannot ever reassimilate or merge again. But culture, without genes or genotypes, and floating through and out from phenotypes, is protean in its sources. It can differentiate and reintegrate; it can assimilate or be assimilated by other culture or exist alongside; a new item can develop from within a society or from others without.

What could these early anthropologists do to develop an evolutionary history of man's culture overnight? They more or less knew the end of the story—the human successes of their day. They could ima-

[5] It is of interest that Lubbock, although professionally a banker, was also a competent naturalist. He drew inferences from original observations and experiments on social insects, which have stood up better than his speculative interpretations of cultural origins.

gine a beginning, which, if there had been any notable evolution, would have had to be quite different, and which they called "the origin." The word was fashionable: Darwin himself wrote the *Origin of Species* instead of the *Evolution of Organisms,* as Huxley points out. Naturally, the beginning of the evolution of culture and custom was posited as having been as different as possible from achieved cultures: it was the contrastive opposite, or the condition of cultureless animals. Then, all that lay between was filled in by plausible speculation, more or less propped by such available evidences as were favorable. It was much the same process of thought as that by which the Greeks at the threshold of their philosophy-science explained the origin of the world: the "first" was water—no, fire, for everything changes—no, it was air—better yet, mind—or wait, nothing had ever grown, for logic breaks down before motion. Like these early Greeks, the early anthropologists did not actually know anything of the course of actual events nor of the processes which had shaped their course. There was really not much they could do but conjecture plausibly. It is in general a sign of ignorance when we profess to seek or know the origin of any group of phenomena without knowing something of its history: the true origin, being farthest from us, is likely to be the last discovered. Probably we tend most to manufacture origins when we have least actual knowledge of post-origin developments. Ignorance of history breeds myth.

Later Phases of Anthropology

However, something did emerge from the tremendous interest that was stimulated in a group of intelligent men such as the evolutionistic anthropologists were. Their answers were mostly of little worth, because they asked the wrong questions; but they founded a science. They did this by occupying a phenomenal area which rested on the organic but extended beyond it, and which therefore had a degree of autonomy, including processes of its own. A central field became defined in these first dozen years: it was that of institutions and custom. Tylor first called this subject matter by a general name, "civilization," in the subtitle of his 1865 book, and Lubbock followed him. By 1871, Tylor shifted to the less-implication-laden "culture" which he took over from Klemm, defined clearly, and used to delimit his field. He also got away from the dominating concern with social-political-legal institutions which preoccupied most his contemporaries, and rounded his culture off by giving due attention in their own right to religion, technology, and even language. Tylor accepted autogenic development, but also recognized external cultural influence; and he was the first to try to ascertain by empirical counts which features of culture intercorrelated or "adhered" to one another. With his stocktaking booklet of 1881, he even crystallized the name "anthropology" for the field of natural science he had staked out.

True, Tylor was also influenced by the century-old stream of social thought concerned with improvement and progress. But this influence was superadded to his natural science thinking—somewhat as Darwin accepted use and disuse effects—without unduly warping his empiricism and curiosity.

A second phase of "classic evolutionistic anthropology" began around 1890 and centered around Frazer, who has influenced the emotional thinking of more non-anthropological readers—including Freud—than any other anthropologist, but who is virtually outside the current of the present-day science. Frazer was a classical scholar by training, and his active following was mainly classical, literary, and British: Andrew Lang, Gilbert Murray, Jane Harrison, Robert Graves; Hartland, Crowley, and Briffault also belong. This group was perhaps more concerned than the first with explanations by endogenous or spontaneous psychology than with generic human evolution.

It is clear that these so-called evolutionistic anthropological movements stemmed from the commotion that Darwin produced in opening up possibilities of more than static interpretation, but that they were scarcely influenced by Darwin's specific work or thought, nor, if they had tried to interpret the history of human culture, would they have had the organized data to do it with. Later anthropologists have had to begin all over again. We have by now learned several things: not to compare ethnocentrically; not to use data torn out of context, nor loose fragments of structure; that good description must be in terms of the described society's and culture's own nature, structure, and function, not in terms of any merely logical scheme—just as in biology; that structure cannot be analyzed too searchingly; that it is wise to lean heavily on the findings of archaeology because incomplete as these are they are always solider than reconstructive conjectures.

The Need for Classification

All in all, during the past half-century, anthropologists have been most interested in pursuing microdynamic work, in analyzing out the structure, and the changes of structure and function, of relatively small societies and cultures, within spans mostly of less than a lifetime, the communities involved being mostly subnational of the order of tribes, villages, or localized minorities. This restriction of scope satisfies the need to work holistically which most anthropologists feel, and yet it makes possible the rather reliable determination of the sociocultural mechanisms involved. But it also leads to spotty and diverse results which are slow in adding up to larger findings, whether of mechanism or of the general course of growth.

Macrodynamic inquiry into culture is being consciously or directly pursued by few anthropologists and by few historians. It is being kept in vision by some; much more than that is probably not feasible at present. The organized corpus of comparable knowledge is as yet too scant and discontinuous: it is much as if Julian Huxley and George Gaylord Simpson had tried to do their present work two hundred years ago. True, there is a rather large total aggregate of cultural description extant; but very few works share the same specific aim. One stresses psychology, another the unique particularities of its culture; one attempts to test a hypothesis of greater or less range, another follows a conventional pattern of topics; some accounts are oversensitive to cultural style, others insensitive, and so on. The situation is made more difficult by the fact that anthropologists still tend to value personal expertise, technical virtuosity, cleverness in novelty, and do not yet clearly recognize the fundamental value of the humble but indispensable task of classifying—that is, structuring—our body of knowledge, as biologists did begin to recognize it two hundred years ago.

I am not speaking in distress. I believe the need will be felt—is being felt; and once it is recognized, classificatory organization may begin suddenly to be supplied on a scale now undreamed of. Historiographers may participate—in fact may well take the leading part. The profession of historiography has only to gain by the recent movement, now well under way, to include historians of so-called exotic areas, of science, of the arts, which were formerly left wholly to Sinologists, scientists, critics, art theorists, and such. This in addition to economic and social history, and "history of ideas," which are of longer standing as admitted specialities within history. The time may even be reasonably near when general or universal history will be attempted by scholars for scholars, as in the days of Voltaire and Herder, but with far greater resources, and perhaps co-operatively rather than through the idiosyncratic genius of some Toynbee or Barnes working single-handed.[6] Purely political history, and national history, will always continue, because they will always be wanted. But there is no reason why historians should not also try gradually to take over the whole history of man, which when thus generalized inevitably becomes the history of human culture.

[6] It is a curious aberration of the moment that most our present-day general histories are aimed specifically at eighteen-year-olds who have just entered the university and who in America at least are as good as untrained in seeing significance in any historical phenomena.

Huxley has clearly recognized both the break in mechanism between organic and cultural evolution (he sometimes calls the latter "noetic" or "psychosocial," with neither of which terms do I quarrel), and at the same time the fact that cultural development has largely taken over the determination of what will happen on this planet both of life and to culture.[7] When he first presented these views to a group of anthropologists at the Wenner-Gren Foundation about 1951, he aroused mainly unrest and opposition: I do not know precisely why, unless the audience mistakenly feared that ethnocentrism and challenge to the autonomy of culture were rearing their heads. Similarly, when in my 1948 *Anthropology* I revived the old problem of progress in culture, and cited four ways at least in which progress seems substantiable by objective evidence, the reception in the profession was cool if not negative: Redfield alone, so far as I know, has spoken up in general agreement. The prevalent attitude was evident again in the International Congress at Philadelphia in 1956 when Russian visitors raised the issue. I confess I am puzzled by the negativism; I hardly like to attribute it to a fear of words—to a dread that because Spencer's and Morgan's nineteenth-century "progress" was ethnocentrically slanted, any acceptance of progress is ethnocentric. To be sure, a property of accumulativeness widely and loosely attributed to culture is connected with the idea of progress; but we have lately begun to distinguish differences in the degree and kind of accumulation in different cultural fields, and there is every reason to think that this problem too can be handled evidentially without loose thinking.

SUMMARY

I have tried to show some of what Darwin in 1859 released in effects outside of biology. Through an unusual combination of circumstances, of which religious dogma was one but only one, biology had remained confined in a static framework for most of the century before, while geology had very nearly broken through to recognition of a long-range history of the earth's crust, in all but formal declaration; and social thought, though speculative and unorganized as to data, had accepted progressive evolution as a principle, though still within a short-range time span. Astronomy in widening its compass beyond the cyclic recurrences within the solar system had embarked both upon a historic approach and a long-term one. Recognition of diachronic

[7] "Man's Place and Role in Nature," pp. 79–97 of *The Unity of Knowledge,* ed. Lewis Leary, 1955; also, "Evolution, Cultural and Biological," pp. 3–25, in *Yearbook of Anthropology—1955,* ed. Wm. L. Thomas, Jr. The respects in which biological and cultural evolution are importantly alike and importantly unlike are skillfully summarized by Huxley in two paragraphs at the turn of pp. 24 to 25 in the second article.

change in biology was therefore overdue; so that when Darwin combined the genius of his insight with masterful use of the long structural-taxonomic accumulation, the principle of natural selection broke through to an explosively revolutionary recognition in biology; and this revolution was immediately extended to the underlying earth sciences and the overlying sciences of man and culture, especially anthropology, which, rooting in the natural sciences, was directly affected as historiography and sociology were not.

While anthropology acquired direction and cohesion as a field of science in the dozen years following 1859, its first findings were largely misdirected and sterile because it possessed no corpus of classified knowledge comparable to that of biology, and had to fall back mainly on speculation. In this it went further astray by short-circuiting a genuinely historical, gradualistic approach into an illusorily diachronic quest for ultimate origin. Tylor, in extricating the concept of culture as the distinctive subject-matter of anthropology, and by a certain sobriety and balance of his speculations, which left room for historical considerations, however salvaged the science, which toward 1900 began to steer into the courses it has since followed. However, anthropologists as a group have been unnecessarily slow in recognizing the necessity of organizing the body of their knowledge into a systematically comparable form, as observational geology and biology have done so thoroughly and fruitfully.

Both the microdynamic and the macrodynamic approach are justified and profitable in the field of culture, and in the end ought to supplement each other. The macrodynamic emphasizes the time factor more and deals with longer time-spans. It thus connects naturally and easily with biological evolution on the one hand and with human historiography on the other. Since biologists have come to recognize that with the evolution of specific human faculties cultural factors have been increasingly superimposed on selective genetic ones, and bid fair to be the prime determinants of terrestrial evolution from now on, it would seem that anthropologists should reciprocate in co-operating with efforts to see evolution as one continuing course, though with a cardinal change in the course. When it becomes recognized that culture is in one sense the endlessly complex and unordered behavior of men seen ordered through a lens of long-enough focus, historiography and culture study can join hands instead of viewing each other across a gulf. We shall then be in position to realize such continuities as exist between the evolution of life forms, the manifestations of culture, the past behavior of men, and perhaps their present behavior in mass.

L. S. B. LEAKEY

THE ORIGIN OF THE GENUS *HOMO*

Darwin believed that the most likely continent in which to search for evidence of human evolution was Africa. He did not base this view upon fossil evidence, but deduced it rather from the distribution and nature of the primates still living in Africa: gorillas, chimpanzees, and numerous genera of catarrhine monkeys. His actual words were:

> We are naturally led to enquire where was the birthplace of man at that stage of descent when our progenitors diverged from the catarrhine stock? The fact that they belonged to this stock clearly shows that they inhabited the Old World, but not Australia nor any oceanic Island, as we may infer from the laws of geographical distribution. In each great region of the world the living mammals are closely related to the evolved species of the same region. It is, therefore, probable that Africa was formerly inhabited by extinct apes closely allied to the gorilla and chimpanzee; and, as these two species are now man's nearest allies, it is somewhat more probable that our early progenitors lived on the African continent than elsewhere. But it is useless to speculate on the subject (Darwin, *The Descent of Man,* 1871).

During the past century there have been many occasions when scientists have looked away from Africa to Asia as the potential "cradle of the human race," as some of them have phrased it. This changed opinion was the result of important discoveries of fossil Pongidae in the Siwalik Hills region of India and of human fossils of the very greatest importance in the Far East—the Java and Pekin fossil skulls from the Trinil and Djetis beds and from Choukoutien.

For many years these fossils discoveries seemed to outweigh the evidence from living genera in Africa, despite the fact that there had been some discoveries, such as those of *Propliopithecus* in Egypt and later of *Australopithecus africanus*—first found by Dart (1925) in South Africa—which might have modified this opinion if they had

LOUIS S. B. LEAKEY is Curator of the Coryndon Memorial Museum in Nairobi, Kenya, and President of the Pan-African Congress on Prehistory. Born in Kenya and Cambridge-educated, Dr. Leakey has earned many honors and wide acclaim for his contributions to anthropology, specializing in African tribal cultures and the origin of man. He is a Fellow of the British Academy, and he is also a Fellow of the Royal Anthropological Institute, and of the Geological Society of London.

been given their full value in the arguments that were being advanced by those who discussed man's origin.

But for many years, Dart's *Australopithecus* find was shrugged aside as representing a juvenile pongid and not a hominid, while the true significance of *Propliopithecus* was also ignored. This was a pity, for these discoveries, when added to Darwin's cogent reasoning from the modern distribution of the primates, should have been *more* than sufficient to keep Africa very much in the picture for students of human evolution.

In 1926 the late Dr. Gordon, a retired medical practitioner living at Koru in Kenya Colony, found some fossil deposits on his farm, deposits which later proved to be of Lower Miocene age. Among the material which he collected, and which was sent to London for study, was a fragment of maxilla of a fossil pongid. This discovery was of such significance that when Dr. A. T. Hopwood, of the British Museum (Natural History), came to Kenya to join me on my first expedition to Olduvai Gorge in 1931, he and I went together to Koru to study the site. After a few days I left him there for three or four weeks with some of my native staff to help him. During this time further primate fossils were found which led to Hopwood's (1933) first descriptions of *Proconsul africanus* and *Limnopithecus legetet.*

The African primate story was unfolding: *Propliopithecus,* an early generalized pongid, in the Oligocene of Egypt; from the Lower Miocene of East Africa—*Limnopithecus,* an ancestral member of the Hylobatinae, or gibbons, now confined to eastern Asia, and *Proconsul,* a generalized pongid with some hominid and some chimpanzee characters; and finally, *Australopithecus* from the Pleistocene in South Africa, a creature which Dart (1925) rightly claimed to be a true hominid.

All these finds, plus the presence in Africa in more recent times of gorillas, chimpanzees, monkeys, and many races of man, added up to convincing indications of Africa as a place of major importance for human and primate evolution. Here, indeed, was a sequence that could no longer be ignored, and the pendulum began to swing back to Darwin's original view, with Asia being regarded more and more as only a secondary centre.

After this, events began to move much faster. In Kenya, from 1932 on, my colleagues and I discovered very large numbers of primate fossils representing the genera *Proconsul* and *Limnopithecus.* These were found in association with fossil faunal and floral assemblages that showed these creatures had once lived in open grasslands, between forest galleries, along the rivers which flowed into an ancient Miocene lake.

In 1936 Robert Broom began his long series of discoveries at Sterkfontein of numerous new fossils of *Australopithecus,* first called *"Plesianthropus"* by him in the belief that the fossils represented a distinct genus (Broom and Schepers, 1946; Broom, Robinson, and Schepers, 1950); later, with the help of Robinson, he discovered fossils representing the genus *Paranthropus* at Swartkrans (Broom and Robinson, 1952).

Dart, too, found additional *Australopithecus* material at Makapan, and it became clear in due course that both *Australopithecus* and *Paranthropus* walked wholly upright and were truly hominid (not merely hominoid) in many of their skeletal and dental characters, even though their brain capacity (in terms of actual volume) was small.

In the meantime, the study in Africa of prehistoric Stone Age cultures had been developing very rapidly, and it had become apparent that tool-making man had been on the African continent, developing his material cultures there, throughout the Pleistocene period. There was, indeed, evidence that his cultural remains in Africa dated back to an earlier period than in either Europe or Asia.

The African Fossil Primates

Let us now briefly consider some of this important African fossil primate material in more detail and review its significance for the study of human evolution in the light of present-day knowledge.

THE EARLIEST PONGID

Propliopithecus, from the Oligocene deposits of Egypt, is still represented by only fragmentary material, but what there is seems to indicate that this was a small, generalized primate whose teeth clearly show pongid affinities; thus it belongs to the Hominoidea. This is not a monkey but rather a member of the generalized stock from which both man and apes could have evolved.

It is a very great pity that we have so little data to guide us, but it seems possible—one cannot put it more positively than that—that a group of pongids at this *Propliopithecus* level of evolution in the Oligocene may have been the direct ancestors of both *Proconsul* and *Limnopithecus.* It is possible that the *Propliopithecus* stock could have been ancestral also to another small pongid of the Kenya Miocene, in which a simian shelf had already developed and in which the general morphology, as well as the height of the lower canine relative to the lower molars and the sectorial nature of the third premolar, have resemblances to the conditions seen in recent Pongidae.

There is a most urgent need for an intensive programme of research in the Oligocene deposits of the Fayum, with the express objective of finding more material of *Propliopithecus* and other small primates. When such finds are made, we may confidently expect to have material which will prove of major importance in the study of the early phases of ape and human evolution.

MIOCENE HOMINOIDS

Turning to *Proconsul* from Kenya Colony, the present position can be summarized as follows: The total number of Miocene fossil specimens representing Hominoidea which I and my colleagues have found now exceeds five hundred, of which the majority represent the genus *Proconsul* and a lesser number *Limnopithecus.*

Proconsul as a genus was divided by Le Gros Clark and Leakey (1951) into three species: one of small size, represented by *Proconsul africanus;* one of medium size, represented by *Proconsul nyanzae;* and one nearly equal in size to a gorilla and called *Proconsul major.*

Mandibles of all these species, in which the symphyseal region is well preserved, are known and *in no case* is there the slightest suggestion of a simian shelf. The simian shelf is, of course, a characteristic anatomical feature of the large living apes of today. It also occurs in fossil apes of the *Dryopithecus* group. Its function would seem to be to strengthen the posterior region of the mandibular symphysis, and it may be linked with the development of large, everted lower canines used for tearing bark from trees and shrubs, during which process there is considerable strain upon this region of the lower jaw.

The form of the mandibular condyle of the jaws is known in both *Proconsul africans* and *P. nyanzae* but has not yet been recorded for *P. major.* It is of special interest since in no case does it exhibit the specialized features which characterize this region in the living great apes but instead shows a most remarkable similarity to the morphology seen in *Australopithecus, Pithecanthropus,* and *Homo.*

The facial regions of both the small and the medium-sized species of *Proconsul* are also known, and they are much less prognathous than modern great apes. In the medium-sized species, *P. nyanzae,* the face is even more vertical than in *P. africanus;* also, the ascending ramus of the mandible is correspondingly higher and more vertical and hominid in character.

So far, only one skull of *Proconsul* is known—that of the small species *P. africanus*—but it is of major interest in the study of hominid evolution for several reasons. In the first place, there is no trace whatsoever of any kind of supraorbital torus. Instead, the structure of the frontal bone—over the orbits and the forehead—is smooth and remarkably like that seen in *Homo sapiens*. This is the case although the *only* skull which we have is that of a fully adult individual. Second, the skull reveals some remarkable retentions of cercopithecoid characters, especially in the nasal region and in the temporal bones but also in some other parts of the skull.

The canine teeth of all species of *Proconsul* are large and of general pongid form, but they are not so everted outward and forward as in the modern pongids. The lower incisors in all three species of *Proconsul* are relatively small, while the upper central incisors of the medium-sized species, *P. nyanzae,* are so similar to the type seen in *Homo* that they may be mistaken for human when isolated from their context; they are quite unlike the corresponding teeth of modern pongids. On the other hand, *Proconsul* molar teeth tend to have a strongly developed cingulum, a feature that some anatomists regard as denoting a considerable degree of specialization. However, such a view does not seem to be well founded; it seems rather that the development of a cingulum is an early character seen in the teeth of many primates and one which has been progressively discarded in various branches during evolutionary development.

A certain amount of skeletal material (cf. Napier and Davies, 1959) representing the *Proconsul* genus is now known, and it is clear that these creatures, living in an open grassland, or savannah, type of ecology, were quadrupedal in gait and possessed arm and leg bones more comparable in proportions to those of cercopithecoid monkeys than those of modern great apes. No pelvic bones have so far been secured, so that we do not yet know the structure of that part of the skeleton.

Limnopithecus, another Miocene genus from Kenya, is now known from a good many specimens (cf. Le Gros Clark and Thomas, 1951). It has been divided by Le Gros Clark and Leakey (1951) into two species; one is Hopwood's original *L. legetet,* and the second and larger is *L. macinnesi.* Both species are clearly related to the Hylobatinae, and their general ancestral relationship to the modern gibbons is, I think, never seriously questioned. However, they were quadrupedal animals and had not yet developed the excessively long arms of *Hylobates* and *Symphalangus. Limnopithecus* mandibles exhibit very much the condition seen in *Hylobates* and *Symphalangus* at the present day. Sometimes there is a small simian shelf, but there may be a complete absence of this character. Individual variation is considerable. When a residual simian shelf is present, it is not of the modern pongid type but more like that of some monkeys.

Another genus which was recognized by Le Gros Clark and Leakey (1951) in the Lower Miocene deposits of Kenya was *Sivapithecus,* represented by a species which they called *africanus*. Unfortunately, so few specimens are known that little can be said about this form. If the provisional identification proves to be justified, it will mean that the Indian species of this genus originated in Africa in the Lower Miocene and moved to Asia at a later date.

Perhaps one of the most significant recent discoveries in the fossil beds of Lower Miocene age in Kenya is that of two mandibles of a small pongid, each with a well-developed simian shelf of the type seen in modern apes. Therefore, we now have evidence that this specialization of the modern pongids, which also occurs in some of the fossil pongids of Europe in Upper Miocene and Pliocene times, was already fully developed in at least one branch of the Lower Miocene pongids. Its absence, therefore, in all samples of all three species of *Proconsul,* when taken in conjunction with the presence of other more hominid characters, such as the shape of the mandibular condyles and of the forehead in *Proconsul,* greatly increases the value of this Kenya genus in the interpretation of hominid evolutionary trends. It strongly suggests that *Proconsul* was already developing in a truly hominid, and not merely hominoid, direction and may thus be ancestral to *Homo*.

It is most unfortunate that no fossiliferous deposits of genuine Pliocene age and containing fossil primates have so far been recorded in Africa. It is true that the South African deposits which have yielded fossils representing *Australopithecus* at Taungs, Sterkfontein, and Makapan were *once* claimed to date from the Pliocene on the basis of finds such as Lycyaena and other survivals, but the total faunal assemblage from these sites does not sustain such a remote age, although Pliocene survival forms most certainly do occur. However, the true age of the oldest australopithecine horizon so far found in South Africa is almost certainly Upper Villafranchian, or the upper part of the Lower Pleistocene (cf. Oakley, 1954).

LOWER PLEISTOCENE HOMINIDS

When we critically examine the *Australopithecus* fossils from the Lower Pleistocene of South Africa, we are at once struck by two very different and apparently contradictory features. On the one hand, the teeth very markedly resemble those of man, e.g., the lower milk molars, and differ greatly from those of the great apes (Robinson, 1952; Le Gros Clark, 1952); on the other hand, various features in the facial structure recall the specialized modern great apes, such as gorilla or chimpanzee. It is this combination of characters which has earned the australopithecines of South Africa the name "ape-men," but the term "near-men" is preferable, for the apelike characters are more apparent than real.

Such limbs and pelvic bones of the Australopithecinae as have been discovered so far indicate, moreover, that these hominids walked erect (Le Gros Clark, 1955).

A great deal of argument has taken place on the question of whether or not these South African members of the subfamily Australopithecinae should be regarded as representing a morphological stage through which the primates passed on the way to becoming *Homo,* and even as to whether they can seriously be regarded as ancestral to such extinct cousins of *Homo* as the genus *Pithecanthropus.*

There are those who maintain that even if the geological age of *Australopithecus* and *Paranthropus* specimens were truly Pliocene, as was originally suggested, *neither* of these two genera *could* have given rise to modern man, because each exhibits certain physical specializations of structure (particularly in the general facial architecture and the method of facial-shortening) which could not have given rise to the characteristic human form. However, others maintain that the major reason for excluding the genera *Australopithecus* and *Paranthropus* from the direct line leading to man is simply that the time element seems to be far too short for such a degree of evolutionary change to have taken place. We shall return to this aspect of the problem presently.

Almost all anatomists and paleoanthropologists are, however, in general agreement that the roots of the human stock must be sought in the general primate group which we call by the subfamily name of Australopithecinae. Therefore, it has long been expected that, somewhere in African deposits of Pliocene age, we may one day discover fossil members of the Australopithecinae which will reveal the common ancestors of the known, but specialized, *Australopithecus* and *Paranthropus* genera and of some other hominid in the direct line leading to the genus *Homo.* Most anatomists would also add that there may well be a fourth separate and distinct branch leading to the specialized *Pithecanthropus* genus of the Far East.

As a corollary to this view we must postulate that somewhere, in deposits contemporary with *Australopithecus* during the Lower Pleistocene, there should exist fossils of the same general stock but already showing some characters that directly foreshadow those found in the genus *Homo.* As we shall see presently, such a find has now been made.

A different view is expressed by another section of scientific opinion which believes that the Far Eastern hominids, as represented by the genus *Pithecanthropus* (with two species, *erectus* and *sinensis*), are examples of a definite intermediate stage of human evolution, lying somewhere in between the Australopithecinae and *Homo.*

An alternative explanation, however, is that the Far Eastern genus represents a specialized hominid side branch, which was originally derived from the common Australopithecinae subfamily in Africa but which became extinct through over-specialization and never led to *Homo* in any form. For many, Pithecanthropus is far too specialized to represent a direct line leading to *Homo,* unless it be to the specialized late Neanderthalers.

We must now turn to the recent discovery in East Africa which has shed such a great deal of new light upon the whole of this complex problem of man's ancestry.

The Discovery of Zinjanthropus

On July 17, 1959, my wife, working with me at Olduvai Gorge in Tanganyika Territory, found a fragment of fossil human skull on the slopes of the gorge at site FLK I. Following up this clue, she saw two hominid teeth *in situ* in the cliff face. When excavations were carried out, a nearly complete skull of a hominid and also a tibia were found. These were lying on a living floor upon an ancient camp site, in association with nine stone tools of the Oldowan culture and 176 waste flakes, which had resulted from the manufacture of the tools on the spot. There was also one hammer stone. Associated with these tools and flakes were the fossilized bones of many small creatures, such as rats, mice, frogs, lizards, birds, fish, a snake, and a tortoise, plus the bones of some juvenile pigs and antelope and a juvenile giant ostrich. The geological horizon in which this discovery was made is Bed I of Olduvai. This was formerly regarded as belonging to the lower half of the Middle Pleistocene, but it is now put into the upper part of the Lower Pleistocene on the basis of the evidence of its fossil fauna content, which compares closely with that of Omo and other Upper Villafranchian sites in Africa (cf. Howell, 1959).

Since the other animal bones found upon this ancient living floor were in all cases broken up, while the hominid skull and the associated tibia were not, it seems reasonably certain that the latter belonged to the maker of the tools and were not merely items in the food supply of some other hypothetical hominid who possibly occupied the site.

The Oldowan is a well-known, early Stone Age culture which antedates the great Chelles-Acheul hand-axe culture in many parts of Africa. It was first found at Olduvai Gorge in 1931 and was named after the gorge (Leakey, 1951). It is now usually regarded as the oldest, well-established Stone-Age culture in the world, since the so-called Kafuan culture (formerly believed to antedate it) is no longer accepted by most prehistorians, because it is generally found in deposits where nature, rather than man, could have been the agent of chipping—that is to say, the Kafuan is mostly recorded from gravel and boulder beds (Bishop, 1959; J. D. Clark, 1959). It must be remembered, however, that the Oldowan tools found in the living floor, at site FLK I, associated with the new fossil skull at Olduvai are *not* the *earliest* examples of the Oldowan culture. Still earlier specimens are known to occur in lower levels of Bed I and also in the deposits at Kanam West, the latter datable, on the basis of faunal evidence, to an early part of the Lower Pleistocene.

The new fossil skull from Bed I at Olduvai appears at first sight to have considerable resemblance to the South African genus of the subfamily Australopithecinae known as *Paranthropus*. In particular, the following similarities are striking: the presence of a sagittal crest, the great reduction of the canine and incisor teeth in contrast to very large crowned premolars and molars, the fact that the fourth premolar is larger than the third, and the general shape of the dental arcade.

However, when a more critical examination is made of the *total* picture presented by the morphological characters of the new skull, it is found that there are a far greater number of characters in which it differs markedly from *Paranthropus* (and also from *Australopithecus*). In consequence, a new genus has been set up to accommodate this newly found, early stone-tool-making man. The name that has been given it is *Zinjanthropus boisei* (Leakey, 1959). The generic name *Zinjanthropus* is derived from the Arabic word *Zinj*, meaning "Eastern Africa," and the specific name is in honour of Mr. Charles Boise, who has helped finance the search for early man in Africa for so many years.

One of the most striking facts about *Zinjanthropus* is that, in a number of important characters, this new skull exhibits specializations which are similar to those found in the genus *Homo* and which have, hitherto, been regarded as among the diagnostic characters of *Homo* and of certain extinct types of man, such as *Pithecanthropus sinensis*.

For example, although some male gorillas and some members of the *Australopithecus* and *Paranthropus* genera occasionally develop mastoid processes, these are not markedly pyramidal as in *Homo*, nor do they approximate in size and general morphology the mastoid process of *Homo. Zinjanthropus,* on the other hand, has mastoid processes which markedly resemble those found in *Homo* and in some extinct types of man.

In their general facial architecture, and more particularly in the malar-maxillary region of the zygomatic arches, the South African "near-men," *Paranthropus* and *Australopithecus,* exhibit similarities to the great apes, especially to the gorilla. In this character the structure of the facial bones in *Zinjanthropus* is quite different and is of the type regularly seen in the genus *Homo* and always in that species which we call *sapiens.* There are, moreover, very great differences between the shape of the anterior part of the palate in the South African genera and in the genus from East Africa. In both *Paranthropus* and *Australopithecus* there is a marked flattening of the anterior part of the palate in the region between the pre-molars on either side and the incisors; in *Zinjanthropus,* as in most of *Homo,* this region is much deeper and more evenly arched. The type of palate which we find in *Zinjanthropus* is, indeed, one which is frequently associated with articulate speech, and it may well be that when the mandible is found, it will also carry indications of articulate speech for *Zinjanthropus.*

In the basioccipital region of the new skull the position of the condyles parallels that seen in *Homo* and indicates that the head was carried in a completely upright position. In fact, the condyles seem to be set even more forward on the fossil skull than they normally are in *Homo sapiens.* This peculiarity may well be directly correlated with the abnormal length of face in the new fossil.

Significance of *Zinjanthropus* to Human Evolution

In the characters which have just been listed it may be said that *Zinjanthropus,* in spite of being classified in the subfamily Australopithecinae, already exhibits specializations which foreshadow *Homo,* and therefore it seems reasonable to accept the genus *Zinjanthropus* as being in the direct evolutionary line leading to *Homo.*

Such a view is reinforced by the fact that the Stone Age culture (the Oldowan) which is associated with *Zinjanthropus* at Olduvai is one which is demonstrably ancestral to the great hand-axe culture.

At Olduvai Gorge, Bed I yields remains of the Oldowan culture at several levels, but, at the junction of Bed II with Bed I, the Chellean stage of the hand-axe culture appears for the first time. Excavations (Leakey, 1958) have shown that Chellean Stage 1 is a culture stage of the hand-axe culture in which the vast majority of the specimens are still made in the Oldowan manner, but the presence of a small proportion of true Chellean-type hand-axes indicates that the next evolutionary stage has been entered. Even in Chellean Stage 2 the proportion of hand-axes to survivals of Oldowan forms remains low, but it is relatively greater, and there is no doubt that the Oldowan culture leads directly to the hand-axe culture.

The significance of this lies in the fact that skulls with a morphology which places them in the genus *Homo* are known to occur with later stages of the hand-axe culture at Kanjera in East Africa and at Swanscombe in England, both during the final stages of the Middle Pleistocene period. The Steinheim skull from Germany also probably belongs to this period.

On the other hand, it is necessary to recall that Professor Arambourg (1955, 1956) of Paris claims that the fossil human remains which he found with an Acheulian stage of the hand-axe culture at Ternifine in North Africa belong to a new hominid genus which he has named "*Atlanthropus.*" He has also expressed the view that *Atlanthropus* is very closely connected with the *Pithecanthropus*-type of man from Java and China. If this claim is eventually substantiated, then we will have a situation in which a late stage of the hand-axe culture is associated with the genus *Homo* at Kanjera and Swanscombe and with an extinct genus allied to *Pithecanthropus* in North Africa.

While such a situation cannot be ruled out as impossible, the claims that the North African genus *Atlanthropus* is really a member of the Pithecantropine group needs very critical examination.

The parts of the skull which are generally recognized as exhibiting diagnostic characters in *Pithecanthropus* are (a) the occipital bone, (b) the temporal bone, especially in the tympanic plate region, and (c) the frontal bone. The mandible in this genus is somewhat variable.

Now the fossil remains of *Atlanthropus* which have so far been recovered from the Ternifine site in North Africa do not include any of these three vital diagnostic parts of the skull. There are, however, two mandibles which certainly show *some* resemblances both in their morphology, and in the dentition, to some of the mandibles of *Pithecanthropus sinensis*—the Chinese species of the genus from Choukoutien. However, it is doubtful if there is anything like adequate evidence at present to justify linking *Atlanthropus* with the genus *Pithecantropus,* except that they are both of Middle Pleistocene age.

Unfortunately, we do not know the type of mandible that goes with either the Swancombe or the Kanjera human remains from deposits of the upper half of the Middle Pleistocene, but there is at least an equal chance that *Atlanthropus* belongs with this group, with which I would also associate the Steinheim skull as a regional variation. All three rank as members of the genus *Homo* and do not belong in any way with the over-specialized *Pithecanthropus* side branch of the human stock. An interpretation of the *Atlanthropus* specimens linking them with the Swancombe, Steinheim, and Kanjera finds would appear to be much more in keeping with the evidence available at present, as well as with the evidence derived from a study of the cultural picture.

Returning to *Zinjanthropus,* we have noted that this genus exhibits specializations which tend toward those found in *Homo* and have suggested that this genus may well be directly ancestral to modern man, while the other two genera in the same subfamily may be regarded as divergent branches of the same general stock, which exhibit specializations in directions away from *Homo* and which eventually became extinct. In this connection it is necessary to remember that there are some characters in *Zinjanthropus* which are worthy of special mention and which do not closely agree with either *Australopithecus* or *Paranthropus,* nor yet with *Homo,* nor with any other hominid or pongid. In particular, *Zinjanthropus* shows a widening of the base of the skull in the occipitotemporal region by the development of a peculiar shelf, laterally, beyond the brain case. This serves as a kind of ledge from which to suspend the large mastoids. To a very much less marked degree there is a suggestion of a similar shelf in some *Paranthropus* specimens; the corresponding flange in gorilla is quite different in structure.

In the genus *Pithecanthropus* the development of the base of the skull had reached a stage where the widest part of the whole skull posteriorly was in the mastoid region and not upon the parietals. The parietals have, however, expanded partially, so as to rise vertically from a point more or less over the mastoid processes before bending abruptly inward.

In other words, it is not difficult to see, in the members of the genus *Pithecanthropus,* a special development of the back of the skull derived from the condition seen in *Zinjanthropus,* and it may here be noted that the tympanic plate of *Zinjanthropus,* taken alone and out of its context, could also possibly be regarded as having affinities with the form seen in some skulls of *Pithecanthropus sinensis.*

If we are to accept the possibility that the genus *Zinjanthropus* is directly ancestral to *Homo,* as represented by the Kanjera, Swanscombe, and Steinheim skulls in the Middle Pleistocene, we must briefly consider the question of the time factor.

Is it possible for a hominid as truly primitive as *Zinjanthropus* to have become *Homo* in the interval between the upper part of the

Lower Pleistocene and the later stages of the Middle Pleistocene, perhaps 400,000 years, or about 80,000 generations?

In this connection it seems profitable to recall that when man domesticates a species of wild animal, he seems to create a situation where the natural processes of evolution are accelerated in respect of morphological change although the actual rate of mutation is not increased. When a man became a toolmaker and began to exert control over his natural environment, he made himself in effect a "domesticated animal." It may well be that, upon becoming man, he initiated a major acceleration of his own rate of physical change in addition to his rapid cultural development.

In recent years there has been a strong tendency among paleoanthropologists and anatomists to treat man as though he was somewhat different from other animals and so to expect him to behave differently in respect of his evolutionary pattern. It has become fashionable to see *Australopithecus, Pithecanthropus,* and *Homo* as three successive stages which are often regarded as representing a direct and simple evolutionary sequence. As a result, attempts are made to fit all available fossil hominid material into one or another of these three supposed stages, which are often regarded as having a *time* significance as well as a morphological one. Consequently, it has even been suggested that *Australopithecus* as a genus represents the Lower Pleistocene hominids; *Pithecanthropus,* the Middle Pleistocene stage; with *Homo* the last and Upper Pleistocene development.

Such a simplified picture is wholly contrary to known facts and it is also at complete variance with what we know happens in other animal groups.

While *Zinjanthropus,* which exhibits certain specializations toward *Homo,* is contemporary with *Australopithecus* from Taungs and Sterkfontein in the Lower Pleistocene period, there is no doubt that the slightly later genus *Paranthropus,* while also a member of the subfamily Australopithecinae, belongs to the Middle Pleistocene and lived at a time when *Pithecanthropus* was thriving in the Far East. *Paranthropus* was, in fact, only a little earlier than known members of the genus *Homo* (as represented by the Steinheim, Kanjera, and Swanscombe variants) in Europe and Africa. The little-known fossil *Telanthropus* from Swartkrans, described by Robinson (1953), may well also prove to be a contemporary species of the genus *Homo* rather than a distinct genus, as suggested at present. It must also be noted that some scientists regard *Telanthropus* as synonymous with *Paranthropus,* but this seems unlikely.

The Need for Classification

Since there can be no doubt that, as we find more and more evidence of early fossil man, the picture will become more complex rather than simplified, there is a clear need to try to establish some sort of taxonomic order in the Hominoidea. This super-family of mammals has all too often been treated as though it were not subject to the ordinary rules of zoological taxonomy and nomenclature.

Properly worded and very carefully thought-out subfamilial, generic, and specific diagnoses need to be prepared, preferably by a small international committee set up for this purpose. This should result in relegating to synonymy a number of misleading names. Such a committee, while it must include anatomists and paleoanthropologists, should also have among its members people with a good knowledge of zoological and paleontological taxonomic procedure.

It is always difficult to decide just what points in a fossil specimen should be given generic value, but it is certain that much could be done to bring the present position more in line with normal zoological practice without going to the extreme of treating all known fossil men and "near-men" as belonging to the genus *Homo,* all living and extinct great apes as representing the single genus *Pongo,* and all living and extinct gibbons as a single genus, *Hylobates.* This would carry the process of simplification too far, but some serious attempt to achieve internationally acceptable taxonomic nomenclature for all the Hominoidea is urgently needed.

In Conclusion

As we enter the second century of the study of evolution, we may summarize the present position in respect of the origin of man as follows:

The tentative suggestion made by Darwin in 1871 that the most likely home of the origin of man was on the African continent has been more than substantiated by fossil evidence ranging from the Oligocene to the Pleistocene.

The picture which is emerging shows that the stock which led to man had many divergent sub-branches, and there are still immense gaps in our knowledge. We may confidently expect that these will gradually be filled in the ensuing hundred years. In particular we need a great deal of new evidence from the Oligocene deposits of North Africa and perhaps also of South Europe, in order to shed light upon the very early stage of the evolution of the Hominoidea.

There is also a major need to discover and explore systematically African deposits of truly Pliocene age, containing fossil land mammals. This is also true of deposits of the Pleistocene period, antedating the deposits which are now yielding *Australopithecus, Paranthropus,* and *Zinjanthropus* specimens. In such beds an early and much more generalized member of the Australopithecinae is likely to be discovered, perhaps along with fossils immediately ancestral to the gorilla and the chimpanzee.

There is also a major need to search for good Pliocene and early Pleistocene fossil beds in the region which lies between Africa and China for evidence of hominids (probably of *Australopithecus* affinities) moving eastward to become the ancestors of the genus *Pithecanthropus.*

Another urgent need is to study, in more detail, the primates in the secondary centre of pongid evolution in North India, an area which may be expected to shed much light upon the origin of the orangutan and upon other cognate problems.

Finally, the search for and the study of the Upper Villafranchian members of the subfamily Australopithecinae, that is to say, *Australopithecus, Paranthropus,* and *Zinjanthropus* must be intensified, since they give promise of throwing more and more light upon the critical stage when the prehuman primates were "crossing the Rubicon" to become true men.

The foregoing material is based on a lecture delivered on 25 November, 1959, at the University of Chicago's Darwin Centennial Celebration and prepared subsequently for inclusion in this volume. I wish to express my gratitude to Dr. F. Clark Howell, who has helped me meet this deadline by compiling the brief bibliography that follows.

Selected References

Arambourg, C. 1955. "A recent Discovery in Human Paleontology: *Atlanthropus* of Ternifine (Algeria)," *Am. J. Phys. Anthrop.,* n.s. XIII, 191–202.

———. 1956. "Une 3ème Mandibule d'"*Atlanthropus*" Decouverte à Ternifine," *Quaternaria,* III, 1–4.

Bishop, W. 1959. "Kafu Stratigraphy and Kafuan Artifacts," *South African J. Science,* LXII, 117–21.

Broom, R., and J. T. Robinson. 1952. *Swartkrans Ape-man, Paranthropus crassidens* (Transvaal Museum Memoirs, No. 6). Pretoria. 123 pp.

Broom, R., and G. W. H. Schepers. 1946. *The South African Fossil Ape-*

men: The Australopithecinae (Transvaal Museum Memoirs, No. 2). Pretoria. 272 pp.

Broom, R., J. T. Robinson, and G. W. H. Schepers. 1950. *Sterkfontein Ape-man, Plesianthropus* (Transvaal Museum Memoirs, No. 4). Pretoria. 117 pp.

Clark, J. D. 1958. "The Natural Fracture of Pebbles from the Batoka Gorge, Northern Rhodesia, and its Bearing on the Kafuan Industries in Africa," *Proceedings of the Prehistoric Society,* XXIV, 64–77.

Darwin, C. 1871. *The Descent of Man.* London. (New York: Modern Library Edition, 1949).

Dart, R. A. 1925. "*Australopithecus africanus:* The Man-ape of South Africa," *Nature,* CXV, 195–99.

Hopwood, A. T. 1933. "Miocene Primates from Kenya," *J. Linnean Soc. London,* XXXVIII, 437–64.

Howell, F. C. 1959. "The Villafranchian and Human Origins," *Science,* CXXX, 831–44.

Leakey, L. S. B. 1951. *Olduvai Gorge. A Report on the Evolution of the Hand-axe Culture in Beds I–IV.* Cambridge: Cambridge University Press. 164 pp.

———. 1958. "Recent Discoveries at Olduvai Gorge, Tanganyika," *Nature,* CLXXXI, 1099–1103.

———. 1959. "A New Fossil Skull from Olduvai," *Nature,* CLXXXIV, 491–93.

Le Gros Clark, W. E. 1952. "Hominid Characters of the Australopithecine Dentition," *J. Roy. Anthrop. Inst.,* LXXX, 37–54.

———. 1955. "The Os Innominatum of the Recent Pongidae with Special Reference to that of the Australopithecinae," *Am. J. Phys. Anthrop.,* n.s. XIII, 19–27.

Le Gros Clark, W. E., and L. S. B. Leakey. 1951. *The Miocene Hominoidea of East Africa* (Fossil Mammals of Africa, No. 1). London: British Museum (Natural History). 117 pp.

Le Gros Clark, W. E., and D. P. Thomas. 1951. *Associated Jaws and Limb Bones of Limnopithecus macinnesi* (Fossil Mammals of Africa, No. 3). London: British Museum (Natural History). 27 pp.

Napier, J. R., and P. R. Davies. 1959. *The Forelimb Skeleton and Associated Remains of Proconsul africanus* (Fossil Mammals of Africa, No. 16). London: British Museum (Natural History). 69 pp.

Oakley, K. P. 1954. "Dating of the Australopithecinae of Africa," *Am. J. Phys. Anthrop.,* n.s. XII, 9–28.

Robinson, J. T. 1952. "Some Hominid Features of the Ape-man Dentition," *Journal Dental Assoc. South Africa,* VII, 1–12.

———. 1953. "*Telanthropus* and its Phylogenetic Significance," *Am. J. Phys. Anthrop.,* n.s. XI, 445–501.

S. L. WASHBURN
AND
F. CLARK HOWELL

HUMAN EVOLUTION AND CULTURE

In *The Descent of Man* Darwin (1871) outlined the structural-functional complexes which distinguish man from the anthropoid apes. He attributed the differences in skull structure to the great size of the brain of man. He noted that "nearly all the other and more important differences between man and the quadrumana are manifestly adaptive in their nature, and relate chiefly to the erect posture of man; such as the structure of his hand, foot, and pelvis, the curvature of his spine, and the position of his head" (p. 515). This led Darwin to speculate that:

> As the progenitors of man became more and more erect, with their hands and arms more and more modified for prehension and other purposes, with their feet and legs at the same time transformed for firm support and progression, endless other changes of structure would have become necessary. . . . The early male forefathers of man were, as previously stated, probably furnished with great canine teeth; but as they gradually acquired the habit of using stone, clubs, or other weapons, for fighting with their enemies or rivals, they would use their jaws and teeth less and less. In this case, the jaws, together with the teeth, would become reduced in size, as we may feel almost sure from innumerable analogous cases" (1871, p. 435).

Darwin recognized the great importance of language and its relation to the size of the human brain. He stressed the importance of reason, social life, and the moral order in the selective process which led to man. He believed that the anthropomorphous apes—gorilla, chimpanzee, orang, and gibbon—constituted a natural group and that "some ancient member of the anthropomorphous sub-group gave birth to man" (p. 519).

A large number of primate fossils have been discovered since 1871.

SHERWOOD L. WASHBURN is Professor of Anthropology at the University of California and is well known for his writings in physical anthropology and his comparative studies of the primates.

F. CLARK HOWELL, Associate Professor of Anthropology at the University of Chicago, specializes in human evolution and paleoanthropology.

Only the Neandertal skull cap and the Naulette lower jaw were known to Darwin. Although Darwin did mention the discoveries of Boucher de Perthes in the Somme River in connection with the antiquity of man (1871, p. 390), almost the whole science of prehistoric archeology has developed since the writing of *The Descent of Man*.[1] As did Huxley (1863), Darwin had to build his theory primarily from the comparison of living forms, since a fossil record was lacking and there was an absence of archeological specimens associated with human fossils. However, he did caution that "we must not fall into the error of supposing that the early progenitors of the whole Simian stock, including man, was identical with, or even closely resembled, *any existing ape or monkey*" (1871, p. 520).

At that time it was impossible to do more than speculate about the order of the actual events in human evolution. Perhaps one of the main reasons why Darwin did not trust his own theory of natural selection further but tried to support it with various subsidiary theories was simply that he did not have sufficient data to see how well it applies in accounting for human origins.

Darwin did express the hope that the necessary fossils of man would be found; surely he would have been delighted to see the extent to which his prophecy has been fulfilled. Equally surely, with his usual caution, he would have reminded us that many of the problems raised by him nearly a century ago are still unanswered. On which continent did man originate? How rapidly did the evolution of man proceed? Why did man lose his body hair? What was the nature of the social life of early man? If progress continues at the present rate in human evolutionary studies, it will be another hundred years before answers are provided to many of the queries posed by Darwin.

This paper will present an interpretation of man's evolution during the Pleistocene. Our contention will be that the actual events follow very closely the order postulated by Darwin, except that the human brain seems to have been a secondary development which evolved later rather than the primary factor in the differentiation of ape and man. The authors recognize that certain points in the interpretation are debatable. However, it is felt that there is sufficient new information to

[1] *The Descent of Man* had three rather separate purposes: (1) to prove that man is one of the Quadrumana and should not be placed in a separate order; (2) to discuss the way in which an ancient anthropomorphous ape was changed into man; and (3) to account for many features of man by the theory of sexual selection. This paper is concerned only with certain aspects of man's origin, and it is assumed that the time has long since passed when proof would be needed for the general fact of human evolution. Today much of what Darwin attempted to explain by the theory of sexual selection or by the effects of use, disuse, or correlation of parts would be attributed to natural selection.

make an attempt at synthesis that is both interesting and particularly appropriate in a volume which honors Darwin and his contributions.

Sequence of Evolutionary Events

A schematic view of the order of events in human evolution during the Pleistocene is presented in Table 1. Certain of these materials may be summarized, as follows.

1. Neither fossils of the Hominidae nor stone implements are known from the terminal Pliocene nor from the basal Pleistocene (earliest Villafranchian [2] stage).
2. The oldest known bipedal Hominidae, capable of using and manufacturing stone implements, were certain australopithecines (genus *Australopithecus*) of late Villafranchian age.
3. A variety of fossil men (genus *Homo,* probably with several species) appears during the earlier Middle Pleistocene associated with types of stone implements that were prepared according to clearly defined traditions of manufacture. Some tools of the Chelles-Acheul tradition (hand-axes, cleavers, and various flake tools) are distributed over Africa, southwestern Europe, and most of India (but not east of the Ganges flood-plain). The Far East, including southeastern Asia, is characterized by a different tradition, the chopper/chopping-tool tradition (Movius 1944, 1948).
4. Anatomically modern man (*Homo sapiens*) appeared very late in the Pleistocene and always with a great variety of implements in stone (and other materials) which changed very rapidly in comparison with those of earlier traditions.

AUSTRALOPITHECINE STAGE

Bipedal locomotion.—Darwin and many workers after him stressed development of bipedal locomotion as a factor in differentiating man from ape. This process freed the hands, made possible the use and manufacture of tools, and led to reduction in the size of the teeth and the facial skeleton. The australopithecines in general represent such a stage in human evolution. Portions of the pelvis of four different individuals from three different sites are now known.[3] This is most

[2] An Old World Villafranchian fauna, so-named after the type locality of Villafranca d'Asti in the upper Po River drainage, comprises essentially later Tertiary forms of mammals but also includes the first appearance of the modern genera *Equus, Elephas* (*Archidiskodon*), and *Bos* (*Leptobos*), and sometimes, *Camelus.*

[3] Two major kinds of australopithecine are known, an earlier form (*Australopithecus*) and a later form (*Paranthropus*), variously regarded as generically, subgenerically, or specifically distinct. The locomotor skeleton is best known in the former, but a portion of the hip bone and the upper ends of two femurs are also known in the later form.

TABLE 1

TEMPORAL AND SPACIAL DISTRIBUTION OF PLEISTOCENE HOMINID SKELETAL REMAINS.

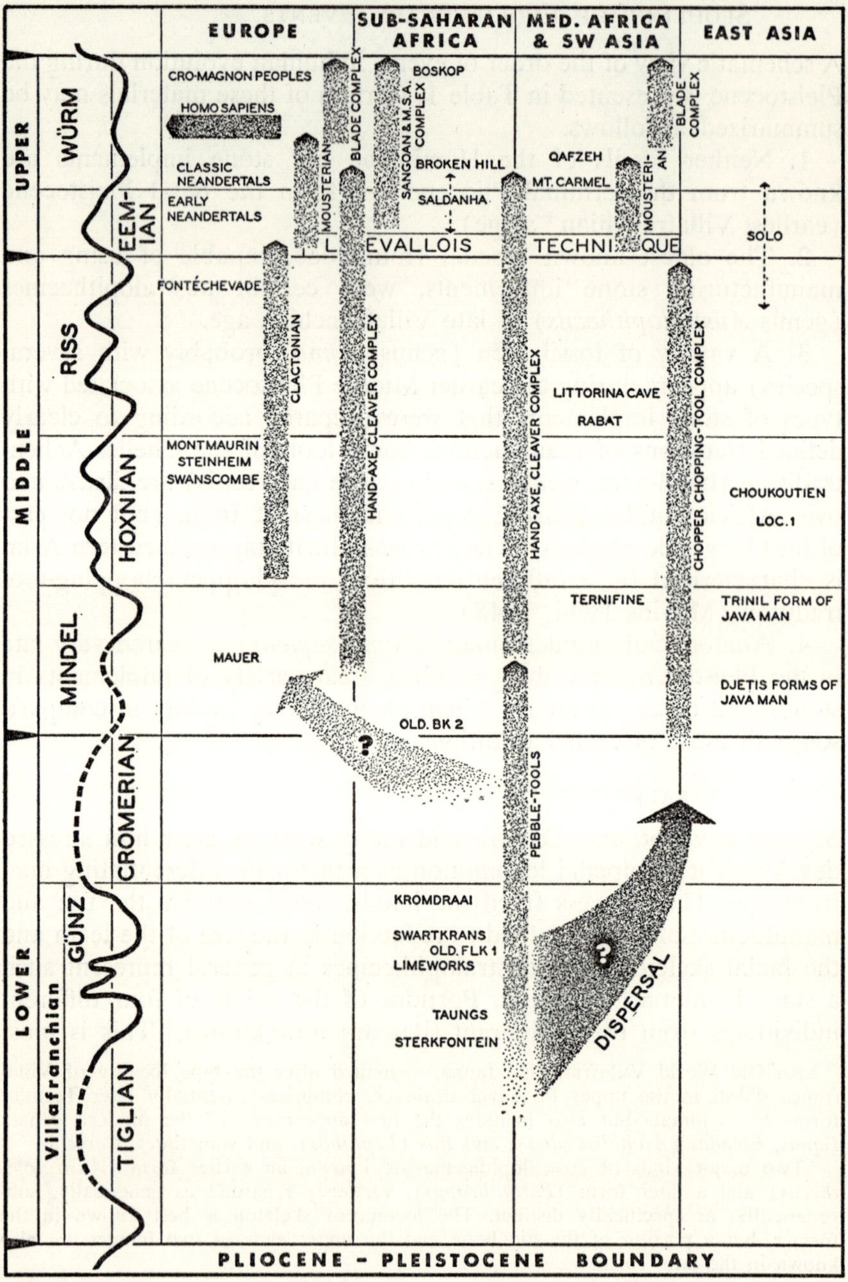

important for the interpretation of locomotion since the pelvis of a bipedal hominid is very different from that of either apes or monkeys (cf. Washburn, 1950; Mednick, 1955). The australopithecine ilium is short, broad, and conforms very closely to the human pattern (Le Gros Clark, 1955 *a;* Howell, 1955). It should be stressed that the human ilium is a very specialized and peculiar bone and that, among all known mammals, the only ilium approximating man's is that of the australopithecines. It is in just those special features related to bipedal locomotion that the ilia of man and the australopithecines are alike. As nearly as we can tell from the ilium and the small fragments of the proximal end of the femur, the hip musculature must have functioned as in man; that is, *gluteus medius* functioned as an abductor and *gluteus maximus* as an extensor. Likewise, judging from the femur, the *quadriceps femoris,* so important in human bipedal locomotion, must have been large (the authors are grateful to Dr. J. T. Robinson for permission to examine these specimens, still in part unpublished). Unfortunately, only fragments of the limb bones are preserved, and these are so short that reconstruction of proportions is impossible.

On the basis of the geological evidence and the associated fauna it is evident that the australopithecines lived in open grassland country, either somewhat drier or not unlike that of southern Africa at present (Brain, 1958). This habitat is very different from that of the living arboreal apes and supports the interpretation of locomotion based on the morphology of the pelvis.

Reduced dentition.—Darwin suggested that the reduction in size of the canine teeth of man was the result of tool use. The canines of all the australopithecines are small and of human conformation (Robinson, 1956). These fossils support Darwin's contention, which in turn suggests that the use of implements may very well be considerably older than the earliest known artifacts. Perhaps, as Darwin suggested, tool use is both the cause and effect of hominid bipedalism, and the evolution of erect posture occurred simultaneously with the earliest use of tools. Primitive apes that were bipedal part-time and were capable of employing objects as implements must have lived in the woodlands or on the margins of forested areas (cf. Oakley, 1954; 1956). The ancestral populations of such apes were possibly like *Proconsul,* a form fully capable of bimanual locomotion through the trees in riverine gallery forests, although it probably also spent much of its waking life in the adjacent, open grasslands. It seems most probable that the australopithecines represent a subsequent evolutionary stage in which such creatures are already essentially bipedal, fully plains-living, and utilizing selected raw materials for the manufacture of stone tools. The preceding anthropoid or proto-hominid stage may very well have occurred at the beginning of the Villafranchian stage; if so, there may be no need to postulate a distinct, fully hominid line prior to the Pleistocene. The great gap in the hominid fossil record is now terminal Pliocene and early Pleistocene. And in Africa at least, sites of such age are extremely rare.

Cranial capacity.—The brains of the australopithecines were surprisingly small. The capacity in the best preserved specimens ranges between approximately 450 and 600 cc. Prior to the discovery of the australopithecine pelvis, the small size of the brain led a number of scientists to dismiss the australopithecines as fossil apes (Keith, 1931; Weidenreich, 1948; and Zuckerman, 1951, 1954). Some remarks on the evolution of the human brain will be made later on, but it is worth noting here that many scientists, including Darwin, have viewed the large brain as the essential factor in the diagnosis of man. For a time some workers even denied that Pekin man was an implement-maker because of the size of his brain. However, at least some australopithecines, with less cranial capacity than that of Pekin man, were already fashioning implements from stone. Consequently, it would appear that after acquisition of bipedalism and the beginnings of tool use new selection pressures led to the expanded brains of the Middle and later Pleistocene.

Earliest evidence of culture.—The oldest stone implements, referred to the Pre-Chelles-Acheul industrial stage, have been known for some time from open sites; they have been found in Africa both north and south of the Sahara and are of Lower Pleistocene age, according to the associated Villafranchian fauna (Howell, 1954). Such evidence is not forthcoming from Villafranchian sites in either Europe or Asia, nor have any such extra-African sites yielded hominid skeletal remains. Hence it is likely that hominids were still restricted to continental Africa for this extended period (Howell, 1959 *a*).

Remains of australopithecines are associated with tools at Sterkfontein, Swartkrans and Olduvai Gorge (Bed I).

At Sterkfontein no Pre-Chelles-Acheul implements have been found in the lower and main australopithecine-bearing pink breccia. The first such association with fragmentary australopithecine remains has come from the overlying red-brown breccia (Robinson and Mason, 1957; also Oakley, 1957). Over two hundred specimens are now known, entirely fresh and unweathered and largely in raw materials (quartz, quartzite, diabase, and chert) which do not occur naturally in the cave deposits, but must have been collected and brought to the site (Robinson, 1958). These include pebble-choppers, core-choppers, rough but retouched flakes, and quantities of fractured and utilized pebbles. An incomplete, longitudinally split and pointed bone implement, worn smooth at one end by use, has also been reported (Robinson, 1959).

A new discovery by Dr. and Mrs. L. S. B. Leakey (1959*a, b;* also this volume) has proved australopithecine tool-making abilities and determined the makers of the Oldowan implements. At an open-air occupation site (FLK) in Olduvai Gorge Bed I, where the Oldowan industry was first recognized, a beautifully preserved hominid skull with excellent associations has been recovered. This new australopithecine (termed *Zinjanthropus boisei* Leakey) is very similar to the large and later form (*Paranthropus*) so abundantly represented at Swartkrans. Stone implements, including choppers and chopping-tools, flakes and flake implements, and cores are associated on the partially-excavated habitation horizon, an old land surface between clays (below) and sands (above), some three meters below the top of Bed I. Fossil remains of animals are also present, including broken bones of numerous rodents, immature larger mammals (e.g., pigs, antelopes), various reptiles and birds.

A number of attempts have been made previously to reconstruct the life habits of the australopithecines (Broom and Schepers, 1946; Dart, 1953, 1955). The most useful general framework for such a reconstruction has been given by Bartholomew and Birdsell (1953). In southern Africa the remains of australopithecines occur in ancient calcite-cemented soils infilling caves. They are typically found in association with a varied but characteristic mammalian fauna of the open grasslands, including very numerous antelopes (mostly medium to small varieties), numerous carnivores (including various hyaenids and sabretooths), suids, equids, and other primates (largely baboons); a microfauna includes numerous rodents, insectivores, hyracoids, and lagomorphs. Very large mammals are rare.[4] The bones are generally broken; however, some well-preserved skulls do occur, and there is a fairly high proportion of skull fragments.

The central question has been whether the australopithecines were the hunters or the hunted (Washburn, 1957). The vast majority of the animal bones show no clear use as tools, but merely represent the remains of meals by animals. Did these creatures already have the human hunting habit? Or were they scavengers? Hyena coprolites occur in the cave deposits, and it seems extremely likely that some broken bones represent remains of their meals (since some hyenas accumulate bones as part of their eating habits). It is by no means certain,

[4] The Limeworks Cave site is an exception. At least five rhinos, a hippo, six chalicotheres, and six giraffids, as well as a fragment of proboscidean molar, have been recorded (Dart, 1957). Some fractured stones from here (Brain *et al.*, 1955), first thought to be implements, are inconclusive and may well be natural (cf. Oakley, 1956 *a*).

however, that this accounts for all the bone accumulations. The southern African sites [5] were worked extensively for lime by commercial firms so that much necessary evidence of detailed stratigraphy and associations has been lost.

The extraordinary new discovery at Olduvai Gorge has clarified some of these important questions concerning australopithecine behavior. It affords clear-cut evidence that these primitive hominids were to some extent carnivorous and predaceous, augmenting their basically vegetal diet with meat, particularly from small animals and the young of some larger species. It is very unlikely that the early and small-bodied australopithecines did much killing, whereas the later and larger forms, which probably replaced them, could cope with small and/or immature mammals. There is no evidence to suggest such creatures were capable of preying on the large herbivorous mammals so characteristic of the African Pleistocene.

Indirect evidence on the habits of the australopithecines will ultimately be gained from their distribution and classification. Unfortunately, several such matters are still under debate. Robinson (1953 *b*, 1955) has attributed to the australopithecines several hominid jaw fragments from the Putjang beds of Java.[6] This hominid falls fully outside the range of variation of the other Javanese "pithecanthropine" forms, whether from the same Putjang beds or the later Kabuh beds. The scant evidence available supports Robinson's conclusion that the dental and jaw morphology of this hominid is generally similar to that of certain australopithecines. It is premature, however, to say that this was certainly a *Paranthropus*, since the cranial morphology is still wholly unknown.

In the southern African sites several types of hominids are represented. These have been treated as genera (*Australopithecus*, *Paranthropus*, and *Telanthropus*) by Robinson (1954 *a, b, c*). The new australopithecine from Olduvai Gorge indicates a still broader dis-

[5] This is especially the case for Limeworks Cave (Makapansgat) where essentially all the australopithecine and associated mammal fauna have been salvaged from the dump heaps of breccia left from the earlier quarrying activities. And yet it is mainly this site which has been employed to support various arguments about australopithecine behavior. Some of the best evidence bearing on the food-getting habits of the australopithecines is that from the Taungs cave, as Dart (1926) pointed out years ago.

[6] This hominid, referred to *Meganthropus palaeojavanicus* von Koenigswald, is represented by several lower jaw fragments collected in 1936 and 1941, by another found in 1939 and provisionally referred to the same species, and by isolated teeth (von Koenigswald, 1949). A more complete specimen, which includes the symphysial region and roots of the anterior teeth as well as the full right premolar-molar series, was found in 1952 at the same Sangiran locality (Marks, 1953). Another hominid, referred to *Pithecanthropus modjokertensis*, is also present in the upper Putjang beds; coexistence is indicated if these forms are present in the same horizon (whether this is so is not clear).

tribution of the larger and later form. Such differences are not found among later fossil hominids. The existence of morphological differences of this magnitude within the same geographic region, and supposedly within relatively short intervals of time, is the strongest evidence that the way of life of these forms was also different from that of later hominids. If these creatures had been fully effective hunters, as was the case among later Pleistocene peoples, the presence of two species from a single site would be most improbable.

Since the australopithecines occupied the same ecological position as the men of the Middle Pleistocene, it is no surprise that they were in due course completely replaced by early members of genus *Homo*. However, the earliest men may very well have been contemporary with the latest australopithecines for a short time.[7] This situation was quite different from that of such forms as the chimpanzee or gorilla, which could coexist with men in a different ecological zone, or the baboon, which could easily escape from man until the development of firearms.

THE EARLY *Homo* STAGE

The early Middle Pleistocene, perhaps even the terminal Villafranchian, seems to have been a time of hominid dispersal into extra-African areas (Howell, 1959 *b*). Hominids were more widely distributed following the end of the Lower Pleistocene (Cromerian interglacial stage). Their stone implements or fragmentary fossilized skeletal remains occur in deposits of early Middle Pleistocene age in Sundaland of southeastern Asia and in southwestern Europe. Paleogeographic conditions would appear to have been particularly favorable for such expansion and the occupation of new habitats; there was maximum continentality as a result of uplift plus greatly lowered sea levels accompanying the expansion of the first continental ice sheets in North Temperate latitudes. The morphological evidence suggests the attainment of advanced evolutionary status for such early men. There were also undoubtedly behavioral differences, including enhanced cultural capacities, in comparison with the australopithecines of the Villafranchian.

Structural advances.—The bipedal hominids of the Middle Pleistocene are larger-brained and indicate major structural and behavioral advances beyond the australopithecine stage. However, there is no

[7] Coexistence seems clearly indicated at the site of Swartkrans. There a later form of australopithecine (*Paranthropus*) is evidently associated with another, distinctive hominid (*Telanthropus*), known only from fragmentary remains (Broom and Robinson, 1950; Robinson, 1953 *a*); the latter bears a quite close resemblance to certain early Middle Pleistocene men (cf. Howell, 1959 *b*).

morphological evidence to preclude their having evolved from an australopithecine group. The locomotor skeleton, or what is known of it, of all the subsequent Pleistocene hominids is remarkably similar to that of anatomically-modern human races. In fact, the femur of Java man (from the Kabuh beds) or that of Pekin man is considerably more like the femur of *Homo sapiens* than it is like that of European (early or classic) Neandertal peoples (cf. Weidenreich, 1941 *a*). This fact would suggest that all variation in limb-bone morphology manifest in the latter half of the Pleistocene may be racial variation rather than illustrative of any major evolutionary trend.

In comparison with the australopithecines the brains of Middle Pleistocene men were very much larger, and the masticatory apparatus, particularly the molar dentition, was much reduced.

Cultural advances.—Fossils of Middle Pleistocene peoples are associated with well-planned stone implements and weapons manufactured according to complex and clearly-defined traditions. These peoples, especially in Africa and Europe, where the evidence is best, hunted and killed large herbivorous animals in some quantity. There is no longer any question that men of the Middle Pleistocene—and even the earlier Middle Pleistocene—were capable and effective hunters. The earliest evidence has been largely provided by the very careful and extensive excavations of Leakey (1957, 1958) in two Chellean occupation horizons at Olduvai Gorge, west of the Crater Highlands in northern Tanganyika. One such horizon (BKII), on an old land surface adjoining a small, natural clay-filled depression representing a former marsh, has provided a rich assemblage of very early Chellean implements: choppers and chopping-tools, utilized flakes and flake-tools, rough polyhedral stones, a few bifacially-worked, picklike hand-axes, and many natural pebbles and stones, all brought to the site. These implements were found in association with articulated animal remains, sometimes whole skeletons but usually only skulls (and horn cores), vertebral columns, and limbs. Most of the bones and the skulls show clear signs of having been broken open to obtain the marrow and brains. This situation provides the earliest and clearest evidence that the primitive folk of the Chellean were meat-eaters, and were sufficiently skilled hunters to kill giant pigs, sheep, oxen, baboons, horses, hippos, and elephants.

One of the strongest indirect arguments that the australopithecines carried on at least a limited amount of hunting is that the hunting of such larger game in the Middle Pleistocene must have been preceded by the development of a taste for meat, probably through scavenging and the occasional killing of lesser game. The earliest use of fire is reported from the cave lair of Pekin man (at Choukoutien Locality 1).

There is also evidence of fire on some open sites of broadly similar age, e.g., in the Swanscombe Middle Gravels of the Thames Valley (Wymer, 1958), and at Cagny-la-Garenne in the Somme Valley (F. Bordes, personal communication). This very important development (cf. Oakley, 1955; 1956 *b;* 1958) plus the increased dependence on the use of implements—of stone as well as other raw materials—probably accounts for the altered selection pressures that resulted in smaller molar teeth.

Distribution of Middle Pleistocene hominids.—There are still relatively few hominid fossils from the Middle Pleistocene. The best preserved early Middle Pleistocene remains, including portions of the cranium, are from Java, either from the earlier Putjang beds (Djetis fauna) or the later Kabuh beds (Trinil fauna). In the former horizon at least two distinctive hominids are present. Only hominid mandibles and teeth are represented in either western Europe (Mauer) or northwestern Africa (Ternifine) at a comparable time. However, slightly later, at a time broadly corresponding to the middle of the Middle Pleistocene of Europe, a good population sample is present from northern China (Choukoutien Locality 1). Two skulls, one only half preserved, provide the only evidence in Europe. Only fragmentary jaws and teeth are known from the end of this time in northwest Africa (Morocco).

There appear to be rather marked differences between the fossil men of the Far East and Europe during the Middle Pleistocene, although the evidence is admittedly scanty and incomplete. These differences may be evidence of two distinctive phyletic lineages worthy of species status. Some workers, like Vallois and Le Gros Clark, although dealing with the problem in a quite different manner, would give such differences full generic rank. Howell (1959 *b*) regards the evidence as indicative of differential dispersal within the early hominid lineage (cf. Table 1). The Asian lineage is conceived of as representing an early, essentially subtropical, primary dispersal, linked with the spread of chopper/chopping-tool tradition, of late Villafranchian age, and persisting in the Far East; it apparently did not extend into Europe mainly due to the high sea levels of the Lower Pleistocene (the Calabrian and especially, the main Sicilian transgressions). The other dispersal, whose center of origin remains unknown but which may very well have been largely equatorial African, was later and was probably linked with the spread of the Chelles-Acheul hand-axe industries of the Middle Pleistocene.

By last interglacial times in Europe there appear larger-brained hominids, already of distinctive morphology, such as those from Ehringsdorf and Saccopastore, which are already recognizable as the populations ancestral to the later classic and other Neandertal peoples of the early Last Glacial in Europe. The last of these more ancient human groups indicate a considerable variety of raciation during the earlier Upper Pleistocene (as evidenced by Solo in Java, Broken Hill and Saldanha in southern Africa, the various European Neandertals, the southwest Asian Neandertals, and very likely many more not yet discovered). There is little evidence of any great structural modifications over the immediately preceding forms. However, there was considerable progress in the technique of implement manufacture as well as in the diversity of stone implements and the development of special tool types.

Homo sapiens STAGE

During the middle of the Last Glaciation, commencing some thirty-five to forty thousand years ago, men structurally like ourselves (*Homo sapiens*) spread rapidly over much of the Old World.[8] This form of man, particularly in Europe and southwestern Asia where the evidence is most complete, is associated with greatly accelerated change, apparently due to both the cumulative effects of culture and the specific biology of *Homo sapiens*. Human material culture changed rapidly and there were extraordinary regional differences, undoubtedly related to specific ecological adaptations. Expanded human populations with enhanced hunting capabilities and specialized equipment, crossed large bodies of water and populated Australia. Such populations

[8] The time and place of the origin of *Homo sapiens* is still uncertain and under debate. The only fossil hominid remains which are possibly of this type, and which are earlier than the main interstadial amelioration of the European Last Glacial, are so fragmentary that their interpretation is open to reasonable doubt. The authors regard the Swanscombe cranial remains as of the same form as the broadly contemporaneous Steinheim skull (cf. Howell, 1959 *b*). Consequently the evidence for the presence of an essentially anatomically-modern population, coexistent with proto-Neandertal folk in Europe during the Middle Pleistocene is poor indeed. However, such a point of view is maintained by Vallois (1949, 1954) and Heberer (1949, 1950).

The evidence for such a population is only somewhat better during the European Last Interglacial stage (or final phases of the preceding glacial). It hinges on the interpretation of the cranial fragments from the cave of Fontéchevade (Charente), recently published in detail by Vallois (1958). Vallois has clearly demonstrated, from very fragmentary evidence, the extent to which this form differed from contemporaneous early Neandertal peoples, with which such a form apparently coexisted. However, it seems certain that the prevalent and presumably dominant human form during the earlier Upper Pleistocene was an essentially primitive type with large brow ridges and other very distinctive morphological features. This form, so far as is known, was completely replaced by *Homo sapiens* some thirty-five thousand years ago. There is scarcely any debate over the main evolutionary stages or over which form was prevalent. There is, however, uncertainty over the immediate precursors of early *Homo sapiens*.

conquered the sub-arctic, living in front of caves or shelters, or in huts and larger semi-subterranean communal dwellings, with food storage facilities, in the open loess lands. Such groups, from an unknown Asian source, peopled the New World. Art made its appearance, in both naturalistic and stylized forms, with engraving, carving (in both bas relief and the round), painting (both monochrome and, subsequently, polychrome) and personal decoration and adornment. Within a span of some twenty thousand years there was more technological change than in the preceding half million years. This is, of course, the record of culture as we know it. It is the evidence of the presence of the restless creator, *Homo sapiens.*

According to the view of evolution of the Hominidae outlined here, at least three major evolutionary stages are recognizable during the course of the Pleistocene: (1) an australopithecine stage, in some cases associated with the earliest stone implements ("pebble-tools"); (2) an early human (*Homo*) stage, during which there was wide dispersal, development and maintenance of significant traditions of stone-implement manufacture, but very slow cultural change; (3) an anatomically-modern (*H. sapiens*) stage associated with major cultural innovations and rapid technological change. It is necessary to stress that such stages were not static, but that change occurred within each and that each represented evolutionary transformation from the preceding stage. Moreover, the relationships of particular stages to one another are not necessarily simple. Thus, although some members of the earlier and more generalized australopithecines may have been directly ancestral to ancient human forms, some populations of the latter may very well have replaced later and more specialized australopithecine populations. Similarly, populations such as those represented in the southwest Asian Upper Pleistocene caves of Oafzeh and Mt. Carmel (Palestine) may have evolved into *Homo sapiens,* but *H. sapiens* forms may have then replaced numerous other populations (such as some Neandertal folk, the Broken Hill-Saldanha peoples of sub-Saharan Africa, and the Solo people of southeastern Asia).

As technical efficiency increases, the structural diversity among hominids at any given time level decreases. Among those at the australopithecine stage, at least three distinct groups are known from the present geographically restricted material (or four groups, if the Javan form, referred to *Meganthropus,* is included). Presumably there must have been even more, especially among later representatives of the group and especially if such populations eventually inhabited much of the tropical Old World. The differences between such forms as Pekin man and Steinheim man, or between the Solo people and European Neandertal populations, suggest very considerable isolation and, correspondingly, distinct races or even, in some cases, species. It is especially interesting that *H. sapiens,* although inhabiting a far wider area and very diverse ecological zones, shows only racial differences.

Interpretation of Human Evolutionary Sequence

If the foregoing is a reasonable statement of the order of human evolutionary events, then an interpretation can be suggested and examined to see how closely it parallels Darwin's theories.

The australopithecines were plains-living, bipedal hominids. Their locomotor-skeleton morphology, in particular, their pelvic structure, is far more similar to that of later human forms than is their cranial morphology. (This same contrast is evident in the femur and skull remains of Java man and leads to the same conclusion.) The australopithecines merely belong to an earlier evolutionary stage in which the contrast between the locomotor skeleton and the skull was even greater. It is interesting that although Weidenreich (1948) regarded the australopithecines as essentially apes because of the size of their brains, he was always convinced that bipedalism evolved prior to any major changes in the brain and skull (Weidenreich, 1913; 1941 *b;* 1947). However, the australopithecine canine teeth were of essentially human size and form and thus depart widely from the canine teeth of the anthropoid apes (Robinson, 1956). Moreover, the evidence would now suggest that some such forms were already stone-implement makers. Darwin postulated that the reduction in the size of the canines of the "early male forefathers" was the consequence of employing stones and clubs for fighting (1871, p. 435). This interpretation may be correct, and it now appears reasonable that forms ancestral to the australopithecines were probably the earliest to employ and even, to some extent, to manufacture stone and other implements. Since the canine teeth were already small and of the human type, the substitution of the use of implements for canines in fighting (and defense in general) must have taken place still earlier. It has been suggested that only after the development of weapons for protection could anatomically unprotected and slow-moving hominids have lived successfully in the African plains (Dart, 1949; 1953).

Selective Value of Bipedalism

Presumably the proto-hominids living on the edge of forested areas or in transitional woodland-grassland zones were neither fully erect nor efficiently bipedal. This would support Darwin's suggestion that "the free use of arms and hands, partly the cause and partly the result of man's erect posture, appears to have led in an indirect manner to other modifications of structure" (1871, p. 435).

It may well be that the basis for the hominid adaptive radiation was *both* bipedalism and the use of implements. Each is so intimately, causally associated with the origin of man that it is perhaps meaningless to speak of one without consideration of the other. Even partially liberating the forelimbs from locomotor functions could have led to some use of implements, and this would have been of such survival value that enhanced bipedalism and further implement use would have evolved together. It must be recalled that the hands which began to be freed with the beginnings of bipedalism were those of an intelligent, manipulative, curious, and playful ape. Such an animal is eminently pre-adapted to implement use and manufacture, as numerous studies have shown (Yerkes, 1943).

What were the factors which initiated and made biologically advantageous the shift from incipient, occasional, inefficient bipedalism to an habitual, efficient, terrestrial bipedalism? Extraordinarily little is known about the necessary background, including environmental situations and the particular structures, physiological mechanisms, or behavior patterns, which formed the basis for the transition of some proto-hominid group to an early (australopithecine) grade. Certainly much more might be learned of these basic, important behavioral and structural preadaptations from field studies of African-ape behavior in particular ecological situations.

SIGNIFICANCE OF DENTAL REDUCTION

Darwin discussed the reduction of the dentition and the facial skeleton as if it had been a single and relatively simple event. The fossil record clearly indicates that the actual events were considerably more complex. Whereas the whole canine-premolar complex is reduced in the australopithecines,[9] and hence closely comparable to the typical

[9] The large upper canine of monkey or ape meets the anterior surface of the sectorial (unicuspid) first lower premolar, and the large lower canine fits into the space (diastema) between the upper canine and lateral incisor when the jaws are closed. The whole complex (well illustrated by Le Gros Clark, 1955 *b*), including the correlated facial structure, masticatory musculature, and associated cranial superstructures (crests), evolved together and is a morphological entity whose parts should not be considered separately. Reduction in size of the human canine clearly affected the crown considerably more than the root dimensions, since the root is still long in the australopithecines and even in the men of the Middle Pleistocene. One argument which has favored the interpretation that considerable canine reduction must have occurred is the relatively long-rooted canine present even in modern representatives of *Homo sapiens.*

human condition, the molar teeth are large and, in many respects, primitive in their morphology. The later form of australopithecine (from the geologically younger sites of Swartkrans and Kromdraai) has still larger molar (and premolar) teeth compared with the earlier form (from the sites of Sterkfontein, Makapan and Taungs). In correlation with such large teeth the australopithecine face is massive; there is, however, some reduction of the anterior portion related to the incisor and canine teeth. In all Middle Pleistocene and later human forms the entire face is very markedly reduced in size. The evidence would favor the conclusion that those selection pressures which maintained large canine teeth seem to have changed long before those which favored large molar teeth. One interpretation of this situation is that implements were first and primarily employed for protection, hence as weapons, and only to a lesser extent in the food quest. Or perhaps, and this seems extremely likely, their earliest use was in obtaining food, but not preparing it since extracting food from the earth would greatly enhance the food supply, but abrasive roots and tubers would increase tooth wear and favor selection for large molar teeth.

In the Middle Pleistocene there is abundant evidence that men fashioned excellent stone implements in standardized forms by well-defined traditions, knew the use of fire, used fire-hardened wooden spears, and killed some large animals. It seems likely that there was some preparation of food, and, with some variation, a trend began toward reduction in molar dentition which continued throughout the remainder of the Pleistocene. However these events may be interpreted ultimately, the authors feel that different rates of evolution in various parts of the hominid dentition, combined with a study of dental wear, masticatory muscle size, and careful appraisal of the associated implements and animal bones will lead toward a much richer understanding of the habits of early man. Even now it seems certain that Darwin was essentially correct in attributing decreases in the dentition, facial skeleton, and masticatory muscles to the use of implements by hominids. This point of view has been criticized as essentially Lamarckian (Vallois, 1955), but there is no suggestion that such changes were the consequence of the inheritance of acquired characteristics. As implements and tools assumed a greater importance and performed functions formerly carried out by muscles, bone, and teeth, it seems clear that selection pressures would have shifted. A correlation between form and function may be produced by selection, and the use of implements and tools has altered the whole life of man, resulting in changed selection pressures leading toward many more changes than those already briefly discussed here.

THE EXPANDING BRAINCASE

When the existing anthropomorphous apes are compared with living man, one of the greatest differences is in the size of the brain. Thus, the area of cerebral cortex in the chimpanzee is only about one-fourth that of the surface area in man, although in the former the convolutions are well marked and have a disposition similar to man's and the cyto-architecture is nearly identical in both. Not only is the brain of man approximately three to four times as large but it is the particular pattern of cortical development, and relations with diencephalic structures, which makes human life as we know it possible. Language, memory, motor skills, foresight, complex social organization, and art are all specific human attributes related to the structure and function of the brain. Although almost nothing was known of the latter when he wrote, Darwin stressed the human brain, its mental powers, and the attributes of reason, language, moral sense, sense of beauty, religion, and social virtues. In fact, far more of *The Descent of Man* is devoted to such topics than to the origin of human morphology. Considering this fact, and their original close agreement regarding man, it is indeed rather strange that Darwin and Wallace eventually disagreed on this very point (cf. Eiseley, 1958).

All verified human fossils discovered up to the 1920's, with the exception of the earlier Middle Pleistocene form from the Kabuh beds in Java, had large braincases. With the exception noted, all such remains were of Upper Pleistocene age. The brains of these peoples, the form and proportions of which could be studied from endocranial casts, were not particularly different in size from those of living *Homo sapiens*. For this reason, and because many scientists accepted the validity of either faked or wrongly-dated human skeletal remains as evidence that large-brained, anatomically-modern man (*Homo sapiens*) existed throughout the Pleistocene, man was often defined on the basis of brain size. It was also often implied that such forms had discovered culture as we know it bit by bit. It would now appear, however, that the large size of the brain of certain hominids was a relatively late development and that the brain evolved due to new selection pressures *after* bipedalism and consequent upon the use of tools. The tool-using, ground-living, hunting way of life created the large human brain rather than a large-brained man discovering certain new ways of life. The authors believe this conclusion is the most important result of the recent fossil hominid discoveries and is one which carries far-reaching implications for the interpretation of human behavior and its origins.

The approximate changes in gross size of the hominid brain, as measured by cranial capacity (in cubic centimeters), during the Pleistocene are summarized below.

TABLE 2

REPRESENTATIVE CRANIAL CAPACITIES *

Homo sapiens	1200–1500 cc.
Java and Pekin man	900–1100 cc.
Australopithecines	450–550 cc.
Chimpanzee	350–450 cc.

* From Washburn (1959).

The figures cover the most common capacities and indicate moderately well the general tendency for an increase in size. In nearly all cases the samples are so small and the difficulties of reconstruction are often so great that these approximate figures are adequate. The significance of the figures is essentially clear and would not be changed by any minor alterations. The important point is that size of brain, insofar as it can be measured by cranial capacity, has increased some threefold *subsequent to* the use and manufacture of implements. Erect, bipedal hominids, capable of making and employing tools, existed early in the Pleistocene when the brain was no larger than that of the existing apes. It can be maintained (if the gorilla is excluded from camparison because of great body size) that the australopithecines had slightly larger capacities on the average, but the difference, if any, is small. It could also be argued that if the australopithecines had been employing and manufacturing implements sufficiently long for the canine complex to be reduced by new selective pressures, then portions of the cerebral cortex may already have increased to some extent. However, in the absence of hominoid skulls of any sort from the upper Pliocene or earliest Pleistocene it is clearly impossible to ascertain certainly whether australopithecine cranial capacities were already augmented.

In the expansion of an apelike proto-hominid brain into a hominid brain, there was not equal expansion of the various parts. For example, the area of the cortex associated with the hand and thumb is greatly expanded compared to the bulk of the motor and sensory areas (Penfield and Rasmussen, 1952). This is a reflection in the cortex of motor skills related to tool use. In the same manner as altered selection after tool use favored shorter fingers and a larger, fully mobile thumb,[10]

[10] It has been argued (cf. Straus, 1949) that the hands of existing apes are too specialized and different in the size of the thumb and the pattern of both extrinsic and intrinsic musculature to provide an ancestral model for the human hand. The essence of this argument is the implicit assumption that the hands of the first implement-wielding hominids were essentially like that of anatomically-modern man. There are, of course, very marked differences among the existing African and Asian

so a much enlarged area of cortex was favored to guide the skilled hand of the implement maker. From the fossil record of the hominids it is evident that there were differential modifications in the form and proportions of the brain. Various functional areas, as delimited by modern investigations through electrical stimulation, were affected unequally and at different times, including those relating to motor skills (particularly the hand), patterns of memory, and the elaboration of thought. Notable in man is the extent of cortex which is associational compared with that which represents motor and sensory projection areas.

The evidence of the cerebral cortex strongly supports the conception that cortical expansion followed the acquisition of tool use. The areas of the cortex associated with human vocalization are very large and the control of the flow of speech is situated in various areas in the dominant hemisphere. Vallois (1955, p. 2108) notes that the brains of apes are essentially symmetrical while that of man is larger on the dominant side. This asymmetry is customarily ascribed to handedness, but hand preference is also present in monkeys with symmetrical brains. Hence it is possible that the asymmetrical condition is correlated with speech and its control by the dominant hemisphere (cf. also Anthony, 1958, and discussion thereafter). Should this prove to be the case, it might be possible to determine whether some of the early men whose skulls are well preserved were capable of speech. In any case it is clear that the reason a chimpanzee cannot learn to talk is that the necessary special cortical areas are either not present or not sufficiently differentiated. Those areas of the cortex associated with persistent motivation, memory, anticipation, and imagination are greatly expanded in the human brain. These abilities are essential to complicated social life. In the future it may become

---

apes in the proportions and morphology of the hand, including length and width of the hand, length (including free length) of the thumb, and also in degree of rotation of the thumb (cf. Schultz, 1936). There is considerable variation in the intrinsic volar thumb muscles, with a general tendency toward weak development or even absence of such muscles. It is well known that the long flexor tendon to the thumb is frequently deficient in orangs (95 per cent) and in gorillas (nearly 80 per cent), but this condition is seen in only about 50 per cent of the chimpanzees (Straus, 1940; 1942). Obviously, however, if the hominid hand evolved from that of an ape, it was from a primitive form and the hand undoubtedly was unlike that of existing species in a number of features. The very well-preserved hand of the early Miocene hominoid *Proconsul* is of considerable interest, since the morphology, including the motions permitted of the hand and wrist, was essentially lower catarrhine (Napier and Davies, 1959). Also, in the case of the primitive middle Miocene hylobatid, *Pliopithecus,* the hand skeleton differed fundamentally from that of the living, and obviously highly specialized, gibbon or siamang (H. Zapfe, 1958). The essential point is that once tools were used, the hand would have been modified by new selection pressures. It is consequently futile to look for human hands among pre-tool-using apes.

possible to demonstrate that this social brain is the outcome of new selection pressures which came with increasingly complex society. The general pattern of the human brain is in many ways like that of a chimpanzee or even a monkey; its uniqueness lies especially in its great size and in the enlargement of particular areas. From the immediate point of view this brain structure makes possible a complicated technical-social life; but from the long term evolutionary point of view, it was altered selection pressures of the new technical-social life which gave the brain its peculiar size and form (Washburn and Avis, 1958; Washburn, 1959).

At the time Darwin wrote there was great emphasis upon instinct. Subsequently, especially among sociologists and anthropologists, the main emphasis shifted to learning. It is now possible to see some of the interrelationships between learning and the structural base. A person can learn any language, but only a human being is capable of learning language. An ease of learning is built in, so to speak (no pun intended), and man can learn thousands of words as easily as a chimpanzee can learn a few sounds. Language must have been of such great importance to our distant ancestors that those capable of learning easily were greatly favored by selection, until learning ability became a distinctive human characteristic. However, this inherited ability in no way determines which language will be learned, and *both* inherited ability and learning are essential for speech, human nature is a product of man's evolution and man does easily the things for which he has been prepared by situations now long past. The crowded, industrialized world in which we live bears little relation to the sparsely populated world of ancient hunters and gatherers. Yet it was in that world in which early man evolved, and much of what men are required to do today is made difficult because selection has not yet had sufficient time to alter the biological base to make the learning easy.

Summary

If comparisons are made now between ape and man and if one speculates upon the reasons for the differences, much the same conclusions are arrived at as those reached by Darwin. The brain and the behavior which it makes possible are all-important. Erect bipedalism frees the hands and facilitates the use of tools. What is new and truly significant is the hominid fossil record that reveals something of the actual stages through which man has evolved and that affords an understanding which comes only from the history of the actual events. The uniqueness of modern man is seen as the result of a technical-social life which tripled the size of the brain, reduced the face, and modified many other structures of the body. When one looks upon the skull of *Homo sapiens,* with its great braincase and small face, one gazes upon the results of cultural and natural selection as they are ossified in the bone.

REFERENCES

ANTHONY, J. 1958. "La Réalisation du Cerveau Humain," pp. 79–88 in *Les Processus de l'Hominisation. Colloquis Internationaux,* Sciences Humaines: Paris: Centre National de la Recherche Scientifique.

BARTHOLOMEW, G. A., JR., and J. B. BIRDSELL. 1953. "Ecology and the Protohominids," *Amer. Anthrop.,* LV, 481–98.

BRAIN, C. K. 1958. *The Transvaal Ape-man-bearing Cave Deposits.* Memoir No. 11, Transvaal Museum, Pretoria.

BRAIN, C. K., C. VAN RIET LOWE, and R. A. DART. 1955. "Kafuan Stone Artifacts in the Post-Australopithecine Breccia at Makapansgat, *Nature,* CLXXV, 16–18.

BROOM, R., and J. T. ROBINSON. 1950. "Man Contemporaneous with the Swartkrans Ape-man," *Am. J. Phys. Anthrop.,* n.s. VIII, 151–55.

BROOM, R., and G. W. H. SCHEPERS. 1946. *The South African Fossil Apemen. The Australopithecinae.* Memoir No. 2, Transvaal Museum, Pretoria.

DART, R. A. 1926. "Taungs and Its Significance," *Natural History,* 315–27.

———. 1949. "The Predatory Implemental Technique of *Australopithecus,*" *Am. J. Phys. Anthrop.,* n.s. VII, 1–38.

———. 1953. "The Predatory Transition from Ape to Man," *International Anthrop. and Linguistic Rev.,* I, 201–218.

———. 1955. "Cultural Status of the South African Man-Apes," *Ann. Rep. Smithsonian Institution, 1955,* 317–38.

———. 1957. *The Osteodontokeratic Culture of Australopithecus prometheus.* Memoir No. 10, Transvaal Museum, Pretoria.

DARWIN, C. 1871. *The Descent of Man.* ("Modern Library") New York: Random House.

———. 1958. *Darwin's Century. Evolution and the Men Who Discovered It.* New York: Doubleday & Co.

HEBERER, G. 1949. "Die Unmittelbaren Vorfahren des *Homo sapiens,*" *Universitas,* IV, 1465–1477.

———. 1950. "Das Präsapiens-Problem," pp. 131–62 in *Moderne Biologie.* Berlin.

HOWELL, F. C. 1954. "Hominids, Pebble-tools and the African Villafranchian," *Amer. Anthrop.,* LVI, 378–86.

———. 1955. "The Pelvic Remains of *Australopithecus,*" *Am. J. Phys. Anthrop.,* n.s. XIII, 396. (Abstract).

———. 1959*a*. "The Villafranchian and Human Origins," *Science,* CXXX, 831–44.

HOWELL, F. C. 1959*b*. "European and Northwest African Middle Pleistocene Hominids," *Current Anthropology,* I, in press.

HUXLEY, T. H. 1863. *Evidence as to Man's Place in Nature*. London: Williams and Norgate.

KEITH, A. 1931. "New Discoveries Relating to the Antiquity of Man. London: Williams and Norgate.

LEAKEY, L. S. B. 1957. "Preliminary Report on a Chellean I Living Site at BK. II, Olduvai Gorge, Tanganyika Territory," pp. 217–18 in *Third Pan-African Congress on Prehistory, Livingstone, 1955,* J. D. CLARK, ed. London: Chatto & Windus.

———. 1958. Recent Discoveries at Olduvai Gorge, Tanganyika. *Nature,* CLXXXI, 1099–1103.

———. 1959*a*. "The Newly-discovered Skull from Olduvai: First Photographs of the Complete Skull," *Illustrated London News,* CCXXXV, 288–89.

———. 1959*b*. "A New Fossil Skull from Olduvai," *Nature,* CLXXXIV, 491–93.

LE GROS CLARK, W. E. 1952. "Hominid Characters of the Australopithecine Dentition," *J. Royal Anthrop. Inst.,* LXXX, 37–54.

———. 1955*a*. "The Os Innominatum of the Recent Pongidae with Special Reference to that of the Australopithecinae," *Am. J. Phys. Anthrop.,* n.s. XIII, 19–27.

———. 1955*b*. *The Fossil Evidence for Human Evolution.* Chicago: The University of Chicago Press.

MARKS, P. 1953. "Preliminary Note on the Discovery of a New Jaw of *Meganthropus* von Koenigswald in the Lower Middle Pleistocene of Sangiran, Central Java," *Indonesian J. for Nat. Science,* CIX, 26–33.

MEDNICK, L. W. 1955. "The Evolution of the Human Ilium," *Am. J. Phys. Anthrop.,* n.s. XIII, 203–16.

MOVIUS, H. L., JR. 1944. "Early man and Pleistocene stratigraphy in southern and eastern Asia." (Papers of the Peabody Museum of American Archaeology and Ethnology, XIX.) Cambridge: Harvard University Press.

———. 1948. "The Lower Palaeolithic cultures of Southern and Eastern Asia," *Trans. Am. Philos. Soc.,* n.s. XXXVIII, 329–420.

NAPIER, J. R. and P. R. DAVIES. 1959. *The Forelimb Skeleton and Associated Remains of Proconsul africanus* ("Fossil mammals of Africa," No. 16). London: British Museum of Natural History.

OAKLEY, K. P. 1954. "Skill as a Human Possession," pp. 1–37 in *History of Technology, Vol. I,* C. J. SINGER *et al.,* eds. Clarendon: Oxford University Press.

———. 1955. "Fire as Palaeolithic Tool and Weapon," *Proc. Prehist. Soc.,* XXI, 36–48.

———. 1956*a*. "The Earliest Took-makers," *Antiquity,* XXX, 4–8.

———. 1956*b*. "The Earliest Fire-makers," *Antiquity,* XXX, 102–107.

———. 1957. "Tools Makyth Man," *Antiquity,* XXXI, 199–209.

———. 1958. "L'utilisation du Feu par L'homme," pp. 135–48 in *Les*

Precessus de L'hominisation. Colloquis Internationaux, Sciences Humaines. Paris: Centre National de la Recherche Scientifique.

PENFIELD, W. and T. RASMUSSEN. 1952. *The cerebral cortex of man. A clinical study of localization of function.* New York: The Macmillan Co.

ROBINSON, J. T. 1953*a*. *Telanthropus* and its phylogenetic significance, *Am. J. Phys. Anthrop.,* n.s. XI, 445–501.

———. 1953*b*. *"Meganthropus,* Australopithecines and Hominids," *Am. J. Phys. Anthrop.,* n.s. XI, 1–38.

———. 1954*a*. "Prehominid Dentition and Hominid Evolution," *Evolution,* VIII, 324–34.

———. 1954*b*. "Phyletic Lines in the Prehominids," *Zts. fur Morphol. u. Anthrop.,* XLVI, 269–73.

———. 1954*c*. "The Genera and Species of the Australopithecinae," *Am. J. Phys. Anthrop.,* n.s. XII, 181–200.

———. 1955. "Further Remarks on the Relationship between 'Meganthropus' and Australopithecines," *Am. J. Phys. Anthrop.,* n.s. XIII, 429–45.

———. 1956. *The Dentition of the Australopithecinae.* (Transvaal Museum, Pretoria, Memoir No. 9.)

———. 1958. "The Sterkfontein Tool-maker," *The Leech,* XXVIII, 94–100.

———. 1959. "A Bone Implement from Sterkfontein," *Nature,* CLXXXIV, 583–85.

ROBINSON, J. T. and R. J. MASON. 1957. "Occurrence of Stone Artifacts with *Australopithecus* at Sterkfontein," *Nature,* CLXXX, 521–24.

SCHULTZ, A. H. 1936. "Characters Common to Higher Primates and Characters Specific for Man." *Human Biol.,* XI, 259–83; 425–55.

STRAUS, W. L., JR. 1940. "The Posture of the Great Ape Hand in Locomotion, and its Phylogenetic Implications," *Am. J. Phys. Anthrop.,* XXVII, 199–207.

———. 1942. "Rudimentary Digits in Primates," *Quart. Rev. Biol.,* XVII, 228–43.

———. 1949. "The Riddle of Man's Ancestry," *Quart. Rev. Biol.,* XXIV, 200–223.

VALLOIS, H. V. 1949. "L'origine de *l'Homo sapiens,*" *C. R. des séances de l'Acad. des Sciences, Paris,* CCXXVIII, 149–51.

———. 1954. "Neandertals and Praesapiens," *J. Royal Anthrop. Inst.,* LXXXIV, 111–30.

———. 1955. "Ordre des Primates," pp. 1854–2206 in *Traité de Zoologie. Anatomie, Systématique, Biologie.* Tome XVII. Paris: Masson et Cie.

———. 1958. "La grotte de Fontéchevade. 2e Partie. Anthropologie," *Archives de l'Inst. de Paléont. humaine, Paris,* XXIX, 1–164.

VON KOENIGSWALD, G. H. R. 1949. "The Discovery of Early Man in Java and Southern China," pp. 83–101 in *Early Man in the Far East,* W. W. HOWELLS, ed. (Studies in Physical Anthropology, No. 1.) Amer. Assoc. of Phys. Anthropologists.

WASHBURN, S. L. 1950. "The Analysis of Primate Evolution with Particular Reference to the Origin of Man," *Cold Spring Harbor Sym. on Quant. Biol.,* XV, 67–78.

WASHBURN, S. L. 1957. "Australopithecines: the Hunters or the Hunted?" *Amer. Anthrop.,* LIX, 612–14.

———. 1959. "Speculations on the Interrelations of the History of Tools and Biological Evolution," pp. 21–31 in *The Evolution of Man's Capacity for Culture,* arranged by J. N. SPUHLER. Detroit: Wayne State University Press.

WASHBURN, S. L. and VIRGINIA AVIS. 1958. "Evolution of Human Behavior," pp. 421–36 in *Behavior and Evolution,* A. ROE and G. G. SIMPSON, eds. New Haven: Yale University Press.

WEIDENREICH, F. 1913. "Uber das Huftbein und das Becken der Primaten und ihre Umformung durch den aufrechten Gang," *Anat. Anz.,* XLIV, 479–513.

———. 1941*a*. "The Extremity Bones of *Sinanthropus pekinensis,*" *Palaeontologia sinica,* n.s. D, No. 5 (Whole Series No. 116).

———. 1941*b*. "The Brain and its Role in the Phylogenetic Transformation of the Human Skull," *Trans. Amer. Philos. Soc.,* n.s. XXXI, 321–442.

———. 1947. "The Trend of Human Evolution," *Evolution,* I, 221–36.

———. 1948. "About the Morphological Character of the Australopithecine Skull, pp. 153–58 in *Robert Broom Commemorative Volume.* Cape Town: Royal Society of South Africa.

WYMER, J. 1958. "Further work at Swanscombe, Kent," *The Archeological News Letter,* VI, 190–91.

YERKES, R. M. 1943. *Chimpanzees: a Laboratory Colony*. New Haven: Yale University Press.

ZAPFE, H. 1958. "The Skeleton of *Pliopithecus* (*Epipliopithecus*) *vindobonensis* Zapfe and Hürzeler," *Am. J. Phys. Anthrop.,* n.s. XVI, 441–57.

ZUCKERMAN, S. 1951. *"An* Ape or *the* Ape?" *J. Royal Anthrop. Inst.,* LXXXI, 57–68.

———. 1954. "Correlation of Change in the Evolution of Higher Primates," pp. 301–52 in *Evolution as a process,* J. HUXLEY, A. C. HARDY and E. B. FORD, eds. London: George Allen and Unwin, Ltd.

CESARE EMILIANI

DATING HUMAN EVOLUTION

Ever since it became apparent that man did not originate by a sudden act of creation, but rather by the normal process of speciation, one of the primary goals of paleoanthropological research has been to date the various stages of human evolution. The different ways by which such dating may be attempted and the complex problems arising therefrom have been discussed by Oakley (1953). This author carefully distinguishes between relative and absolute dating methods, and he lists four different (and often complementary) methods under each one of these two categories. The four methods listed by Oakley for absolute dating of fossil hominids are the following: (1) direct age determination of the bones; (2) age determination of the source bed by determining the age of material (charcoal, shells) associated with the bones; (3) age determination of the source bed by correlating such bed with a deposit of known age; (4) age determination of the source bed by correlating such bed with certain geophysical parameters (the Milankovitch curve, for instance), under the assumption that these parameters have a bearing on the history of the Pleistocene. The fourth method is actually an integral part of the third, insofar as it has been used to establish Pleistocene chronologies or, in other words, to estimate the ages of deposits with which source beds are to be correlated. Its value is limited to the validity of the assumption upon which it is based.

The first and second methods have been applied extensively by means of radiocarbon techniques. I will not discuss radiocarbon here, which is a topic familiar to everybody today, but I will rather deal with current possibilities of dating fossil hominids beyond the range of radiocarbon. These possibilities fall largely under Oakley's methods 2 and 3.

Important steps in the evolutionary process leading to the develop-

CESARE EMILIANI teaches marine geology at the Marine Laboratory of the University of Miami while continuing his research on paleotemperatures and isotope geology. A native of Bologna, Italy, Dr. Emiliani came to the University of Chicago in 1948, where he earned his Ph.D. and became associated with the studies of Harold C. Urey.

ment of modern man took place between a million and fifty thousand years ago. Unfortunately, this time interval is perhaps the most difficult one to date. Suitable material formed at earlier and later times can be rather accurately dated by various methods based on radioactive isotopes, such as U^{238}, U^{235}, Th^{232}, Rb^{87}, K^{40}, and C^{14}. Twenty induced or secondary radioisotopes with half-lives ranging from 30,000 to 2×10^6 years are known. Some of these are artificial; others exist in nature, being formed either by the disintegration of long-lived parent isotopes or by nuclear bombardment in continuous natural processes; none is a leftover of the primordial formation of the elements (being too short-lived). In order for any of these radioisotopes to be useful for dating, it is necessary that either the radioisotope itself or its parent element be incorporated into the material to be dated at the time that this material was formed. It is also essential that no subsequent losses or additions have occurred. In the particular case of human evolution, the material most desirable to date is bone. Unfortunately, bone is physically and chemically very active, so that fossil bone is likely to be contaminated with all sorts of additions and losses. A recent attempt by Sackett (1958) to date bones using the U^{238}/Th^{230} ratio has given uncertain results, because uranium and thorium seem to migrate quite freely into the bone material.

If bone does not seem suitable for dating purposes beyond the range of C^{14}, other material closely associated with fossil bones or artifacts may be more appropriately used. Dating of shells occurring in caves (food refuse, using the U^{238}/Th^{230} or the Pa^{231}/Th^{230} ratios, seems feasible, and measurements should become available in the near future.[1] These two dating methods might be usable for shell material formed during the time interval between 10,000 and 200,000 years ago. For the time interval between 200,000 and 2×10^6 years ago it does not seem likely, at present, that suitable methods of dating human evolution *directly* will be developed in the near future. Man, however, developed during a period of the earth's history when repeated glaciations occurred, so that human and prehuman fossils which can be closely correlated with glacial and interglacial events may be indirectly dated if such events are dated.

Studies of Pleistocene history have been based mainly upon the continental record in glaciated areas and especially upon till and loess sheets, fossil soils, and terminal moraines. On that basis, Calvin

[1] A few measurements by Sackett (1958) indicate that modern shells, while containing varying amounts of uranium, may not contain thorium in measurable quantities. The presence of uranium and the absence of thorium in freshly deposited shells are, respectively, necessary and convenient for dating shells by the U^{238}/Th^{230} method.

(1897), working in the North American mid-continent, recognized five successive glaciations separated by four interglacials. In Europe the classic work of Penck and Brückner (1909) recognized four glaciations separated by three interglacials. The discrepancy between the European and North American classifications was believed eliminated when the Iowan stage of Calvin was reduced to a substage of the last glacial age, the classical Wisconsin. However, recent C^{14} measurements by Rubin (Ruhe, Rubin, and Scholtes, 1958) have shown that the Iowan deposits of Iowa are much older than the earliest classical Wisconsin deposits. It is now believed that, as Calvin proposed, five major glaciations occurred, with the last two glaciations (Early and Main Würm in European terminology, Early and Main Wisconsin in North American terminology) separated by a "cool" interglacial.

The above difficulty and many others encountered by geologists in their attempts to unravel the history of the Pleistocene epoch on the basis of the continental record stem from the fact that that record, especially in glaciated areas, is always discontinuous and fragmentary. A cursory look at the enormous literature is sufficient to make anyone marvel at both the complexity of the record and the ingenuity of the scholars who have coped with it. The task of these scholars would have been incomparably easier if some stratigraphic section covering the entire Pleistocene were available, showing, for instance, a complete sequence of alternating tills and soils. Unfortunately, such a section seem to be available nowhere in the glaciated areas.

In order to obtain a complete section of sediments covering the whole Pleistocene, one has to turn to deposits formed under water in periglacial or non-glacial areas or in the ocean. Complete sections of Pleistocene deposits may exist on the bottom of certain lakes, such as, for instance, the lakes of northern Florida. Coring of such lake beds and pollen analysis of the cores may yield important information on the climatic changes caused by glaciation, and, if the sections of lake deposits are complete, the climatic pattern of the Pleistocene might be reconstructed rather easily. Work along these lines has been and is being done by, among others, Sears and Clisby (1955), Clisby, Foreman, and Sears (1957), and Maarleveld and Van der Hammen (1959) on the lake beds of Mexico City, San Augustin Plains (New Mexico), and the Sabana de Bogotá (Colombia).

While lake sediments offer interesting opportunities, the best sections of continuous Pleistocene sediments have been found on the ocean floor. About 40 per cent of the ocean floor is carpeted with a type of sediment called "*Globigerina* ooze." This sediment consists essentially of clay with imbedded test of planktonic protophyta and

Foraminifera, the latter representing 30 per cent or more of the dry weight of the bulk sediment. *Globigerina* ooze accumulates on the ocean floor at rates ranging from a few to several centimeters per thousand years. Following the development of the piston corer by Kullenberg (1947), deep-sea sediments can be sampled as far down as 20 meters below the surface of the ocean floor. Thus *Globigerina*-ooze sections covering the whole Pleistocene time have become available in recent years. The study of these sections, especially by the modern method of oxygen isotopic analysis developed by Urey and co-workers (Urey *et al.*, 1951; Epstein *et al.*, 1951, 1953), has shown that the surface water of the ocean underwent numerous and apparently periodic temperature variations in the recent past. These temperature oscillations are best evidenced by deep-sea cores from the equatorial and North Atlantic and adjacent seas (Emiliana, 1955*a, b,* 1958), a probable result of the fact that the Atlantic Ocean, being surrounded by large ice sheets at its northern end during the glacial ages, was subjected to more marked temperature variations than were the other oceans (cf. Emiliani and Geiss, 1959). The temperature curves obtained from various deep-sea cores were combined into a single, generalized temperature curve (Fig. 1), which is believed to have more than regional significance. Radiocarbon measurements on the upper portions of some of the deep-sea cores (Rubin and Suess, 1955, 1956) have proved that the last temperature minimum of the deep-sea cores corresponded in time to the Wisconsin glaciation. Extrapolations of the radiocarbon data, together with inferences from ionium-radium data by Urry (1949) on other deep-sea cores, provided a basis for a tentative time scale of the temperature variations (Fig. 1). According to this time scale, the Early Würm glaciation occurred about 65,000 years ago. This figure has been substantiated recently by a radiocarbon date from the Groningen laboratory (Haring *et al.*, 1958). A reliable, absolute time scale for the deep-sea stratigraphy back to 200,000 years ago, based on the changing Pa^{231}/Th^{230} ratio (Rosholt, 1957; Sackett, 1958), should become available in the very near future. Preliminary measurements by Rosholt (personal communication) appear to substantiate the time scale of Figure 1. Altogether, one may presently venture to say that the time scale of Figure 1 is probably correct within 10 per cent all the way back to 300,000 years ago.

The older portion of the generalized temperature curve is based on two deep-sea cores from the Caribbean, both of which are less than 10 meters long. Longer, undisturbed cores of *Globigerina*-ooze facies are not yet available from the Atlantic and adjacent seas, so that a clear picture of the temperature variations before 300,000 years ago

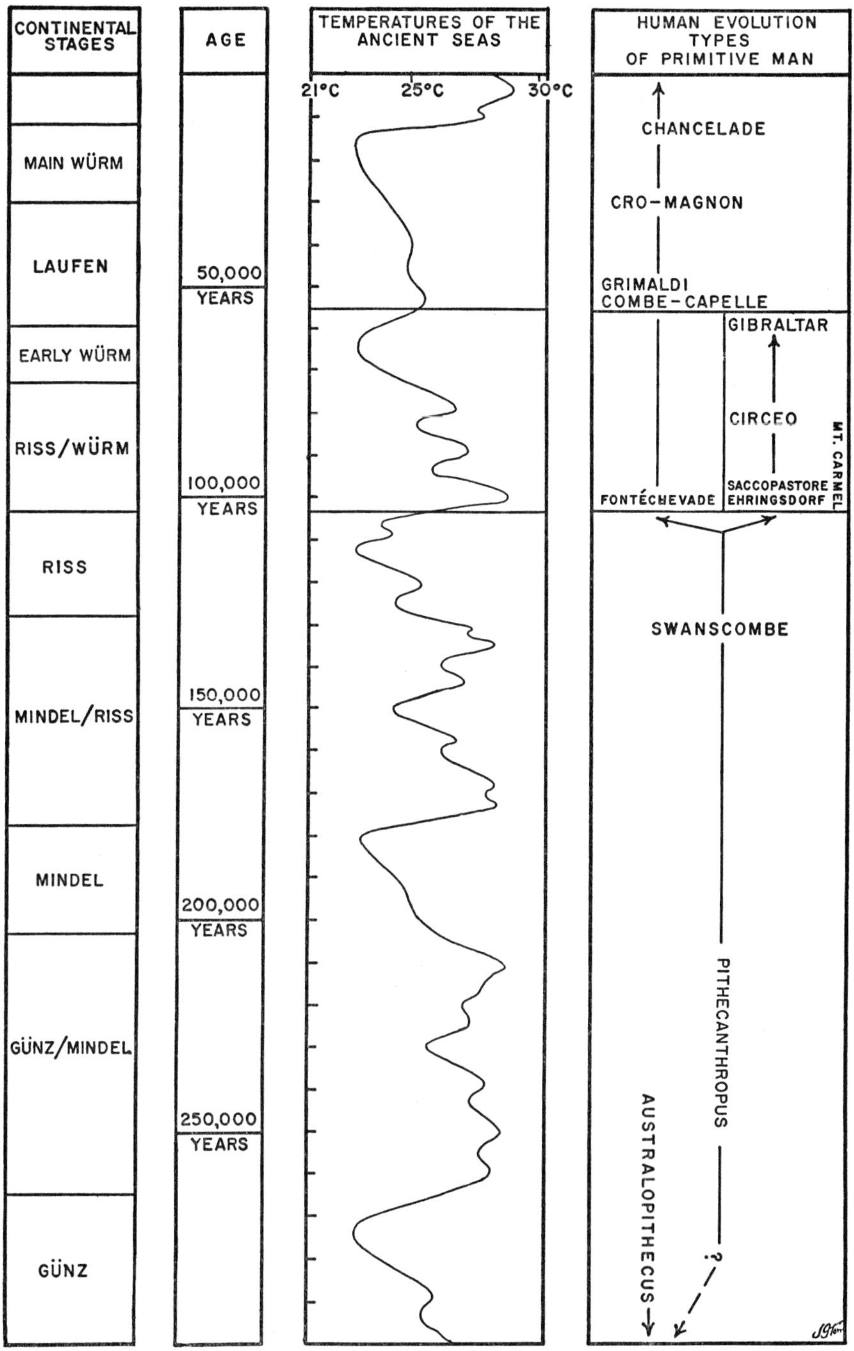

FIG. 1.—Time scale of temperature variation in the Pleistocene, based on the method of analyzing deep-sea sediment. Tentative dates are then attributed to the human and prehuman fossils according to the glacial and interglacial events with which they are associated.

cannot yet be presented. Some deep-sea cores from the Pacific Ocean extend back in time much more than 300,000 years. The surface temperature variations of the Pacific Ocean, however, were much smaller than those of the Atlantic and adjacent seas, so that the Pacific cores are not so suitable for a clear reconstruction of the temperature variations during the Pleistocene epoch. In spite of this difficulty, isotopic analysis of a long core from the eastern equatorial Pacific (core 58 of the Swedish Deep-Sea Expedition, 1947–48) clearly revealed a broad trend of decreasing temperature from the Late Pliocene into the Early Pleistocene (Emiliani, 1955*a,* Fig. 11).

The percentage of $CaCO_3$ varies remarkably throughout the Pacific core in question, as well as in other cores from the same general area (Arrhenius, 1952). This variation is believed to have been caused, ultimately, by climatic changes. In particular, high-carbonate layers are thought to have been deposited during glacial ages, and low-carbonate layers during interglacial ages (Arrhenius, 1952; cf. also Emiliani, 1955*a*). Since the temperature cycles revealed by the deep-sea cores from the Atlantic and adjacent seas apparently lasted about 40,000 years each, a similar duration would obtain for each carbonate cycle of the Pacific cores. The Plio-Pleistocene boundary was tentatively placed at about 610 cm. below the top of core 58, because this level immediately follows a high-temperature level and immediately precedes a marked carbonate maximum (Emiliani, 1955*a*). About 15 carbonate cycles occur above this level, so that a duration of about 600,000 years may be calculated for the whole Pleistocene epoch. Officially, however, the Plio-Pleistocene boundary is not defined on the basis of the deep-sea sea stratigraphy, but on the Late Cenozoic stratigraphy of Italy (International Geological Congress, 1950). Specifically, this boundary is placed at the time when certain northern species of marine invertebrates entered the Mediterranean. It is not known at present whether this event corresponds in time to the level of core 58 mentioned above, so that no estimate can be given for the age of the officially established Plio-Pleistocene boundary.

As previously mentioned, the temperature variations of the Pleistocene are best studied in the deep-sea cores from the Atlantic and adjacent seas. The stratigraphic record of these deep-sea cores has been divided in stages identified by positive integers, following a system introduced by Arrhenius (1952). No. 1 is the present, high-temperature age; No. 2 is the preceding low-temperature age; No. 3 is the high-temperature age preceding age 2; etc. Thus odd numbers refer to high-temperature ages and stages, even numbers to low-temperature ages and stages. A tentative correlation between temperature stages and the glacial and interglacial stages of the continents is shown in Figure 1. This correlation is supported by C^{14} age measurements on both continental and pelagic material, reaching back to about 70,000 years and, therefore, including the Early Würm glaciation. A reasonable extrapolation, together with a comparison between the generalized temperature curve of the deep-sea cores (Fig. 1) and the continental record worked out by Brandtner in Austria and Moravia (Brandtner, 1954, 1956, and personal communication), strongly suggest that stage 5 is the last interglacial (Riss/Early Würm), and stage 6 at least part of the Riss glaciation. The correlations for older stages are more uncertain. If the suggested correlation is correct all the way back to stage 14, the Günz glaciation would be dated at about 280,000 years ago. Although this age may seem relatively young to Pleistocene scientists accustomed to older estimates which placed the Günz glaciation at about 600,000 years ago (cf. Zeuner, 1945, 1952), it is nevertheless in agreement with such older estimates if these are corrected for the now generally accepted shorter duration of the postglacial time (Emiliani, 1955*a*).

Although neither the chronology of the deep-sea deposits nor their correlation with the glacial and interglacial events of the Pleistocene epoch are yet well established beyond about 100,000 years ago, that is, beyond the last interglacial, the chronology and correlations of older deep-sea deposits afford time estimates for continental events which are undoubtedly superior to previous estimates.

Some important human and prehuman fossils and groups of fossils are rather surely correlated with glacial and interglacial stages of the continental stratigraphy. These, as discussed above, may be correlated, with various degrees of confidence, with the stages of the deep-sea deposits. The latter can be dated, or ages can be reasonably estimated. The above relationship results in estimates of the ages of selected human and prehuman fossils (Emiliani, 1956). Although this method of dating human evolution may seem rather devious, it nevertheless provides age estimates which, at the present time, would be otherwise impossible to make. This method is based on the following three requirements: (1) the stages of the deep-sea stratigraphy must be dated; (2) the stages of the continental glacial stratigraphy must be correlated with the deep-sea stratigraphy; (3) the fossil hominids to be dated must be correlated with the continental glacial stages.

The first requirement is strictly fulfilled only back to 70,000 years ago, by radiocarbon measurements. Older deep-sea deposits are tentatively dated by extrapolations and inferences, as discussed above. The second requirement is strictly met again only back to 70,000 years ago, by radiocarbon measurements. Beyond this age, the work by Brandtner (1954, 1956, and personal communication) on the loess and soil profiles of Austria and Moravia rather strongly suggest that the correlation shown in Figure 1 is probably correct at least back to the Riss glaciation. The third requirement is met only by the selected fossils shown in Figure 1. Some fossil hominids, in fact, have not yet been surely correlated with given stages of the Pleistocene continental stratigraphy. These are omitted from Figure 1.

Some interesting, though tentative, conclusions may be drawn from Figure 1. Modern man (*Homo sapiens sapiens*) appears already to have been present on the earth 50,000 years ago. The Fontéchevade remains are about 100,000 years old, and the Swanscombe skull bones about 125,000 years old. If these are assigned to an ancestral subspecies of *H. sapiens sapiens* (cf. Vallois, 1958), the latter appears to have originated between 100,000 and 50,000 years ago. *Pithecanthropus* extended through the Günz/Mindel interglacial. If *Pithecanthropus* is ancestral to *Homo,* the first Homininae may have appeared sometime between 200,000 and 125,000 years ago. The older limit is uncertain, however, because the early Homininae may have been contemporaneous with the late Pithecanthropinae. A similar relationship exists between the Australopithecinae and the Pithecanthropinae. If the former are ancestral to the latter, speciation might have occurred substantially earlier than the time when the late Australopithecinae were in existence.

Homo sapiens neanderthalensis may have originated between 125,000 and 100,000 years ago. In any case, the Neanderthals were already in existence 100,000 years ago. Recent paleotemperature analyses of marine shells associated with the neanderthaloid jaw discovered in the deep layers of the Haua Fteah cave of Cyrenaica (McBurney, Trevor, and Wells, 1953) have unmistakably given fully interglacial temperatures, which result in indirect dating of the jaw at about 100,000 years. This date is probably correct within 10,000 years, because the high temperatures shown by the shells existed in the Mediterranean for only a relatively short time around 100,000 years ago. The Neanderthals appear to have existed for only about 50,000 years, a time interval which may represent a total of only about 2,000 generations.

The Pleistocene glaciations undoubtedly had a profound significance for the evolution of man. One may even venture the opinion that, if glaciations had not occurred, modern man might have failed to develop. *Pithecanthropus,* which was already in existence before the beginning of the major Pleistocene glaciations, was, in fact, a remarkably evolutionary type. He was apparently quite capable of taking care of himself and of disposing suitably of his enemies. If glaciations had not created uncomfortable conditions, *Pithecanthropus* might have developed into a large, interbreeding population, occupying the whole earth. Glaciation favored migrations and the splitting-up of populations into small groups. Conditions for efficient evolutionary processes were thus created. These resulted, ultimately, in the production of modern man. Today, modern man is a very large, interbreeding population. No further evolution seems likely, except by artificial means or by isolation on planets outside the solar system.

Bibliography

ARRHENIUS, G. 1952. *Sediment Cores from the East Pacific.* Repts. Swedish Deep-Sea Exped. 1947–1948, Vol. V, Fasc. 1.

BRANDTNER, F. 1954. "Jungpleistozäner Löss und fossile Böden in Niederösterreich," *Eiszeitalter u. Gegenwart,* IV–V, 48–92.

———. 1956. "Lössstratigraphie und paläolitische Kulturabfolge in Niederösterreich und in den angrenzenden Gebieten," *ibid.,* VII, 127–75.

CALVIN, S. 1897. "Synopsis of the Drift Deposits of Iowa," *Amer. Geologist,* XIX, 270–72.

CLISBY, K. H., FOREMAN, F., and SEARS, P. B. 1957. "Pleistocene Climatic Changes in New Mexico, U.S.A.," *Veröff. Geobot. Inst. Rübel, Zürich,* XXXIV, 21–26.

EMILIANI, C. 1955*a*. "Pleistocene Temperatures," *Jour. Geol.,* LXIII, 538–78.

———. 1955*b*. "Pleistocene Temperature Variations in the Mediterranean," *Quaternaria,* II, 87–98.

———. 1956. "Note on Absolute Chronology of Human Evolution," *Science,* CXXIII, 924–26.

———. 1958. "Paleotemperature Analysis of Core 280 and Pleistocene Correlations," *Jour. Geol.,* LXVI, 264–75.

EMILIANI, C., and GEISS, J. 1959. "On Glaciations and Their Causes," *Geol. Rundschau,* XLVI (1957), 576–601.

EPSTEIN, S., BUCHSBAUM, R., LOWENSTAM, H., and UREY, H. C. 1951. "Carbonate-Water Isotopic Temperature Scale," *Bull. Geol. Soc. America,* LXII, 417–25.

———. 1953. "Revised Carbonate-Water Isotopic Temperature Scale," *ibid.,* LXIV, 1315–25.

HARING, A., DE VRIES, A. E., and DE VRIES, H. 1958. "Radiocarbon Dating Up to 70,000 Years by Isotopic Enrichment," *Science,* CXXVIII, 472–73.

INTERNATIONAL GEOLOGICAL CONGRESS, 18TH SESSION, GREAT BRITAIN. 1948. *Recommendations of Commission Appointed to Advise on the Definition of Pliocene-Pleistocene Boundary,* Rept., Part 9, p. 6. London, 1950.

KULLENBERG, B. 1947. "The Piston Core Sampler," *Svensk. Hydrol. Biol. Komm., Skr.,* Tredje Ser., Hydr., Vol. I, Part 2.

Maarleveld, G. G., and van der Hammen, Th. 1959. "The Correlation between Upper Pleistocene Pluvial and Glacial Stages," *Geol. en Mijnbouw,* n.s., XXI, 40–45.

McBurney, C. B. M., Trevor, J. C., and Wells, L. H. 1953. "The Haua Fteah Fossil Jaw, *Jour. Roy. Anthropol. Inst.,* LXXXIII, 71–85.

Penck, A., and Brückner, E. 1909. *Die Alpen im Eiszeitalter.* Leipzig: Tauchnitz.

Rosholt, J. N., Jr. 1957. "Quantitative Radiochemical Methods for Determination of the Sources of Natural Radioactivity," *Anal. Chem.,* XXIX, 1398–1408.

Rubin, M., and Suess, H. E. 1955. "U.S. Geological Survey Radiocarbon Dates. II," *Science,* CXXI, 481–88.

———. 1956. "U.S. Geological Survey Radiocarbon Dates. III," *ibid.,* CXXIII, 442–48.

Ruhe, R. V., Rubin, M., and Scholtes, W. H. 1958. "Late Pleistocene Radiocarbon Chronology in Iowa," *Amer. Jour. Sci.,* CCLV, 671–89.

Sackett, W. M. 1958. "Ionium-Uranium Ratios in Marine Deposited Calcium Carbonates and Related Materials." Thesis, Washington University, St. Louis, Mo.

Sears, P. B., and Clisby, K. H. 1955. "Pleistocene Climate in Mexico," *Bull. Geol. Soc. America,* LVI, 521–30.

Urey, H. C., Lowenstam, H. A., Epstein, S., and McKinney, C. R. 1951. "Measurement of Paleotemperatures and Temperatures of the Upper Cretaceous of England, Denmark, and the Southeastern United States," *Bull. Geol. Soc. America,* LXII, 399–416.

Urry, W. D. 1949. "Radioactivity of Ocean Sediments. VI. Concentration of Radioelements in Marine Sediments of the Southern Hemisphere," *Amer. Jour. Sci.,* CCXLVII, 257–75.

Vallois, H. V. 1958. *La Grotte de Fontéchevade.* Part II: *Anthropologie.* (Inst. Pal. Humaine, Arch., Mem. 29.)

EDGAR ANDERSON

THE EVOLUTION OF DOMESTICATION

There is a paradoxical inverse relation between the studies of evolution and of domestication in 1859 and in 1959. Darwin collected facts about domestication to buttress his studies of evolution. His monumental *Variation of Plants and Animals Under Domestication,* written in 1868, led directly to such precursors of genetics as Bateson's *Materials for the Study of Variation* (1894) and de Vries's *Species and Varieties: Their Origins by Mutation* (1905). Our resulting understanding of variation has been so illuminated as to produce a whole corpus of experimental and observational data dealing with the processes of evolution. This understanding now illuminates in *its* turn the problem of domestication from which Darwin drew his facts and insights.

Unlike many of his colleagues and disciples, Darwin realized increasingly that variation was more than a simple axiom. To him, as to us, it was a fit subject for study and experiment. A century later we are coming to realize that domestication itself must also be patiently analyzed by scientific methods. Genetics has greatly clarified our understanding of variation; the applications of genetics to the biological problems of systematics have provided us with a whole body of analytical techniques pertinent to the study of domestication. What are the differences between germ plasms of a cultivated plant and its wild precursors? We do not yet have an answer for a single domesticated plant nor for any animal. A number of critical investigations are actually under way, a few have been partly completed. As a whole they tell us little more than that the problems of domestication, seen in the light of genetics in 1959, are as much of a challenge and an opportunity as were the problems of variation in 1859.

Let us begin with a short survey of pertinent information derived from some of these modern studies.

EDGAR ANDERSON is Curator of Useful Plants at the Missouri Botanical Garden and Engelmann Professor of Botany at Washington University, St. Louis. His work in plant genetics has earned him many honors, including the Linnean Society's Darwin-Wallace Medal in 1958. He is a past president of the Botanical Society (1952) and of the Society for the Study of Evolution (1958). His major publications include *Introgressive Hybridization* (Wiley, 1949) and *Leaves, Stems, and Measurements* (Michigan, 1960).

Some Recent Findings in Major and Minor Crops

WHEAT

The cytological discoveries of Kihara (1924) and Sax (1922) opened up the entire field of the origins and relationships of the world's wheats. We now know that (1) wheats are polyphyletic, and (2) they come from crosses involving at least three species and at least two genera.

The major crosses took place at different times and in different places. Their evolution under domestication is a process rather than an event, a process whose important early stages are still matters for experiment. Kihara, one of the world's greatest cytogeneticists, has spent a lifetime of basic research on the problem, as have other able workers in Russia, Western Europe, North America, and Mexico. While the major outlines of wheat classification are now clear, resting in part on an understanding of germinal architecture at the microscopic and submicroscopic levels, there are important portions of their classifications (not to mention their history) that are still frankly provisional. Conclusions as to the origins of agriculture based on little more than published knowledge of the cereal grains and their histories will be obsolete before they are published.

SUGAR CANE

The pioneering studies of Venkatraman (1938) and his colleagues, followed by Parthasarathy's brilliant cytological analyses in 1948, have revolutionized our notions of the origin of sugar cane and the nature of the problem. It is now established that the so-called "Noble canes," frequently assumed to be the wild relatives from which the cultivated strains arose, are one of the results of domestication rather than a contributing cause. Recently Grassl (1959) has been able to demonstrate experimentally that they probably arose from an early domesticated sugar cane that was carried to New Guinea, hybridizing there with a giant reed which was being grown as a protective stockade for houses and gardens. It is now certain that the sugar canes are a vast, highly-polyploid complex involving several genera of grasses. Their domestication has been going on gradually for millennia and has involved millions of square miles. It is significant that when the veracity of Venkatraman's reports of intergeneric hybrids with sugar canes was still being debated by scholars, hard-headed businessmen in sugar companies were already trying them out in breeding programs as a possible source of disease resistance.

POTATOES

Vavilov and his colleagues (reported in Bukasov, 1930) and Hawkes in 1944 showed that our cultivated potatoes spring from a vast polyploid network of wild, weed, domesticated, and feral species and races, stretching from the southern Rocky Mountains through Mexico and Central America to the highlands of South America. Scores of species in this complex have been given valid botanical names, yet even on the basic level of classification, though general outlines are clear, much is still frankly provisional.

COTTON

Brilliantly investigated by Hutchinson (1947) and his colleagues for a quarter-century and by other able research groups in North Carolina and Texas, cottons are now known to be polyphyletic, involving both European and New World species. Their evolution under domestication is a process extending over thousands of years and perhaps involving other genera.

MAIZE

The problem of maize classification and history was reopened (Anderson and Cutler, 1942) by the demonstration that nineteenth-century classifications of maize were provisional, incomplete, and largely artificial. A few acrimonious publications in this field had obscured far more important facts, but in the past twenty years our knowledge of maize has been revolutionized (Mangelsdorf and Reeves, 1959).

The disputes are largely about details. There is general agreement on most of the main points and increasing reliance on the special techniques developed for executing the studies.

By international cooperation—involving the governments of Mexico, Colombia, and the United States, the Rockefeller Foundation, and experts from Brazil, Peru, and several American laboratories—the multitudinous strains of maize in the New World have been collected, stored under refrigeration, studied, and provisionally classified. The enormous technical labor necessary is being supplied largely by Mexican and Colombian agronomists and to a lesser extent by commercial corn-breeding companies. More than a dozen monographs in Spanish and English have dealt with outlines of the classification (Brown, 1953; Brieger *et al.*, 1958; Hathaway, 1957; Roberts *et al.*, 1957; Wellhausen *et al.*, 1952, 1957). Related studies have been started in Japan on Asiatic varieties of maize and have been in part published monographically (Kihara, 1956).

It is now documented, and generally agreed, that the evolution of maize is a process continuing over some thousands of years and almost certainly involving more than one genus.

OTHER CROP PLANTS

There have been other significant studies of major crop plants: the work on the cultivated and weed hot peppers (*Capsicum*) (Heiser and Smith, 1953); the patient monographic studies of New World squashes and pumpkins, both archeological and modern (Cutler and Whitaker, 1956, and in press; Whitaker and Bird, 1949; Whitaker *et al.*, 1957); and studies being prepared for publication at the moment—on the cultivated tomato (Rick) and a survey of the varieties of manioc (Roger). This work has enlarged our understanding of these crop plants and of the people who grew them. Some of the most illuminating studies have involved the taxonomy, history, and genetics of crops of minor economic importance. Heiser (1955), studying the evolution of wild and cultivated sunflowers (*Helianthus annuus, sensu latiore*), has demonstrated among other things that domestication is a long-continued process in which cultivated strains have repeatedly entered into the ancestry of the weed and "camp-follower" races, just as these have played important roles (note the plural) in the development of the cultivated sorts. He has demonstrated that the wild annual species, the weeds, the "camp-followers," the cultivated seed and oil varieties, the ornamental varieties, all form a complicated interwoven network of descent and evolution involving a millennium, at the very least, covering most of the southern United States, with important and as yet little known chapters in northwestern Mexico.

Jonathan Sauer (1959), working with the ancient grain amaranths, an almost forgotten set of crop plants of ancient high cultures in the New World, revolutionized our understanding of these important plants. From pre-Columbian and early post-Columbian documents he demonstrated that they were closely associated with Aztec, Mayan, Incan, and other New World cultures as ceremonial plants, as well as important food source. Working with their classification, he subjected their spiny-chaffy inflorescences (they are close kin to the hairy pigweeds) to minute and detailed microscopic examination. Through this laborious process, he was able to demonstrate (1) that they are closely related to the New World weed amaranths and (2) that two species had arisen in cultivation, one centered in the Inca area, the other in the valley of Mexico, with variants in southern Mexico and the Mayan areas.

With this background he turned his attention to the Asiatic grain amaranths which for some centuries have been a staple crop in Kashmir and other Himalayan borderlands. Though they have been commonly assumed to stem from Asiatic weed amaranths, Sauer demonstrated beyond all reasonable doubt that they are identical with the cultivated species developed in the Aztec and Inca areas and still grown there as minor crops. His work was done with such painstaking attention to all relevant detail (including a comprehensive technical monograph on the New World weed amaranths) that even E. D. Merrill (1954) went out of his way to give it a clean bill of health in one of his characteristic diatribes against all published evidence for pre-Columbian contacts between the Old World and the New. The bulk of Jonathan Sauer's researches has been published, but the bearing of his conclusions will be more apparent when he publishes his work on modern collections made among African and Asiatic peoples.

Such modern monographs of crop plants as those just cited have demonstrated that domestication is a different problem than it was previously said to be. Cytogenetics (and its god-child, biosystematics) have enabled us to cut some of the brush out of the way, but these techniques are laborious and time-consuming, and many crops of key historical importance still await detailed study. For the present, one who has been probing in this area can do little more than indicate the dimensions of the problem of domestication, outlining kinds of evidence that might lead to helpful insights, suggesting the kinds of questions we should be preparing to ask ourselves. Let us first put on record some guideposts to future advance that are justified by the studies of the past two decades.

Reasonably Well-Established Hypotheses

THE ORIGIN OF A CULTIVATED PLANT IS A PROCESS, NOT AN EVENT

Multiple origins are already known with certainty for some crop plants. Reticulate relationships with a crop's precursors and with the weeds derived from it are provisionally documented. Reticulate relationships between substrains are the rule rather than the exception. There is a distinct possibility that this kind of continuing complexity may be the norm rather than the exception among cultivated plants.

WEEDS AND CULTIVATED PLANTS

The weed-to-cultivated-plant relationship is no longer something that can be taken for granted (Anderson, 1952). Pointing to a closely related weed as the probable precursor of a cultivated plant poses two additional problems: (1) the origin of the weed and (2) the nature of the relationships between the weed and the plant; this does not eliminate the original problem of the origin of the cultivated plant. We have already documented cases of weeds which arose from cultivated species and vice versa. In lettuce we can point to present-day use of weed lettuce in breeding new cultivated varieties as well as to frequent introgression from cultivated varieties into weed lettuces, greatly increasing *their* variability.

THE EARLIEST AGRICULTURES ARE NOT EUROPEAN

We know that agriculture did not originate in Europe. Therefore, we should deliberately avoid looking at the problem from a European point of view. A clean-crop agriculture, relying largely on cereals and domesticated animals, is (considering the various agricultural patterns of the world) a specialized system. Agriculture most probably came from the tropics or subtropics. People from these areas who have had least contact with Europeans grow plants in groupings for which I know of no names in European languages.

One of the commonest patterns is a plot immediately adjacent to the house in which are grown a variety of plants used for a variety of purposes. In Asia, Africa, and Latin America such a plot will be protected with a fence or stockade (frequently of living plants). It will contain trees, shrubs, vines, and annuals. It will be simultaneously an orchard, a vegetable garden, a dump-heap-compost-heap, and a medicinal garden. Maize and sorghum will frequently be raised there; other cereals rarely if at all. More often than not, few or none of the plants will be in straight rows. I have mapped examples of such plots in Costa Rica, in Guatemala (Anderson, 1950), in Honduras (1954), in Colombia, Ethiopia, and India. In none were more than a bare majority of the plants in rows. Planting and harvesting (insofar as climate permitted) were continuous processes running throughout the year. The plants might be grouped as actively discouraged weeds (very few of these!), permitted weeds, encouraged weeds, and cultivated plants. There might be fruits and vegetables, seed and root staples, and plants with uses other than nourishment: fiber plants (including thatch and plants for brooms); utensil-producers such as the calabash tree, gourds, etc.; poisons, stimulants, vermifuges, and other drugs; condiments; cosmetics; plants used in brewing to help sterilize utensils; and plants used in ritual.

In the plots that I have sampled, the most curious and most variable feature has been the ornamentals. There are either (1) many flowers and other ornamentals, even when the people are poor and there is no market for the blooms, or (2) there are virtually none at all, even though—as in Ethiopia—the soil and climate are excellent and the people intelligent, highly-skilled agriculturists. As we shall see below, these two extreme types exhibit very definite distribution patterns for the world as a whole and show obvious ethnic affinities.

If we do not even have names in European languages for such plots and their subtypes, we are scarcely ready to talk intelligently about—or even plan intelligently to study—the origins of domestication. We need first of all to know what kinds of concepts are applied to such plots by their owners. Malinowski (1935) and his followers have shed much light on such problems but, so far as I am aware, have taken very much for granted either the presence or absence of flowers and ornamentals. Since in my experience this is, in both hemispheres, the greatest single variable in the whole phenomenon, we are greatly in need of scrupulously careful field work and analysis.

CROPS WITH MULTIPLE USES HAVE SPECIAL SIGNIFICANCE

The longer a crop has been in cultivation, everything else being equal, the more likely it is that various *kinds* of uses will have been found for it. Under these conditions we may expect it to include special strains bred for particular uses. Sorghums, for instance, include annual forage grasses, perennial forage grasses, syrup sorghums, grain sorghums, broomcorn, even popping sorghums used in making confections. Turmeric, which the Western world knows only as a condiment, is widely used as a body paint, a depilatory, in ritual, and as a food color. Hemp has specialized varieties for fiber, for drugs (marijuana and hashish), and for oil seeds, as well as rampant weedy forms. Squashes and pumpkins are used for rattles, dishes, oil seeds, ornaments, vegetables, and confections. It is probable that all three of these groups (sorghums, squashes, and turmeric) have been in cultivation much longer than our major cereals.

CLASSIFICATION OF A CULTIVATED PLANT INVOLVES SUCCESSIVE APPROXIMATIONS

An estimate in the light of all the knowledge which can be brought to bear on the subject is the best we can do today. With data collected as a result of new insights gained from this estimate, we can make a better one tomorrow and so advance through a series of increasingly accurate approximations. While in general this is somewhat true of monographs of a wild-growing species, the taxonomic complexity of classifying cultivated plants is of another order of difficulty from those of our wild-growing species. The degree to which classifications of cultivated plants must proceed iteratively, and on the whole slowly, toward a final goal has not been generally appreciated even by those with monographic experience with wild plants.

Problems for Investigation

EARLIEST DOMESTICATION OF NON-FOOD PLANTS

Primitive peoples in various parts of the world use plants for body paint, for living stockades to protect themselves from each other and from wild animals, as poisons, for chewing, for fatigue drugs, and for purposes of ritual. I would suppose that most if not all these usages might have been characteristic of very early man. Certainly living stockades, or plants which take root readily even under barbaric treatment and which are useful in providing privacy and security, are widely used among primitive peoples. Some of our most charming tropical ornamentals probably originated in this way and have a long history of domestication. Body paints seem to be very ancient; when we look into the cultivated plants which are used as body paint, we find plants such as the turmerics and their relatives, which on quite different considerations must have been domesticated very early.

I have only a few observations of my own with regard to the antiquity of coffee as a fatigue drug. All known sources of caffeine were discovered by primitive peoples. None have been added by biochemical and botanical surveys. How early in man's history did he start cultivating such plants as coffee? Before he started cultivating coffee, how much earlier and in what ways did African man first begin to affect the course of the evolution of these fatigue-relieving plants? Though they most certainly have been used in some ways by prehominids,* at what point did this begin to affect the population dynamics of wild coffees? How soon did this induce disturbances of their barrier systems, so that stronger introgression between species became common and laid the foundations for the origins of polyploid proto-domesticates or semi-domesticates? How soon in his career would man have started to play a role in the spread of these living artifacts? These are the kinds of questions for which we may hope to find answers, now that we see the importance of the questions.

Before we can get into any such position of authority, we shall have to work hard at a real understanding of the wild, semi-wild, and cultivated coffees of Africa. The basic taxonomic work has not yet been finished. There is not a modern monograph of the genus nor any portion of it. We now have only the best guess which can be made, in the light of the haphazard and fragmentary collections in European and American herbaria. Cytology has given us new insights in this

* In Ethiopia monkeys are one of the most difficult plagues to deal with in the coffee forests. Since they eat the berries and since the Ethiopian coffees are somewhat weedy, monkeys may well be one of the agents which help to spread the plant.

genus as in many others. We now know that both *Coffea arabica* and *Coffea robusta* are tetraploids. Rauk has recently established the fact that the semi-wild coffees of Ethiopia are tetraploids. The diploids from which these tetraploids most probably arose are clustered in central Africa. When did the tetraploids make their way northward, and in what ways was the process affected by man and by prehuman hominids?

FIELD-SAMPLE STUDY OF SMALL GRAINS

Early in the maize program, referred to elsewhere in this essay, it became apparent that to understand this crop effectively we must, as in modern field genetics, take thc interbreeding group as a unit. Accordingly field-sample and field sample-analysis methods were gradually worked out and standardized. When I was in India, I was impressed by the prevalence of mixed crops, for instance, a mixture of grains sown in wide strips, with mixtures of three varieties of mustards (belonging to two genera) grown in rows between the grain strips. Indian wheat experts described to me the conspicuous within-field variability of aboriginal grain fields along the Himalayan border. I then began to suspect that field-sample methods might be necessary in understanding the grains raised in ancient centers of variability, such as those defined by Vavilov.

It was not until I became associated with the Oklahoma-Ethiopia agricultural group that there was a chance to test this hypothesis. The results are significant not only for Ethiopian agriculture and for the understanding of modern wheats and wheat breeding, but also for the study of agricultural origins. Ethiopia is one of the chief centers of variability for the hard wheats—the tetraploid groups which include the primitive emmers and the durum wheats used in macaroni and other *pastas*. All of the hard-wheat fields sampled in Ethiopia proved to be mixtures in which hard wheats were usually predominant, though many of them included bread wheats (presumably hexaploid). Random samples were made from various hard-wheat fields. Though they were all heterogeneous, it was found by careful investigation that they had been sown as mixtures and grown as mixtures. They were reaped as mixtures, harvested and threshed as mixtures, baked and brewed as mixtures, even though frequently the same farmers were growing bread wheats in fields scrupulously devoid of other grains.

Among the hard wheats there was always great diversity, black, gray, and red forms being almost universal, in addition to straw-colored ones. Very rarely there was a plant of emmer, while plants apparently to be classed as durum wheats were noticeably more like emmers than they are in other parts of the world.

The problem was one of such vast proportions and so significant to the use and understanding of wheat, that it was turned over for study to Professor Jack Harlan of Oklahoma, who has been trained in this field. Before that point, however, through the courtesy of Dr. Ernest Sears, I made ear-to-row tests from one random Ethiopian collection. Of the thirty plants tested, several were obviously segregating. Three had such variable siblings as to suggest that the ear collected was a second- or third-generation descendant of a cross between very different wheats.

These observations and tabulations, when we free ourselves from European notions of clean-cereal agriculture, suggest very strongly (1) that it was in such mixtures as this that our polyploid wheats developed; (2) that the variability of Vavilov's centers is due to the fact that these are remote areas in which ancient patterns of grain farming are most likely to survive; (3) that we should probably start thinking not about the origin of wheat or the origin of barley but about the origin of small grains. In proto-agricultural times may not all fields have been plots where likely plants of various sorts were sown and reaped and used as mixtures?

THE IMPORTANCE OF MINOR CROPS TO THE ORIGIN OF DOMESTICATION

The probable historical importance of such crops has been stressed by Carl Sauer (1952) and others. His argument was that a crop low in yield and difficult to grow would never have been domesticated if there had been something else better at hand. Such crops, rather than one of the world's major domesticates might trace back to the earliest eras of agriculture. Likely candidates for study would be such plants as the minor millets, which belong to several genera—millet being merely a general name for an inferior grain—the jack bean, grains of paradise, perennial rye, tuberoses, and African marigolds, which were domesticated in Mexico in pre-Columbian times and used for ritual. For reasons which are obscure they are used for ritualistic purposes in other parts of the world as well.

WEED POPULATION DYNAMICS

The study of weed population dynamics, as yet scarcely begun (J. Sauer, 1959), is a probable key to many important problems. We need to determine the various intermediate stages between an out-and-out weed and an out-and-out crop plant. Noxious weeds, permitted weeds, encouraged weeds can be found in areas of ancient agriculture when we become conscious of such possibilities.

EMOTIONAL ATTITUDES TOWARD PLANTS

What are people's attitudes toward plants in general, toward food plants in particular and toward ornamental plants? Devoted plant care and emotional identification with certain plants are strong driving forces in some cultures, trades, and castes, and not in others. Malinowski (32) made a brilliant exposition of the role of plants in the life and religion of the Trobriand Islanders. The interest in ornamental plants as such is widely different in different cultures. In Costa Rica, for instance, all classes of society are devoted to ornamental plants of many kinds, including people living just barely above the subsistence level who receive no income from the ornamentals they cultivate so skilfully. In Ethiopia, on the contrary, well-to-do Galla farmers, though devoted to their crops and herds, average less than one ornamental plant per four houses (by actual count) in rural areas of Kaffa province. Before we can investigate the problem of domestication intelligently we must have at least some preliminary orientation in understanding these phenomena.

WHERE DID AGRICULTURE ORIGINATE?

In the light of the above established facts and reasonable hypotheses, in what part (or parts) of the world and at what time (or times) should we be looking for the origin (or origins) of agriculture? The following speculation-in-the-light-of-all-likely-evidence should not be allowed to solidify into a serious working hypothesis. It is meant to suggest some of the points we should start thinking about if we are to find the solid and significant evidence on which a good working hypothesis might be based.

Those who are familiar with the suggestions put forward in Carl Sauer's *Agricultural Origins and Dispersals* (1952) can readily see that they induced me to search in India and Africa for the kinds of facts which have in turn led to a somewhat different interpretation. This, in its turn, will have served a good purpose if it stimulates others to reflection and observation leading to more effective hypotheses. I shall first enumerate a few areas in which careful monographic analysis of existing data would probably lead to decisive evidence.

EVIDENCE OF EARLY TRANSMISSION OF PLANTS FROM AFRICA TO INDIA

The agricultural connections between southern India and Africa are so obvious that numerous authors have called attention to them. It is reasonably certain that a fairly late introduction took place from India to East Africa (Murdock, 1959). The movement I am suggesting on botanical evidence—which will become more convincing when detailed and documented—is in the opposite direction, i.e., from Africa to India. It must have taken place much earlier, preceding the Aryan invasion, judging by the facts that: (1) all these plants bear Sanskrit names; (2) all of them are crops of such minor importance as not to have merited attention had anything as sophisticated as Late Stone Age domesticates been available; (3) all of them were unknown in ancient Egypt, or appear very late in the record. I shall confine myself to five plants which I investigated in Africa and in Ethiopia and have since studied in herbaria and in the readily accessible literature. Each is so little known elsewhere as to require a few words of explanation and description.

1. Guar or cluster bean (*Cyamopsis*).—This legume is widely used in south India as a green manure and as cow feed, though even for this purpose the seeds have to be stewed into a kind of vegetable bouillon and the cattle deprived of other food until they learn to relish it. By picking the seed pods while they are immature, in the manner of okra, and mixing them with other vegetables and strong spices, they can be used as human food. Even then, according to my one educated informant, "they were the sort of thing you had to eat because your grandmother thought it was good for you." In recent years, as the result of basic research on the chemistry of its endosperm, guar has become one of the world's most important food stabilizers and industrial gums. Though it has a Sanskrit name, the botanical evidence suggests that it was domesticated in Africa.

2. *Galla potatoes* (*Coleus sp.*).—In India and parts of Indonesia and in Africa one finds these high-quality tuberous rhizomes, a little larger than one's thumb, in native markets. They have been so little studied that it is not yet certain whether one or several species are involved. Botanically it is more likely that they came from the highlands of Africa than from India or Indonesia, though either or both are possibilities.

3. *Pearl millet*.—This is a common grain in Indian markets. The plants I grew in Honduras from seeds bought in India varied as much from plant to plant as did first back-crosses from hybrids between distinct species. A morphological analysis of existing varieties, utilizing the method of extrapolated correlates, should indicate with precision the species from which each variety was derived.

4. *Ragee millet* (*Eleusine corocana*).—This plant is of great importance to South India since it can produce a harvest even in years when the monsoon fails. It is widespread in Central Africa, where it has been meticulously monographed by Portere; a similar study in

India would allow us to discuss its origin with precision. The closely-related goose grass (*Eleusine indica*), a world-wide weed, is a diploid and a likely candidate for one of two species from which ragee millet —a tetraploid—might have originated as a true-breeding hybrid. An analysis of the variation in *E. corocana* plus precise comparison with *E. indica* would make it possible to draw up a technical description of the other putative diploid parent. In Ethiopia I frequently saw a semi-cultivated variety of *E. indica* in native markets. It had been developed for its long wiry culms, which are used in making the expertly fashioned native sieves. The sieve itself is knit from these stems, producing a fine tough sieve of remarkably even mesh. I know of no reports of such sieves from other parts of the world. If they are unique to Africa, the presence of a specialized use for goose grass there and of a domesticated or semi-domesticated robust strain developed for this use, increases the likelihood that *E. corocana* arose there and then spread to India in pre-Aryan times.

5. *Aframomum ("grains of paradise"; Molucca pepper)*.—This ancient spice has until recently been known only in India, its seed pods and seeds having appeared there in trade since very early times. It tastes something like strong black pepper except that the effect is more localized on the tongue, almost like a series of microscopic needle pricks. It is used in India as a spice and in love potions and other village magic. During the nineteenth century it enjoyed a minor vogue in England but was eventually suspected of having narcotic properties. In Ethiopia I found it universally in Galla markets. There one saw not merely the brown pods of last year's crop, but the fading vermilion ones recently gathered. Through the courtesy of the grandson of the old Sultan, I saw it growing on his properties. It is a semi-domesticate in wooded areas near streams, where it is encouraged to multiply. Though related to the various cultivated and semi-cultivated gingers and turmerics of the Orient, the entire genus is supposedly African. Bunting and Rauk have now begun to clarify the classification of these ancient crops and semicrops. Comparative anthropological field work on their various uses in Africa and Asia should produce more critical evidence.

SIGNIFICANCE OF ATTITUDES TOWARD ORNAMENTAL PLANTS

Before going to Ethiopia, I had made detailed studies of house-orchard-gardens in Guatemala, Costa Rica, and Colombia, as well as more casual observations of the same phenomena in Mexico and India. In all these places I had been increasingly struck by the importance of flowers and other ornamental plants. However, in Ethiopia

I found that among the Kaffa Gallas, though house-side tangles of vines, shrubs, vegetables, condiments, and drug plants were universal, the number of plants grown purely as ornamentals averaged less than one per every four houses in rural districts. It was clear that the use of flowers had been increasing among those Ethiopians who had had most Arab and European contacts. It was less developed among the more negroid Gallas than among the strongly Semitic Amharas. It was apparently greater in the vicinity of Harrar (I made detailed studies in only a single village), where Arab influence is stronger than in the countryside near Addis Ababa.

Reading and consultation revealed that attention to ornamental plants varies widely from tribe to tribe and from region to region throughout Africa and even among negroid peoples in the Caribbean. Those who have travelled widely in Africa, if they have any opinion at all, hold that attention to flowers is strongest in East Africa, where Arabian and Asiatic influences are most predominant, and slightest in West Africa.

From my own observations plus what I could learn from personal interviews and the literature, including a quick look at the cross-cultural files deposited in the Princeton University Library, one point is clear: If we take a world look at floral versus non-floral agricultures, they are concentrically arranged around two poles. The pole of non-floral, seed-crop agriculture is in central Africa. The pole of floral agriculture is in Indonesia, radiating outward to Oceania, to the flowery kingdoms of China and Japan, to India, and even to dry and rocky Afghanistan.

Could it be possible that the Fertile Crescent of the Near East became so through being at an early date the area in which at least two quite different early agricultures were undergoing cross-fertilization? One was characterized, among other things, by a devotion to seed crops and a complete lack of interest in flowers and ornamental plants. It would have spread out from the hominid center in Africa while the other, flower conscious and including many root crops, was spreading out from the ancient center in Indonesia.

Before we can begin to organize even the existing data around this suggestion, a most important question remains unanswered: What was the relation of sacred trees and sacred groves to these two agricultural complexes? Sacred trees and sacred groves are still sacred throughout much of Africa and India. Are they indeed the "first temples," purely religious elements in the cultures of which they are or were a part? Apparently so, in my experience. To what extent are they domesticates or semi-domesticates? They have been so universally ignored that we do not yet have really critical evidence about their botanical classification, distribution, and religious and linguistic affiliations.

This is why the early Mediterranean records, rich with references to such trees as the olive and the fig, are so intriguing and yet not clear; they give promise of yielding important new understandings. We get tantalizing glimpses of earth-mother cults that wax and wane: Demeter changes in importance and in her very nature; Persephone joins her and Triptolemos as well; seeds and flowers, associated symbolically with these cults, strive for status as emblems of accomplishment and power; flowers become symbols of the Lower World. In late Egyptian times people of importance are represented as carrying elaborate, stylized, scepter-like bouquets of the greatest sophistication. At what times and from what direction do seeds, flowers, and trees enter the religions, the myths, the inscriptions, and the artistic legacies of various cultures?

When I began to study some of the primitive cultivated plants that I had seen in India, I began to think along these lines. After my trip to Ethiopia this hypothesis began to take definite shape. Meanwhile, on very different evidence, Murdock (1959) quite independently had come to the conclusion that there had been a separate African origin of agriculture, "most probably to be credited to the Mande peoples around the headwaters of the Niger in the extreme western part of the Sudan, less than 1,000 miles from the shores of the Atlantic ocean." This is in the same area which had been indicated by the plants in which I was interested, though I knew nothing about the peoples concerned.

Murdock and I operated by wholly different methods, using on the whole quite different evidence. Though he knew about some of the plants I have cited, he produced a long list of plants quite unknown to me: fonio (*Digitaria exilis*), earth pea (*Voandzia subterranea*), geocarpa bean, (*Kerstingiella geocarpa*), fluted pumpkin (*Telfairia occidentalis*), ambary (*Hibiscus cannabinus*) and yergan (*Cucumeriopsis edulis*). That we should have reached such similar conclusions independently and have independently seized upon the significance of such seemingly unimportant plants, helps to strengthen both hypotheses, to my way of thinking.

EVIDENCE FROM PERENNIAL RYES

The suspicion that there may have been an independent center of domestication in Africa lends new significance to the study of perennial ryes. There are two of them, both montane species: *Secale africanum* in South Africa, and *Secale montanum,* surviving here and there around the Mediterranean, widely distributed in the Middle

East. Both are wild species, tending toward extermination under heavy grazing. A cultivated (or semi-cultivated) variety of *S. montanum* has been reported in one localized area but has been very little studied.

These perennial ryes are sharply set off from the common cultivated annual rye, *Secale cereale,* by structural differences in at least two chromosomes and in other ways. A weedy form of rye is found in the Middle East and the theory is now generally accepted that the weed rye, spreading northward along with other cereals, eventually gave rise to rye, the crop plant, becoming progressively at an advantage over its fellow cereals when grain agriculture advanced across the sandy plains of northern Europe.

It was formerly naïvely assumed that, since weed ryes were known and had probably given rise to rye, the crop-plant, the origin of rye had been studied in sufficient detail. We now know that the origin of a weed as successful as this is a more difficult (and frequently more important) problem than the origin of the crop plant from the weed. Fortunately the origin of weed rye has now been studied in detail by Ledyard Stebbins and his former student Howard Stutz, using refined taxonomic, cytological, and genetic techniques. The main points are clear. Crosses between *S. montanum* and its distant relative *S. sylvestre* were fertile enough to permit introgression between the two species, thus creating weed ryes of at least two sorts, *Secale vavilovi* and *S. cereale.* Stutz (1957) has presented convincing data and it is now reasonably certain that rye originated in some such way, probably in early Neolithic times somewhere in the Near East, where we now find increasingly detailed evidence of the development of agriculture as we know it in Europe and the Middle East.

However, as we have seen, when we look for them, we can find tantalizing fragmentary evidences of far earlier agricultures, proto-agricultures, in Africa and in Indonesia. The beginnings of African proto-agriculture must have been well under way while the Arabian and Saharan deserts were still much moister regions than they have been since the early Neolithic. In the light of Stutz's detailed investigations, there is a distinct possibility that perennial ryes grew at various points between South Africa and the Near East in some kind of semi-domesticated state that we can as yet only imagine.

Perennial sorghums are still known in these same areas, as well as weed, grain, syrup, fuel, and broomcorn annuals. We must therefore keep our minds hospitable toward evidences of an African proto-agricultural complex involving perennial grasses, functioning at least in part as cereals.

Where and when did men first begin to domesticate plants? In terms of germ-plasm structure and population dynamics, what happens to a plant when it is domesticated? This is a fundamental problem on which we have some pretty theories and excellent mathematical models but virtually no facts. Only one thing is clear. It is time we put to rest the dogma about early agriculture bequeathed to us by nineteenth-century botanists, anthropologists, and prehistorians. We learned them in our salad days and easily forget what guesses in the dark they originally were. Now is the time to follow the lead of Darwin and lay such myths aside. We should take fresh looks at the existing evidence and start thinking about the most effective questions to ask, the most likely areas in which to find critical new evidence.

The studies resulting in this report would have been impossible without generous and understanding support from the Guggenheim Foundation, the National Science Foundation, and Oklahoma State University.

BIBLIOGRAPHY

ANDERSON, EDGAR. 1950. "An Indian Garden at Santa Lucia, Guatemala," *Ceiba* I, 97–103.

———. 1952. *Plants, Man and Life.* Boston: Little, Brown & Co.

———. 1954. "Reflections on Certain Honduran Gardens," *Landscape* (Summer, 1954), 21–23.

ANDERSON, EDGAR, and HUGH C. CUTLER. 1942. "Races of *Zea Mays:* I. Their Recognition and Classification," *Annals Missouri Bot. Gard.,* XXIX, 69–88.

BUKASOV, S. M. 1930. "The Cultivated Plants of Mexico, Guatemala and Colombia," *Bull. Appl. Bot., Genetics, and Plant Breeding,* Suppl. 47, pp. 261–73.

BATESON, WILLIAM. 1894. *Materials for the Study of Variation.* London: Macmillan Co.

BROWN, W. L. 1953. "Maize of the West Indies," *Tropical Agriculture,* XXX, 141–70.

BRIEGER, F. G. *et al.* 1958. *Races of Maize in Brazil and Other Eastern South American Countries.* (National Research Council Pub. 593.)

CUTLER, R. C. and T. W. WHITAKER. 1956. "*Cucurbita mixta* Pang, its Classification and Relationship," *Bull Torrey Club,* LXXXIII, 253, 260.

———. 1960. "Origin and History of the Cultivated Cucurbits of the New World," *Amer. Antiquity,* in press.

DARWIN, CHARLES. 1868. *The Variation of Plants and Animals under Domestication.* 1897 ed.; 2 vols. New York: Appleton and Co.

DE VRIES, HUGO. 1905. *Species and Varieties: Their Origin by Mutation.* London.

GRASSL, C. O. 1959. "Introgression between *Saccharum* and *Miscanthus* in New Guinea and the Pacific Area," *Proc. 9th Inter. Bot. Cong.,* II, 140.

HATHAWAY, W. H. 1957. *Races of Maize in Cuba.* (National Research Council Publication 453.) National Academy of Sciences.

HAWKES, J. G. 1944. "Potato Collecting Expeditions in Mexico and South America. II. Systematic Classification of the Collections," *Imp. Bur. Pl. Breed. and Genetics,* 142 pp.

HEISER, CHARLES B., JR. 1955. "The Origin and Development of the Cultivated Sunflower," *American Biology Teacher,* XVII, 162–67.

HEISER, CHARLES B., JR., and PAUL G. SMITH. 1953. "The Cultivated Capsicum Peppers," *Economic Botany,* VII, 214–27.

HUTCHINSON, J. B., R. A. SILOW, and J. G. STEPHENS. 1947. *The Evolution of Gossypium.* London: Oxford University Press.

KIHARA, H. 1924. "Cytologische und Genetische Studien," *Mem. Coll. Sci., Kyoto Imperial University,* Ser. B, I, 1–200.

———. 1956. *Land and Crops of Nepal Himalaya.* Kyoto: Kyoto University Press.

MALINOWSKI, BRONISLAW. 1935. *Coral Gardens and Their Magic.* New York: American Book Co.

MANGELSDORF, PAUL C., and R. G. REEVES. 1959. "The Origin of Corn," *Botanical Museum Leaflets, Harvard University,* XVIII, 329–400.

MERRILL, E. D. 1954. "Botany of Cook's Voyages," *Chronica Botanica,* XIV, 161–384.

MURDOCK, GEORGE P. 1959. *Africa.* New York: McGraw-Hill Book Co.

PARTHASARATHY, N. 1948. "Origin of Noble Sugar Cane," *Nature,* CLXI, 608.

ROBERTS, L. M., *et al.* 1957. *Races of Maize in Colombia.* (National Research Council Publication 510.) National Academy of Sciences.

SAUER, CARL O. 1952. *Agricultural Origins and Dispersals.* (Bowman Memorial Lectures, Ser. II). New York: American Geographical Society.

SAUER, JONATHAN D. 1959. "Coastal Pioneer Plants of the Caribbean and Gulf of Mexico." Mimeographed Report on Project NR-388-047, Geography Branch, Office of Naval Research, pp. 1–20.

SAX, KARL. 1922. "Sterility in Wheat Hybrids," *Genetics,* VII, 513–52.

STUTZ, HOWARD C. 1957. "A Cytogenetic Analysis of the Hybrid *Secale cereale* x *Secale montanum, Genetics,* XLII, 199–221.

VENKATRAMAN, T. S. 1938. "Hybridization in and with the Genus *Saccharum.*" Presidential Address, in *Proceedings of the Twenty-fifth Indian Science Congress, Calcutta,* pp. 1–18.

WELLHAUSEN, E. J., *et al.* 1952. *Races of Maize in Mexico.* Publication of the Bussey Institution of Harvard University, Cambridge.

———. 1957. *Razas de Maiz en la America Central.* (Folleto Technico No. 31, Secretaria de Agricultura y Ganaderia) Mexico City.

WHITAKER, T. W., and J. B. BIRD. 1949. "Identification and Significance of the Cucurbit Materials from Huaca Prieta, Peru." *American Museum Novitates,* No. 1426, pp. 1–15.

WHITAKER, T. W., H. C. CUTLER, and R. S. MACNEISH. 1957. "Curcurbit Materials from Three Caves near Ocampo, Tamaulipas," *American Antiquity,* XXII, 353–58.

STUART PIGGOTT

PREHISTORY AND EVOLUTIONARY THEORY

When asked to contribute to this symposium, so predominantly concerned with the natural sciences, it seemed to me that a valuable contribution might be a critical one, which would enquire how the study of prehistory and inevitably that of history as well could in any way be related to evolutionary theory at large.

Such a relationship can be approached from two angles. Before we can examine any current concepts, we must know something of their own evolutionary history, and we must therefore properly begin with some discussion of the history of prehistoric studies in relation to the contemporary modes of thought about evolutionary processes. This enquiry will take us back before 1859 and into the eighteenth century, for I think it can hardly be denied that some of the most persistent lines of thinking about prehistoric, and indeed early historic man—concepts still current today—are significant outcomes of European eighteenth-century thought and must be seen in the light of that fact.

After trying to perceive something of the bases of the presuppositions and assumptions on which we construct our picture of prehistory, we may then turn to a necessarily brief examination of the question as to how evolutionary theory may be thought to be applicable to what we can know of pre-literate human societies. But before we go further, we must try to decide what we are talking about. We must seek for a definition of "prehistory," for the word is deceptive in its apparent simplicity.

The logical French make their way out of the difficulty by saying, quite simply, that prehistory is what it means—the phase before history. History begins, potentially at least, with the invention of writing, an event which, so far as we know at present, took place in

STUART PIGGOTT, Abercromby Professor of Prehistoric Archaeology in the University of Edinburgh, has been mainly concerned with the Neolithic and Early Bronze Age cultures of the British Isles in their relation to the European Continent. He has conducted extensive fieldwork and excavations in England, most recently directing the excavations at Stonehenge. His published works include *The Neolithic Cultures of the British Isles.*

certain restricted areas of western Asia somewhere about 3000 B.C., in round figures good enough for our present purpose. Therefore, the French continue, *la préhistoire* ends—everywhere, all over the world —with the beginning of Sumerian or Egyptian history, and we enter the phase of *protohistoire*. This continues, in any given region, until the local appearance of written documents which can properly be regarded as a basis for history in the usual sense.

This is all very well, but its disadvantages can be perceived perhaps more quickly by a New World audience than by one in the Old. Do you really have to call the study of the archaeology of the Eastern Woodlands cultures up to the time of the first European settlement "protohistory"? What of the Eskimos, the Maoris? What of those cultures in East Asia which continue the technique of Lower Paleolithic stone-working into the period of the European Middle Ages? English usage has never been happy about "protohistory," and I think we have been right, if more parochial, to use the term "prehistory" right up to the point where the local historical record starts, whether with Julius Caesar and after him, the Emperor Claudius coming to Britain or with Catlin and Bodmer and Miller crossing the wide Missouri. But though I think we should go on using the word, its usage must not disguise the fact that there are degrees of prehistory, especially in the Old World; for, once historical communities were in existence, all non-literate, non-historic, prehistoric communities were necessarily contemporary in time with regions with written history and frequently in some sort of relationship with them.

Primary prehistory, absolute prehistory—qualifying terms have been used to denote the kind of prehistory wholly antecedent to any written record—the French *préhistoire* once again. After that, you can group everything non-historic into secondary prehistory, as Grahame Clark (1955) has, or, with Christopher Hawkes (1951) you can point to a series of increasingly close relationships between non-literate and literate societies to indicate the degree of remove of each from wholly antehistoric time to one that is truly protohistoric. I shall use "prehistory" here to denote the study of human societies in the past for which no direct literary record exists. I would not say, for instance, that Herodotus makes the Scythians a historical people by describing them in the fifth century B.C., and I would add that a literate civilization whose script cannot be read has to be treated as prehistoric: the classic instance here is that of the Indus Valley civilization and, until yesterday, was that of Mycenaean Greece.

What I would regard as the essential, the unifying, factor in all prehistory in this wider sense of the study of non-literate communities in the past is that our knowledge is based on archaeological evidence, and the techniques of archaeology alone can inform us of the very existence of such communities. This is not to say, of course, that archaeological techniques do not play a large part in the recovery of literate cultures as well, but, for the cultures which we have agreed to call "prehistoric," archaeological evidence of some kind or other provides our entire basis. What follows from this is, I think, of paramount importance and insufficiently recognised: the nature of the evidence dictates the nature of the inferences which can be properly drawn from it. Dr. Slotkin, of the University of Chicago, in a valuable critique of Collingwood's views of our cognition of history, put it neatly when he distinguished the past-in-itself, which we try to discover by various approaches, from the past-as-known, which they reveal (Slotkin, 1948). There can be more than one past-as-known, according to the type of evidence, of techniques, of approach, you employ, and what I want to stress here is that the past-as-known which is based on archaeological evidence is not, and cannot of its nature be, the same as the past-as-known based on evidence which involves the written record in lesser or greater degree. We may deplore the fact, but that is no excuse for disguising or ignoring it. As an analogy, I would remind you that outer space can be examined both by optical and by radio telescopes, but the technique of the employment and the nature of the results obtained are not coincident. What you perceive from Jodrell Bank is not what you perceive from Mount Palomar.

The prehistoric past-as-known, then, must always be different from that known by means other than the archaeological techniques of the recognition, recovery, and interpretation of the surviving material culture from antiquity. This evidence has its own peculiar potentialities and its own restrictive limitiations. Questions posed of it will be meaningful only insofar as they respect the nature of this evidence and do not seek to strain it beyond its proper limits. As Hawkes pointed out a few years ago to an audience in America (1954), it is not difficult to construct a scale of ascending difficulty or unreliability in inference from archaeological evidence on its own. Since an overwhelming proportion of archaeological evidence is, in fact, the product of ancient technology, it is precisely on this aspect of prehistory that archaeological inference is at its most reliable. Following on this, the determination of the rudimentary subsistence economics of the communities under review, while complex in procedure and involving a cooperative approach between archaeologists and the natural sciences, has again a high degree of reliability. But, once we go beyond this point and try to determine such aspects of a prehistoric community as its social structure, its political institutions, its language, its religion, its modes of thought, we are either straining inference to (or beyond) its permissible limits and, even then, obtaining little that is reliable by so doing or making empty platitudes based on analogies which we would be hard put to it to defend as valid. The past-as-known by archaeology, and therefore that of prehistory, is very largely a technological past, not necessarily because technology was overwhelmingly important to the communities we study or that it has any intrinsic validity in itself over other human activities, but because this is the aspect we can best perceive by means of the only evidence which we possess for prehistory.

In order to interpret the observed phenomena recovered by archaeological techniques and, indeed, all phenomena about the human past recovered by historical means in the widest sense, some working hypothesis is needed which will organise the inchoate details into some sort of intelligible pattern. As Frankfort (1951, p. 21) put it, "a viewpoint whence many seemingly unrelated facts are seen to acquire meaning and coherence is likely to represent a historical reality." I would suggest that we adopt a useful piece of terminology hitherto used almost solely in the field of scientific theory, and call such a viewpoint, such a working hypothesis, a *model* of the past (Piggott, 1958a). To go back to Slotkin's phrase for a moment, the observed phenomena are the products of the past-in-itself, the model we use will be framed in accordance with their nature, and it will permit of their arrangement into an intellectual construct which will be the past-as-known within the terms of that particular model.

This somewhat lengthy preamble is, I feel, necessary in order that we can define our terms and use words and concepts in an agreed sense. To understand the relationship of prehistory to evolutionary theory, we must at the outset try to see not only what our perception of the prehistoric or historic past amounts to but how theories about the past, whether evolutionary or not, can come to be applied. I would say that what we have to consider is the construction of various models of the past, some of which may embody evolutionary concepts.

The models which have been devised have naturally been constructed in terms of the prevailing climate of thought at the time of their formulation: the theological and mythological models of the past used in the Middle Ages are early examples of this. A series of cyclical models, concerning themselves predominantly with the historical past, have been elaborated since Vico (see Collingwood, 1946, pp. 63–71) in the early eighteenth century, and their protagonists include Petrie (1922), Spengler (1926), and Toynbee (1933). These are in a sense evolutionary and, indeed, embody notable fallacies of biological analogy, but more detailed consideration is best postponed for the moment, except that we may note that the cyclical model of Toynbee has, particularly in its later parts, a strong theological tinge.

The two most important models of the prehistoric past which have been constructed are both evolutionary in nature, and both are pre-Darwinian in origin. These models may be distinguished as the technological and the economic, respectively, the latter being closely associated with a variant (again evolutionary) which is social-evolutionary. The first model is responsible for well-known Stone, Bronze, and Iron Ages (with their subdivisions); the variant of the second produces the orthodox Marx-Engels view of ancient society.

The technological-evolutionary model, or that of the Three Ages, originated in recognisably modern form, as is well known, as a result of the problems facing Christian Thomsen in arranging the archaeological collections of what is now the National Museum of Denmark before 1836, when the new system was set out in the Preface to the Museum guidebook. There had been anticipations of the scheme in the eighteenth century, as a result of trends of thought which will be discussed below, but it was Thomsen and his followers who popularized the idea and gave it firm roots in European archaeology.

This is not the place to discuss in detail the Three Ages or their subsequent subdivisions into a terminology which is still with us. Daniel (1943) has made such an examination and, in it, stresses certain points which are cardinal in our understanding of this particular model. In the first place, it originated as, and has always essentially remained, a museum man's classification, based on the predominant substances used for edge-tools surviving from prehistoric antiquity. It was originally conceived simply in these terms, though soon to be supported by the evidence of stratigraphy on Danish sites. It was not (as Daniel points out) an inference made from the observation "of existing human communities in various technological stages and arranging these observed stages in a hypothetical order," but its "roots are in the facts of Danish prehistory," however much its acceptance was helped by the prevalent contemporary desire to see "progress" wherever possible. But in its essentials it was purely a technological model, it being left for prehistorians such as Childe to reinterpret "archaeological ages as technological stages" and to redefine the scheme in terms of economic changes inferred from those of technology, so that the model was really altered into an economic-evolutionary one.

The prehistoric past-as-known within the terms of the technological model is, then, that of the Upper Paleolithic, or the Middle Bronze Age, or the Early Iron Age, or whatever subdivided nomenclature is used. The Paleolithic period or the Bronze Age have no existence in the past-in-itself, but only as constructs in the past-as-known in terms of the technological-evolutionary model. Use another model, and you get another name—much of the Anatolian Late Bronze Age is politically the Hittite Kingdom; within the Early Iron Age would equally fall Periclean Athens and Celtic farmsteads on the southern English downland.

Denmark is a region where the technological sequence is, in fact, relatively simple (though not so simple as it seemed in the last century) and presents itself in a form far more clear-cut than in many another European area: perhaps it was no chance that the Three Ages scheme was first invented there. And the further the terminology and concepts contained in the technological model are extended beyond its point of origin, the less meaningful do they become. Stone Ages, Bronze Ages, Iron Ages, hardly form a satisfactory conceptual framework within which to perceive the ancient Orient, for instance, though they do at least share with the European usage of the terms some sense of comparable antiquity. But in south India "Neolithic" cultures continue until late in first millennium B.C., and in East Asia the terms "Bronze Age" and "Iron Age" are used for societies up to A.D. 1400. As strict statements of technological stages, divorced from any chronologican content, these phases are presumably permissible; but, since farther west the same terms are closely bound up with a time scale which is quite discordant, it is difficult to justify the wider application of this particular model and its attendant terminology.

The technological model, then, originating in the early nineteenth century from an objective observation of the museum collections in Copenhagen, was easily developed, in accordance with current modes of thought, into an evolutionary sequence for prehistory, which, in the then state of archaeological techniques, seemed practically the only one which could be justifiably based on the evidence of material culture. But already, since the later eighteenth century in France and in Britain, a theoretical sequence for antiquity involving the social structure and subsistence economics of societies in the past had been sketched out.

Unlike Thomsen's Three Ages, the early stages of thought which were to lead to the construction of the social and economic model of prehistory were not based on observed archaeological fact but were in large part the outcome of the doctrines of optimism, the inevitability of progress, and the perfectibility of man current in the eighteenth century. Americans do not need reminding of the part that this temper of thought played in the founding of the United States, but one of its by-products was that interest in the past of society which one especially associates with the Scottish Primitivists in the second half of the century, themselves profoundly influenced by Rousseau and Montesquieu. The concept of the Great Chain of Being, as Lovejoy (1942) has demonstrated, was by then a well-established concept—a model, in fact—explanatory of the universe and man's place in it, and was modified in the eighteenth century by the introduction of a time element so that it became a process—still, of course, God-willed—in which the links were added one by one to join the lowest to the highest (cf. Piggott 1956). In fact, Erasmus Darwin, Charles's father, had been instrumental in popularizing this new version of the Great Chain. Once this time element had been introduced, it was possible to construct rudimentary evolutionary schemes, not, of course, based on observation and inference, but on a priori theory, yet nevertheless of great importance in preparing the conditions in which such scientific evolutionary theory would find acceptance. Among such schemes were those of the origin and development of human society, and in 1785 Pinkerton defined the successive stages as savage, pastoral, barbaric, and civilized. It was left to the American anthropologist Lewis H. Morgan to restate this basic sequence in a more detailed form in 1877, his system being "based primarily not on archaeological evidence but on the comparative study of modern primitive peoples, the arrangement of these existing economies and societies into an evolutionary sequence, and the projection of this hypothetical sequence into the prehistoric past" (Daniel, 1950, p. 188).

It appeared that this sequence could be legitimately supported by inferences drawn from prehistory as seen in terms of the technological model and that it would, in fact, be possible to construct an alternative, social-evolutionary, model. But it is important to realise that in the form in which it was propounded by Morgan and enthusiastically championed by Engels in 1884 as introducing "a definite order into human prehistory," it was almost entirely based not on direct inferences from archaeological evidence but on second-order inferences drawn from the prehistoric past-as-known as seen in terms of the technological model. It was left to recent archaeologists to elaborate techniques in collaboration with natural scientists whereby archaeological evidence could be recovered and interpreted with the specific end of obtaining information on the subsistence economics of the community it represented: thence forward the model framed in these terms had a greater degree of reliability than a construct based on inferences from inferences.

Prehistory today, then, is basically constructed from archaeological evidence by viewing this evidence in terms of conceptual models of the past which are largely technological and economic, both also containing not only the simple diachronic element implicit in all historical studies but a sequential relationship implying development and cause and effect. Since 1859 all such developmental schemes have naturally been profoundly affected by the evolutionary concepts current in the natural sciences: these have affected not only prehistory but history. But we must be on our guard against importing a concept proper to one discipline into another where it may seem to fit with deceptive conviction. It is fatally easy to confuse the arresting simile, the apt analogy, with the phenomena or process it was originally used to illuminate; the metaphor becomes reality. The biological analogies of birth and growth, maturity and decay, were taken from their context and imported into the cyclical models of history so laboriously contrived by Spengler and Toynbee: "ostensible similes pervade the argument with an implied assurance that they reflect historical situations" in the latter writer. If the situations are indeed historical in the sense that they are documented by a written record as well as by surviving material culture, the detection of the flaws in an argument presented in these terms is less difficult than in situations in prehistory, where the evidence is so much more limited in content and in potentialities. Owing to its imperfections, archaeological evidence is easier to fit into preconceived patterns than is historical, and doctrinaire schemes of social evolution can best be documented by the convenient ambiguities of archaeology.

An even more insidious danger in the interpretation of prehistory and history is making value-judgments involving "postulates which fulfil an emotional need" (Frankfort, 1951, p. 24) in the modern Western world which are then imported into the past or used as criteria to construct a scale of imaginary values. It is this which leads to what Collingwood called "progress created by historical thinking" (Collingwood, 1946, pp. 321–34), and to this I will return. But in the meantime in our study of prehistory we can at least refrain from the temptation (in Frankfort's words) to "project the axioms, habits of thought, and norms of the present day into the past which, as a result, seems to contain little that is unfamiliar to us" (Frankfort, 1951, p. 19). Richmond has made the point even more strongly, with specific application to prehistory, when he talked of "certain phases of archaeological study where, failing the resources of thought contemporary with the material from which to infuse it with life, ideas are imported from present-day experience and ancient man is anachronistically saddled with views which he would have found at best strangely unfamiliar and at worst grossly distasteful" (Richmond, 1950, p. 6).

This is especially necessary to remember when we are dealing with prehistory, in which the viewpoint conditioned by the technological model is so dominant. If we, as a result of uncritically accepting a prevailing climate of thought in which technological innovation and development are set at a premium, then make value-judgments and rate the worth of this or that society in such terms, we are falling into precisely that trap, the perils of which we have just been warned against by philosopher, historian, and archaeologist, as well (I hope) as by our own common sense. Archaeology is and must always remain very largely a descriptive discipline, illustrative of man's material culture in the past. To obtain some sort of coherence of vision, some working hypothesis, which will give our observed phenomena a meaningful relationship, we must construct, for the non-literate past, prehistory. This we can best do in terms of conceptual models specifically adapted to the nature of the evidence at our disposal. These and their validity in proportion to the degree in which they can legitimately use archaeological evidence will each produce a past-as-known within its own framework of reference: the validity of any such model can again be checked by the agreement or discordance between its inferences and those of other models.

If we then look at the evidence on which prehistory is based, with a view to considering whether it is susceptible of providing inferences which in their turn could be used to support or refute any evolutionary theory, we must be struck by its potential ambiguity, an ambiguity due not only to logical difficulties in interpretation but arising out of such factors as the accident of survival of the material constituting the evidence itself or the reliability of archaeologists' excavation reports, which occupy the peculiar position of being virtually primary sources for the construction of prehistory.

The accident of survival, whereby the student of prehistory has so often to deal with the durable remnant of the material culture with which he is concerned, is clearly an incalculable factor which may distort even the basic technological picture obtained by archaeological methods. We have no warrant for assuming that substances resistant to chemical or bacterial decay possessed an intrinsic superiority or significance to their original makers; yet substances such as stone, metal, or pottery will bulk large in archaeological evidence, while perishable artifacts will have vanished or, at best, survived only in exiguous traces. Our knowledge of human handiwork in the Lower Palaeolithic is almost wholly confined to the observations we can make about the processes of stone-flaking employed to make implements, most of which are frankly of unknown purposes; but, had the nature of organic and inorganic materials been different, we might have had to deal with surviving fibres and lost stones, and our technological model would contain not an Old Stone Age but an Old String Age. The fact that the stones do in fact survive need not make us rate them, *on that account,* as the most significant feature of the cultures we are examining.

This may seem obvious, but reputable archaeologists have sometimes failed to appreciate the fallacy inherent in rating prehistoric communities in terms of their surviving material culture. Words such as "degenerate" are taken from their usage to denote an assumed place in a typological series of pots, for instance, and transferred with an emotive and even moral connotation to the makers of the vessels; people with poor and scanty pottery become stigmatized as "poverty-stricken," though their poverty may well have been only in their failure to provide the archeologist with his favorite product, and in one recent study they become "outcasts," driven out by "expanding, more progressive, groups"—the whole of this value-judgment being based on potsherds (*Annual Report,* Univ. London Inst. Archaeol., 1952). This is an extreme case, but other comparable and equally unjustifiable extrapolations are not difficult to find.

The standard of reliability of excavation reports is a factor less appreciated in the construction of prehistory than it should be (Piggott, 1959; Smith, 1955; Hawkes, 1957, p. 1). Excavations are elaborate laboratory experiments that cannot be repeated or checked if they have been total, since they will then have wholly destroyed the site, and the reliability of the information yielded is in direct proportion to the competence of the excavator—a quality as incalculable as it is difficult to assess from the report. I cannot here discuss the illusory view, sometimes held, that a completely objective record is possible, desirable, or even attainable; I would quote only one example of inference based in good faith on what appeared to be a sound record of large-scale excavation on more than one major site—the instance of the Indus cities. Here, owing to inadequate excavation technique, the cities of Mohenjo-daro, Harappa, and the others were believed to be phenomena, unique in the ancient Orient, of urban communities functioning without material evidence of defenses, or of any centralised power embodied in palace or temple within a citadel. Wheeler's later excavations—indeed, Wheeler's trained observation of the sites before any new excavation had taken place—detected and revealed the citadels, and the normal pattern of ancient urban organization was at once apparent, but not before the bogus picture derived from faulty excavation had contributed to more than one theory of social evolution.

Nor does the ambiguity of archaeological evidence for evolutionary theory end here. It is not only the rapid development of prehistoric studies that can render invalid an interpretation within a few years of its making, but the inherent uncertainty attaching to extrapolation from what may survive of material culture to theories of social evolution that they may be thought to reflect or embody. The most determined efforts to construct a sequence of social evolution in prehistory are those which derive, as we have seen, from the eighteenth-century Primitivists and Lewis Morgan, which, in the form in which they are canonical in Communist thought (or should we not use a theological model, and say, rather faith?), are presented as a theoretical construct which can be used to arrange the archaeological evidence or within the framework of which the archaeological evidence can be fitted and adjusted. Pliable and ambivalent in its nature, its interpretation necessarily fraught with ambiguities, it is not surprising that the evidence has appeared to support the a priori scheme; unfortunately, the same evidence could plausibly buttress up alternative sequences.

A very interesting attempt to interpret the archaeological material of a restricted area of northern Europe in Marxist terms was made by Gordon Childe in his *Scotland before the Scots* in 1946. Childe, as I have argued in another place (1958b), felt an urgent necessity to justify the study of prehistory, to make it socially respectable within the temper of thought of the modern world, to demonstrate that it could be "useful"—a so-called social science projected into the remote past. He constantly sought for a conceptual model which would enable him to interpret archaeological evidence in terms of social evolution, and it was clearly necessary to experiment with the Marxist model, which seemed to him at that time to "have produced narratives which seem more historical than a succession of invasions and are yet just as objective and solidly based on observed data." Without inquiring too curiously into what is meant by "more historical" or pausing over the term "objective," we may note here that the apparently convincing narrative of prehistory presented in *Scotland before the Scots* fails to stand up to re-examination of the archaeological evidence at almost every point ten years or so later. To some extent this is the result of new evidence, but it is far more from a critical reassessment of the same material as that used in 1946 without preconceived theories of how (if at all) it could be fitted into a pattern (if any). The current picture is certainly more chaotic, in terms of any "evolutionary" scheme, than the neat sequence which could be so convincingly demonstrated by the application of the Morgan-Engels-Marx model. It still may well be fallacious, but at least in its present form it does not mislead by appearing to present a pattern where none can be justified. Perhaps it is "more historical" for this reason, if for no other.

If, then, we now try to determine whether any uniform evolutionary process can be perceived in prehistory, we must limit the enquiry to a search for a consistent technological development, for only in the field of technology can inference from archaeological evidence be sufficiently reliable, technology in this instance being taken to include means of subsistence and so, at second remove, to some extent the economic structure of the societies that we believe we have isolated on the grounds of their surviving material culture. Our model of the past is necessarily the technological one, which we use, it must be remembered, not because it has any intrinsic superior validity over other models, but because it is best suited to the archaeological evidence on which prehistory is based.

But even within the restricted framework of this model we see a complex picture. Immediately we have to reckon with societies which innovate against those who conserve: in the former we may infer that technological development was socially acceptable, in the latter it was not so valued. We are seeing in the past what anthropologists can see in the present: "In one society technology is unbelievably slighted even in those aspects of life which seem necessary to ensure survival; in another, equally simple, technological achievements are complex and fitted with admirable nicety to the situation" (Benedict, 1934, quoted by Frankfort, 1951, p. 22). The past viewed in archaeological terms alone—the non-literate past of prehistory—may sometimes present a picture which we may suspect of being deceptively uniform. The essential diversity of character between cultures or civilizations perceptible when a written record exists may be hidden or obscured: "The archaeologist must count himself the luckier when literature can inform him of standards contemporary with his material," Richmond wisely observed, in the context already quoted, and Frankfort demonstrated this in his subtle analysis, more than once quoted here, of the contrasting forms of the ancient civilizations of Mesopotamia and of Egypt, which could be thought, on archaeological evidence alone, to display a deceptive parallelism. But even with this proviso, enough diversity and inequality in technological and allied developments can be perceived in prehistoric antiquity to justify us in thinking that, in fact, the disparities were even greater than we can apprehend.

The prehistoric past-as-known must always be of such a nature that the prehistorian, with the best will in the world, can hardly share "the historian's supreme duty of doing justice to each civilization on its own terms" (Frankfort, 1951, p. 24), because he cannot know what these terms were. This is a task difficult enough, in all conscience, in the instance of fully literate civilizations, but it is virtually precluded by the legitimate models wherein the prehistoric past can be perceived. What little can be perceived certainly gives us no ground for thinking that we should try to reconstruct these terms by importing our contemporary standards and scales of social or moral values into antiquity. Nor should we seek to establish general laws of development by improperly employing constructs, legitimately based on the evidence of the pure or natural sciences, to explain observations and inferences based on evidence of a wholly different order. Only by a frank recognition of the nature of our evidence and the limitations of permissible inference from it, can we hope to construct a satisfactory intellectual discipline for our study.

References

BENEDICT, RUTH. 1934. *Patterns of Culture.* Boston: Houghton Mifflin.

CLARK, GRAHAME. 1955. *The Study of Prehistory.* Inaugural lecture, Cambridge University. London: Cambridge University Press.

COLLINGWOOD, R. G. 1946. *The Idea of History.* Oxford: Clarendon Press.

DANIEL, G. E. 1943. *The Three Ages.* London: Cambridge University Press.

———. 1950. *A Hundred Years of Archaeology.* London: Duckworth.

HAWKES, CHRISTOPHER. 1951. *Proc. Prehist. Soc.,* XVII, 9.

———. 1954. *Amer. Anthropologist,* LVI, 155.

———. 1957. *Adv. Sci.,* LIV, 1.

FRANKFORT, H. 1951. *Birth of Civilization in the Near East.* London: Williams & Norgate.

LOVEJOY, A. 1942. *The Great Chain of Being.* Cambridge, Mass.: Harvard University Press.

PETRIE, FLINDERS. 1922. *Revolutions of Civilization.* London: Harper.

———. 1956. *Antiquity,* XXIX, 152.

PIGGOTT, STUART. 1958*a. Antiquity,* XXXII, 78.

———. 1958*b. Proc. Brit. Acad.,* XLIV, 305.

———. 1959. *Approach to Archaeology.* London: Black.

RICHMOND, I. A. 1950. *Archaeology and the After-Life in Pagan and Christian Imagery.* London: Oxford University Press.

SLOTKIN, J. S. 1948. *Antiquity,* XXII, 98.

SMITH, M. A. 1955. *Archaeol. News Letter,* Vol. VI.

SPENGLER, OSWALD. 1926. *The Decline of the West.* English translation. London: Allen & Unwin.

TOYNBEE, ARNOLD. 1933. *A Study of History.* London: Oxford University Press.

UNIVERSITY OF LONDON, INSTITUTE OF ARCHAEOLOGY. 1952. *Ann. Rept.,* VIII, 75.

FRANÇOIS HENRI BORDES

EVOLUTION IN THE PALEOLITHIC CULTURES

In publications dedicated to prehistory, whether monographs, textbooks, or popular works, one often finds the term "evolution" used sometimes to refer to the general evolution of civilizations or human types, at other times attached in adjectival form to the name of a culture ("evolved Mousterian") or a tool type ("Evolved Châtelperron point").

In earlier days it seemed perfectly satisfactory to parallel the evolution of Paleolithic cultures with the physical evolution of man. Thus Heidelberg Man was responsible for the Abbevillian; Neandertal Man for the Mousterian; and *Homo sapiens* for the Upper Paleolithic. This viewpoint, which is probably reasonably correct in the broadest sense, nevertheless today appears to correspond only loosely with reality. At any rate, the evolution of the Paleolithic cultures was formerly considered as being linear: the Abbevillian begat the Acheulian, which in turn begat the Mousterian; then—linked to, or separated from, the Mousterian, depending on the author—came the Upper Paleolithic, with its three great divisions of Aurignacian, Solutrean, and Magdalenian. Each of these principal cultures was itself subdivided into subperiods, and distinction was made, for example, between the Early Mousterian, the Middle Mousterian, and the Upper or "Evolved" Mousterian. In time, the progress of prehistoric research made it apparent that evolution had not been the same everywhere on earth and that the reality was much more complex than had been thought. D. Peyrony, for example, correctly distinguished the Périgordian from the contemporary, but quite different, Aurignacian, and the Mousterian of Acheulian tradition from the Typical Mousterian.

But it is a fairly obvious fact that evolution—which is to say, change plus continuity—must have taken place, at least in certain cultural

FRANÇOIS HENRI BORDES is Professor of Prehistory at the Faculté des Sciences de Bordeaux. He was affiliated with the French National Center for Scientific Research from 1945 to 1956, studying Pleistocene geology and Paleolithic archaeology, and has about 75 publications to his credit. He is currently specializing on the Mousterian cultures.

phyla, since the time of the appearance of *Homo sapiens.* As far as the passage from the Lower to the Upper Paleolithic is concerned, if we reject the idea so dear to the writers of science fiction that modern man is descended from the crew of a Martian spaceship marooned on earth during the Quaternary period, then we must recognize that it took place somewhere, or even in several places, on this globe (Bordes, 1958). So here, too, there must have been evolution, whether it preceded or followed the appearance of *Homo sapiens.* The fact of evolution (in the broadest sense) of the Paleolithic cultures therefore seems established beyond possible doubt. But the problems posed by this evolution and its processes are often poorly understood. One has to delve into the realms of semantics and the analysis of preconceived ideas to understand, for example, exactly what is meant by such terms as "Evolved Mousterian" or "Evolved Châtelperron point." Fundamentally, the question is this: What criteria, other than relatively valueless aesthetic ones, should be applied in deciding whether one industry is more "evolved" than another?

The criterion of stratigraphy can be very useful here. An industry, or a phase of an industry, located stratigraphically after another, may be pronounced more evolved. But this holds good only when the industries belong to the same phylum; there exists such a thing as false evolution, based on errors in interpretation of stratigraphic and typological data. The clearest case is that of the Mousterian.

At the time of the first excavations on the upper terrace of the very complex type site, Le Moustier (Dordogne), there was observed the superposition of La Quina-type Mousterian over the Mousterian of Acheulian tradition. Explorations at the site of La Quina, in Charente, led to the discovery of a considerable number of very fine tools made on flakes, while the flake tools of the Mousterian of Acheulian tradition are often less attractive, or at any rate less striking. The La Quina-type Mousterian was therefore considered as Upper or "Evolved" Mousterian, since it was aesthetically superior to the Mousterian of Acheulian tradition, which contained bifaces or hand-axes (obviously relics of the Acheulian!), and it lay over the Mousterian of Acheulian tradition at Le Moustier. So the sequence Mousterian of Acheulian tradition = Early Mousterian, La Quina-type Mousterian = Upper Mousterian seemed to be well established. This "evolution" seemed even further confirmed by the fact that when thick levels of the Mousterian of Acheulian tradition are found, the upper part clearly contains fewer bifaces than the lower. Therefore, whenever the Mousterian of Acheulian tradition was found without stratigraphic context, it was pronounced "Early," and wherever the La Quina-type Mousterian was found in similar circumstances it was declared

"Evolved." But, in actual fact, the reality is quite otherwise. D. Peyrony (1930) gave the first blow to this theory by isolating the Mousterian of Acheulian tradition as a culture apart from the others, basing his work on the interstratification at the lower site of Le Moustier. In 1931 R. Vaufrey observed that the Mousterians at Ehringsdorf (Germany) in the last interglacial period "already worked their flint as the Upper Mousterians of La Quina were to do later." But these announcements passed almost without notice.

However, there exist sites (e.g., La Gane, near Groléjac, in Dordogne) where the Mousterian of Acheulian tradition comes *after* the La Quina-type Mousterian. The same thing exists at Pech de l'Azé (Bordes, 1954–55) and at Combe-Capelle, both also in Dordogne. In this latter site the so-called "stratigraphic reversal," together with the crude appearance of the industry, caused an absolutely typical La Quina-type Mousterian to be labeled "Tayacian" (Peyrony, 1943). In reality, there is an interdigitation of the Mousterian of Acheulian tradition with the La Quina-type Mousterian and even with other types of Mousterian which need not occupy us here. The point is that it is a question of different phyla evolving in parallel fashion and not of an evolution from one to the other: the Mousterian of Acheulian tradition is sometimes "evolved" or Upper, and the La Quina-type Mousterian is sometimes early.

Finally, if we analyze the two industries from the typological point of view, the theory of evolution from one to the other is completely demolished. The thick side-scraper, which appears as far back as Level 3 at La Micoque, is a tool almost as ancient as the biface. The two industries are quite different: the thick side-scraper, the side-scraper with bifacial retouch, and the true *limace,* for example, do not exist in the Mousterian of Acheulian tradition. In the La Quina-type Mousterian true bifaces and backed knives are nearly or completely absent. And the ordinary or mutually shared tools are in very different proportions. Last of all, the "forerunner" tools, i.e., those which would be developed in the Upper Paleolithic, clearly appear more important in the Mousterian of Acheulian tradition, especially in the final phases, which are nevertheless described as "degenerated" by certain writers (e.g., Vaufrey, 1933) because of the rarity and crudeness of the characteristic bifaces. We have, then, two distinct phyla and not an evolution from one to the other. Yet, if we have to fix on any Mousterian as being more evolved than the others, it is rather the Mousterian of Acheulian tradition.

On other occasions this or that industry is described as being "evolved" on the basis of completely a priori typological considerations. For example, the Pradels write (1954) at the end of their study of the site of L'Ermitage, "The Mousterian at l'Hermitage, containing very rare bifaces, is an evolved Mousterian, as is shown by the *pointes à perfectionnement,* the *limaces,* the blades, the end-scrapers on blades and by the presence of burins." But the truth is that bifaces are rare or even absent in certain of the very ancient industries belonging to the Clactonian cycle dating from the Second Interglacial on; the *pointes à perfectionnement* which are described include, it is true, some which really are stemmed points, but also some asymmetrically stemmed pieces which could easily have resulted from chipping accidents, from repairs, or even from the original shape of the flake and, being without retouch, are of doubtful validity. Also among the *pointes à perfectionnement* are some whose bases have been thinned by the removal of the bulb of percussion; this type, so far from being an innovation, is one which appears as far back as the early Upper Acheulian. As far as the elongated points (which are included in the *pointes à perfectionnement*) are concerned, these occur at the Mousterian site of Combe-Grenal and are especially numerous in Level R, which is by no means the upper layer, since it is covered by twenty-five other Mousterian levels where these points are rare or absent. The *limaces* are known at Ehringsdorf in the last interglacial; blades are abundant in certain Upper Acheulian industries (e.g., at Montguillain and sites in the Somme Valley); finally, burins and end-scrapers are known from the Middle Acheulian at Atelier Commont, in the Somme Valley.

In another publication Miss Garrod (1956) does not hesitate to refer to a "Levalloisian, already rather evolved" in the 15-meter beach deposits at Ras-Beirut (Lebanon) on the basis of 10 worked flints (4 rolled and 6 unrolled *flakes*) reported by Father Fleisch (1946). Further comment here would, I think, be superfluous.

It is evident, then, that it would be far better to renounce this sort of a priori thinking, as well as oversimplified ideas about stratigraphy, and to follow an entirely different course if the real evolution of the Paleolithic cultures is to be understood. First of all, it is essential to establish exact and delicate stratigraphies. Second, the industrial types and subtypes must be clearly distinguished; it will get us absolutely nowhere to mix different phyla, such as the Mousterian of Acheulian tradition and the La Quina-type Mousterian. The distinction which D. Peyrony made between the Aurignacian and the Périgordian was a giant step forward in this direction. Third, the chronological relationships between the various phases of development of a single phylum must be fixed as firmly and closely as possible. Finally, having arrived at this point, we can proceed to follow the *real* evolution of tool types and their respective frequency percentages and thereby come to an understanding of the evolution of the Paleolithic material cultures. When we have established the criteria at the basis of evolution, which, as already mentioned, is constituted by continuity plus change, we can determine the general direction of this change by comparisons with chronologically preceding or succeeding stages in the same phylum.

However, the industrial types should be established not only on the basis of certain "characteristic" objects, or even by a mere qualitative study of the assemblage, but also by a quantitative study through the use of statistics. Real comparisons thereby become possible and fruitful. A study of this kind in the Mousterian of Acheulian tradition, principally at the site of Pech de l'Azé (Bordes, 1954–55), has enabled us to establish inside this industry an evolutionary sequence which seems to be general, at least in the southwest of France. The Mousterian of Acheulian tradition presents two great types which we have called "Type A" and "Type B," the former being the older. Type A is characterized by a moderate percentage (from 25 to 40 per cent) of side-scrapers of all sorts (except La Quina types); by a great variety of tools; by the presence of backed knives (though usually in small numbers only); and by the existence of bifaces, which, although in various proportions, are always important. Type B is characterized by a sharp falling-off in the quantity and quality of side-scrapers (between 5 and 10 per cent) and bifaces; by the strong development of backed knives and denticulated tools; and by the development of certain tools which herald the Périgordian, e.g., backed knives tending toward the Châtelperron knife, burins (sometimes double), end-scrapers, perforators (not always, however), and truncated blades. At Pech de l'Azé one of the layers (Level A) presents an intermediate industry. The same evolution can be followed, though less clearly, at Le Moustier (Levels G and H), where new excavations using finer stratigraphic methods are waiting to be done, and in every site where several levels of the Mousterian of Acheulian tradition have been distinguished.

Now, if we follow the cultures in the same phylum from the Lower Acheulian to the Mousterian of Acheulian tradition, we can observe the evolution of several types of tools and, in particular, of the backed knives. Crude and often only partially backed in the Lower and Middle Acheulian, they become more delicate in the Upper Acheulian but are still usually fabricated on large flakes. In the Mousterian of Acheulian tradition the dominant type is, first, the Abri Audit-type knife—still broad, with a rather narrow back; later other types appear, made on more elongated flakes and, finally, on blades, with the retouched back becoming broader and approaching the Châtelperron type which characterizes the Périgordian I industry. This evolution continues into the Upper Paleolithic. The Châtelperron-type knife itself becomes finer and narrower and then seems to adapt itself to other purposes, becoming the Gravette *point,* which is different from the primitive knives; the knives, however, persist in the same levels. This backed knife, almost unknown in the Aurignacian and Solutrean, continues as a rare form in the Magdalenian and takes on an important new lease on life in the Azilian—on the one hand, under the form of the Azilian point and, on the other, in the *lame de canif* (penknife blade) and another tool shaped like the rough segment of a circle.

The evolution of the cultures themselves can be considered as being of two distinct kinds: (1) normal, or linear, evolution and (2) *évolution buissonnante,* or ramifying, evolution.

The Solutrean in the Périgord region of France is a good illustration of linear evolution. Against the background of a basic tool kit which changes little from the Lower to the Final Solutrean, there occurs a series of characteristic and sometimes prolific tools which succeed or rather carry on from each other. In the beginning there are the *pointes à face plane;* then appear and develop the laurel leaves and, finally, the shouldered points and willow-leaf points. The continuity of the Solutrean, with its common base in different levels and the persistence of the *pointes à face plane* almost to the very end, cannot be doubted; but neither can one doubt the presence of change, for the dominant role in the assemblage passes clearly from the *pointe à face plane* to the laurel-leaf to the shouldered point. This continuity, which has recently been questioned by certain authors on theoretical grounds, has since been clearly demonstrated by D. de Sonneville-Bordes (1958).

On the other hand, according to the same author, the Upper Périgordian seems rather to possess a ramifying type of evolution, although really trustworthy data are too rare to allow us to state this too categorically. After Périgordian IV, with Gravette points, come three phases or branches: Périgordian V^1, with Font-Robert points; Périgordian V^2, with truncated elements; and Périgordian V^3, with Noailles-type burins. These characteristic tools are often very numerous. In the few sites where a refined stratigraphy has been carried out, these sublevels of Périgordian V seem to follow each other chronologically in the above order, but we are still not in a position to say that there is a continuity from one to the other. Although the sequences of Périgordian IV → Périgordian V, with Font-Robert points; Périgordian IV → Périgordian V, with truncated elements; Périgordian IV → Périgordian V, with Noailles-type burins, seems to be beyond all reasonable doubt (Gravette points, more or less numerous, exist in each of these subtypes of the Périgordian V), the continuity of Périgordian V, with Font-Robert points → Périgordian V, with truncated elements → Périgordian V, with Noailles-type burins is not established. It may be that we are dealing here with different branches, each derived from Périgordian IV and evolving independently, and that the appearance of evolution caused by their superposition in two or three sites is due solely to chance. It would seem that, instead of having one characteristic tool "take off" from another without abruptly disturbing the continuity of the earlier tool (as in the Solutrean), the Périgordian V shows a sudden replacement of one tool type by another. But it is wise to remind ourselves that this effect may be only an illusion. Nevertheless, the problem is particularly important because, if it is a case of divergent branches, the Périgordian V, with Font-Robert points, might be considered a possible ancestor of the Solutrean-type retouch, since this technique appears on certain Font-Robert points.

Another example of ramifying evolution is provided by the Acheulian. At the Middle Acheulian level there appears the flint-working technique known as the "Levallois technique," and from this moment two great branches of the Acheulian, together with the industries which derive from them, run side by side: the Acheulian of Levalloisian technique and the Acheulian of non-Levalloisian technique. It seems, however, that *inside* each of these great technical branches there developed, similarly, further typological branches or subbranches, characterized by different types of bifaces, or at least by the dominance of different bifacial types.

In the Final Magdalenian a microevolution takes place, and several special kinds of tools (e.g., Teyjat points, Laugerie-Basse points) appear, but we still cannot say whether this evolution is of the linear or of the ramifying type. In the same way, in certain Final Magdalenian sites there appear geometric microliths, forerunners of the post-Azilian Mesolithic ones, while in other sites the first Azilian points apparently occur alone without accompanying the geometric forms.

Evolution can also take place through a pauperization both of tool types and of the tools themselves. The Azilian is an example of this; it is certainly derived from the Magdalenian, but it undergoes a subtraction or disappearance of a large number of tool types, especially of burins, and the assemblage concentrates around certain very prolific types such as the Azilian points and the small, short end-scraper.

The main handicap we have to face today in studying the evolution of the Paleolithic cultures stems from the methods of excavation and classification traditionally employed in collecting the great majority of the specimens which are found in the museum collections; the old methods allow us only to get at the averages, the rough draft, without the fine detail that we must have to establish the means and direction of the process. A hypothetical example will serve to illustrate this situation.

Suppose that in a given site we have three levels of the Périgordian V industry. At the top is Level C containing a Périgordian V with Noailles burins; under this is Level B with truncated elements; and below comes Level A with Font-Robert points (these layers may be slightly separated from each other by thin sterile layers or, as in most sites where this superposition has been observed, not separated at all). In the old excavations everything was lumped together, and so the collections contain all three types simultaneously. But in more recent and more careful excavations (e.g., that of D. Peyrony at La Ferrassie, 1934), the three levels were excavated; however, even in this case each level was considered to be a homogeneous unit, an entity in itself—in other words, everything from a given level was classified together, and no attention was paid to the possible variations in the *vertical* distribution of the objects in each level. A study of the possible evolution of Périgordian V, based on these collections, is accordingly impossible. On the other hand, if more modern methods are used and the vertical positions are observed within each level, there are three possibilities which can result (see Fig. 1).

1. (Fig. 1, *a*). The Font-Robert points exist in considerable quantity up to the top of Level A and are absent in Level B (truncated elements); likewise, the truncated elements exist in large numbers up to the top of Level B, which, however, contains no Noailles-type burins. The probability, then, of its being a case of quite different cultures replacing each other in this one site is very strong, and, if there has been an evolution, it must have taken place somewhere else. If every site gives a similar vertical distribution, then the probability of an evolution becomes practically nil.

2. (Fig. 1, *b*). The Font-Robert points reach a maximum in quantity in Level A, then decrease and disappear *in* the base of Level B. The truncated elements appear either in the top of Level A or in the base of Level B, attain a maximum, and disappear *in* the base of Level C. The Noailles-type burins take off from here and carry on in Level C. In such a case, there is a strong probability of an evolution from one type to another.

3. (Fig. 1, *c*). In this case there is a perfect "relay" from one type of tool to another, with no overlapping. Evidently, in this case it might

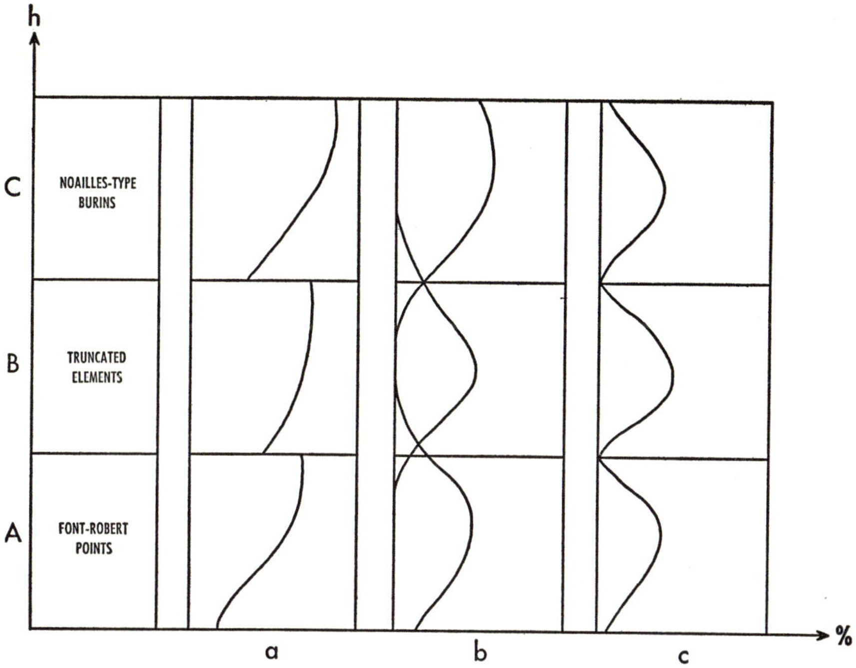

FIG. 1.—Hypothetical results of applying modern methods to a study of three levels of the Périgordian V industry, by noting the vertical distribution of objects at each level. For explanation, see text.

be cutting it rather finely to consider so neatly timed and perfect a replacement as an evolution.

But up to the present time no excavation using such modern methods has been undertaken for such a sequence, and the old excavations give only the averages, so it is impossible to study the evolution of Périgordian V if such an evolution exists. The same observations apply also to the other cultures. Indeed, we are lucky when the former excavators did not confuse archeological and geological levels, and we can be certain that in a given museum series there is really only a single pure culture!

The most heartrending case in point is that of the site of Roc de Combe-Capelle (Dordogne), where Hauser's "excavations" completely destroyed a layer which seems to have contained Font-Robert points at its base and *pointes à face plane* at the top. Might this have been the key to the origin of the Solutrean?

Evolution by means of the appearance and development of a new type seems to have been extremely rapid at times, and is difficult, even impossible sometimes, to follow. We may fairly safely suppose that an invention develops only if its natural and cultural milieu is propitious (it is needless here to present such historical examples as Leonardo da Vinci's invention, or rather attempt at invention, of the airplane and the submarine). When the milieu is not favorable, inventions do not flourish and are quickly forgotten. On the other hand, if the milieu is ready for it and the invention fills a need, it may develop in an explosive manner. It is not difficult to imagine that the propulsor or dart-thrower was rapidly adopted after its invention and that its relative rarity in archeological deposits is due to the fact that the majority must have been made of wood. The dart-thrower, however, depends on the existence of a small dart point, in either stone or bone, which must be both sharp and relatively light; so the Acheulians would probably have had little use for such an invention.

It will always be difficult for prehistorians to discover the exact origin of this or that implement type, therefore, since if an invention did not give an important advantage, it did not go into heavy production and so the chances of finding it are slight. On the contrary, if it was really advantageous, it was immediately copied, and, taking into account the relatively gross picture reflected by archeological investigations, one will get the impression, thanks to the explosive development of this invention, that an advanced stage of its utilization has already been attained. In reality, of course, it may be only the beginning. Such a case is somewhat analogous to what Teilhard de Chardin called *la résorption automatique de la base des phylums* in paleontology. We need not be surprised, then, if certain characteristic tools from certain prehistoric levels seem to follow the law of "all or nothing," that is to say, they are either completely absent or are present in great numbers; this is particularly true when dealing with series collected by the old excavation methods.

These observations are especially pertinent when studying evolution inside a given culture. But if the environment undergoes great changes, it is entirely possible that such an explosive phenomenon might occur and constitute the passage from one culture to another. Thus it is possible that the transformation from the Magdalenian to the Azilian was also more or less explosive; there may have been an abrupt acceleration in the production of certain tool types which reached the point of creating an apparent discontinuity which does not really exist, since the characteristic tools of the Azilian appear in the Final Magdalenian.

Such appearances of characteristic elements, this "foreshadowing" of the succeeding culture, is in my opinion one criterion of evolution (this word, of course, being stripped of any ethical implication) which is safer to apply than any other. The Mousterian of Acheulian tradition, Type B, is "evolved" because it developed the tools which were to characterize the later Périgordian, even though the flint working is inferior in quality to that of earlier stages. In addition, the tools which were characteristic of the Mousterian tended to disappear. So evolution must be judged according to the total and proportions of type frequencies and not only on the appearance of novel forms which sometimes were discontinued.

Can we distinguish any particular direction or tendencies in the general evolution of prehistoric cultures? I believe so. It seems possible to distinguish three great cycles which developed in a parallel fashion, though with astonishing convergences at times, in the inhabited parts of the earth.

The first great cycle is that of the Lower and Middle Paleolithic. Even if there was a common origin or connections between the Abbevillian-Acheulian of Africa, Europe, and Asia, it remains none the less true that later development was distinct. Each of the three continents gave birth to a "Mousterian" stage of evolution—the various Mousterians of Europe and Central Asia, the Soan and Choei-Tong-Keou type industries of South and East Asia, the Stillbay industry of East and South Africa (North Africa seems, at that stage in spite of a later development of the Aterian, to belong to the great European complex, in the widest sense). And, on the other side of the world, the Tasmanian industry (I do not say culture) seems by many traits to have been at a Mousterian stage when its makers were destroyed.

The second cycle is that of the Upper Paleolithic—Mesolithic and concerns peoples who were still exclusively, or nearly exclusively, hunters. The Upper Paleolithic (in the sense of an evolutionary stage) existed in Europe, North Africa (the Typical Capsian), in Asia (China), and in America (Folsom, etc.). In Africa it seems that this stage, if not jumped, was avoided, and after the Stillbayan we find industries which are more Mesolithic than Upper Paleolithic, although we cannot assume an evolution unless we can rule out any outside influence. There exists in Kenya, in the famous "Capsian," an industry of this Upper Paleolithic–Mesolithic type. The true Mesolithic, in the sense of an evolutionary stage, seems to be widespread over the whole world.

The third great cycle is that of the Neolithic, which is characterized much less by pottery and polished stone than by cultivation of the soil, domestication of animals, and food production. It is recognized today that multiple centers of "Neolithicization" have existed throughout the world.

The extraordinary evolution—it has even been called a "revolution"—caused by the Neolithic has brought about a complete change in man's mode of life. But, as I see it, there was another revolution in even earlier times which seems to have passed almost unnoticed: the revolution brought about by the invention of the long-distance projectile-thrower, first the spear- or dart-thrower, then later, perhaps in the Upper Paleolithic, the bow. This revolution, when combined with a steppe environment with its rich herds of animals, produced a milieu so favorable to hunting peoples that we may safely consider the resulting cultures of the Upper Paleolithic as the first great civilization, with its peak or climax in the Magdalenian. For it was indeed a civilization, with everything that word implies in terms of a cultural superstructure: rites, legends, songs (all, unhappily, forever lost), and art, even if based on magic or religion (medieval and Greek art was, after all, nothing else). All of which indicates a relatively dense population, rather well-off materially and capable of providing a certain amount of leisure time to think and create.

I have my thanks to give to M. Philip Smith for his excellent translation of the original French text.

References

BORDES, F. 1954–55. "Les Gisements du Pech de l'Azé (Dordogne)," *Anthropologie,* LVIII, 401–32; LIX, 1–38.

———. 1958. "Le Passage du Paléolithic Moyen au Paléolithic supérieur," *Neanderthal Centenary,* pp. 175–81. Utrecht.

FLEISCH, H. 1946. "Le Levalloisien du niveau + 15 à Râs-Beyrouth (Liban)," *Bulletin de la Société préhistorique française,* XLIII, 299–301.

GARROD, D. 1956. "Acheuléo-Jabroudien et Pré-Aurignacien dans la grotte du Taboun (Mont Carmel)," *Quaternaria,* III, 39–59.

PEYRONY, D. 1930. "Le Moustier, ses gisements, ses industries, ses couches géologiques," *Revue anthropologique,* Nos. 1–3, 4–6.

———. 1934. "La Ferrassie," *Préhistoire,* III, 1–92.

———. 1943. "Combe-Capelle," *Bulletin de la Société préhistorique française,* Vol. XL, pp. 243–57.

PRADEL, L. and J. H., "Le Moustérien évolué de l'Ermitage," *Anthropologie,* LVIII, 433–43.

SONNEVILLE-BORDES, D. DE. 1958. "Recherches sur le Paléolithique supérieur en Périgord." Thesis, Paris.

VAUFREY, R. 1931. "Les progrès de la paléontologie humaine en Allemagne," *Anthropologie,* XLI, 517–51.

———. 1932. "Le Moustérien de tradition acheuléenne du Pech de l'Azé," *Anthropologie,* XLIII, 425–27.

GORDON R. WILLEY

HISTORICAL PATTERNS AND EVOLUTION IN NATIVE NEW WORLD CULTURES

Culture is the means whereby man adapts himself to his natural and social environments and to his pre-existent cultural milieu. The history of this adaptation is the story of cultural evolution. Like biological evolution, the processes of the evolution of culture are selective ones by which the species promotes its survival and fulfilment. The "species," in this case, is man and the social groups in which he lives. The courses which man, society, and culture follow in meeting the challenges of environmental opportunity and in seeking and achieving adaptations are not programed by any laws of inevitability. "Evolution," as Dobzhansky has observed (1958, p. 1096), "is not striving to achieve some fore-ordained goal; it is not the unfolding of predetermined episodes and situations." Nor are these courses, as the same author makes clear, completely the result of pure chance.

Man adapts himself to his natural, social, and cultural settings. It is, however, in his relationship to the natural environment that his adaptive efficiency may be most easily adjudged. Depending upon a combination of the natural resources and the technical culture which he evolves to take advantage of them, he perishes or thrives as a hunter, collector, or farmer. There is abundant evidence from archaeology and history to see how man's technical culture, as directed toward the problems of subsistence, has evolved, more or less steadily, toward greater effectiveness.

The annals of human history also show that as man's technical heritage evolved, it did so concomitantly with significant changes in his social settings. These changes created environmental challenges of another order. The selective-adaptive equation of a better spear to kill more game became complicated by the factor of more people both to kill and to eat the game. Despite the fact that through the ages the spear has also been used by some to dissolve the demographic di-

GORDON R. WILLEY is Bowditch Professor of Mexican and Central American Archaeology and Ethnology, Harvard University. His archaeological researches have resulted in several important publications, most recently *Method and Theory in American Archaeology* (University of Chicago Press, 1958).

lemma, the main trends of social inventiveness have been those which sought ways to accommodate man to man as well as man to nature. This accommodation is expressed in various forms of social and political organization; and, although it is more difficult to evaluate the adaptive efficiency of these social forms than it is to appraise the toolmaking and food-producing aspects of culture, it is possible to trace a general evolution in their history of increasing size and complexity. Developing along with the technological order and the social order is the other aspect of man's culture which is concerned with the realm of ideas. Religion, art, science, world view—it has been summed up in essence by Redfield as the "moral order" (1953). The evolution of these institutions is not well understood; however, their growth appears, in some degree, to be related to both the technological order and the social order. Thus Kluckhohn (1958) states:

> . . . The association of type of economy with a specific kind of social organization and of these and other aspects of culture with one sort of moral order as opposed to another is by no means altogether a random one. Hence one can anticipate that there will ordinarily be some determinable relationships between the size of social groups and characteristics of their value systems.

Perhaps an analogy with biological evolution may again be useful here. In this regard Dobzhansky (1958, p. 1097) has said:

> Genes determine the possibility of culture but not its content, just as they determine the possibility of human speech but not what is spoken. The cultural evolution of mankind is superimposed on its biological evolution; the causes of the former are non-biological without being contrary to biology, just as biological phenomena differ from those of inanimate nature but are not isolated from them.

In this same way it seems probable that the technological and social basis of a culture or civilization provides a platform for and sets limits to its artistic, religious, and philosophic forms, but the forms themselves and their content are the resultants of human genius and the events of particular culture histories.

In attempting to trace the courses and processes of the evolution of New World cultures, we will define, first, the major subsistence and settlement types that are found in prehistoric America. These will then be described and examined within the framework of what are called "basic New World life-patterns." These life-patterns are historical spheres or continuities of great geographic extent and long chronologic persistence. Each is characterized by an underlying natural environmental-subsistence ecology, as well as by particular societal or community types and by cultural institutions. The reconstructions of cultural ecology from natural environmental habitats and tool types, of social units from settlement patterns, and of cultural forms from whatever available clues are, of course, limited by the nature of archaeological data. The quantity and quality of archaeological inference will vary greatly as one moves from the relative security of subsistence and settlement interpretations to those involving the more abstract—and, at the same time, more uniquely and interestingly human—results of men's minds and hands. Inference and its reliability will also vary with the richness or poverty of the record and with the possibilities of injecting into it the reinforcements of ethnohistorical documentation. Nevertheless, any culture history of the native Americas which is to be more than a chronicle account of artifact types must reckon with such inferences.

The extent to which cultural institutions and forms are functionally and causally interrelated with each other and with subsistence and settlement (societal) types is, of course, a major nexus of the problem of human cultural evolution. We cannot pretend to offer many answers in this discussion, but if human beings living in social groups in the natural world are the creators and bearers of culture, it seems likely that such interrelationships exist.

New World Patterns of Culture

SUBSISTENCE TYPES

There are three principal subsistence types native to the New World: big-game hunting, gathering-collecting, and cultivation.

Big-game hunting refers to the stalking and killing of large mammals as a principal source of food and clothing. It was a mode of subsistence widespread in the Americas in Pleistocene times (antedating 7000 B.C.). In later eras it was also followed on a somewhat more restricted geographic scale. It is, obviously, conditioned strongly by natural environmental circumstances.

Gathering-collecting refers to all those means of procuring food from the natural setting except the hunting of big game or the cultivation and domestication of plants and animals. The nature of the foods may be small animals, fish, shellfish, roots, seeds, nuts, and berries. The occasional taking of large game is not precluded, although this is not a specialization. At one time or another gathering-collecting economies characterized most parts of aboriginal America. Although there are clues to their appearance as early as the Pleistocene, they are more typical of the recent geological era. *Gathering-collecting* may be subtyped into (*a*) *gathering,* where food resources were extremely meager, the techniques for obtaining them simple and unspecialized, and the efforts of the people involved devoted almost wholly to the food quest; [1] (*b*) *collecting,* where there are more ample resources and more effective techniques; and (*c*) *intensive collecting,* which is marked by abundant resources, specialized techniques, and the accumulation of food surpluses.

Cultivation of food plants is the only major type of food production in native America.[2] Its origins may begin as early as the close of the Pleistocene, but the pattern was one of only partial economic significance until the second millennium B.C., when it became of primary importance in Nuclear America (Willey and Phillips, 1958, pp. 144–47). Maize was the most important staple crop, and at the time of the discovery of America by the Europeans it was found spread over the southern half of North America and the northern two-thirds of South America. Manioc, another staple, had a more limited distribution. From the points of view of techniques and productiveness, native American *cultivation* types may be considered as (*a*) *incipient,* where either food plants, farming techniques, or both are but partially developed and the produce has only a minor role in the economy; (*b*) *established,* where the primary means of food-getting is farming; and (*c*) *intensive,* where, as a result of favorable soils, climatic conditions, a variety of nutritious plants, and the employment of techniques such as irrigation, terracing, "floating gardens," and fertilizers, a high productivity is assured.

SETTLEMENT PATTERNS

As they reflect the nature of the community, we have classified New World native settlements into five major types: *camps, semipermanent villages, permanent villages, towns-and-temples,* and *cities.*

The *camp* site is small in extent and marked by thin and scattered refuse. There are few or no clues to dwellings or other structures. When found as the only type of settlement for a society and culture, it is reflective of a small, non-sedentary community unit, probably a wandering band of less than one hundred individuals. Such bands roamed over territories of varying extent and followed a hunting or a gathering-collecting subsistence (Beardsley *et al.*, 1956, pp. 135–

[1] This distinction between "food-gathering" and "food-collecting" follows the terminology that Braidwood (1958*b*) has suggested for the Old World. "Food-gathering" pertains to the cultures of the Lower and Middle Paleolithic and to an absence of regional specialization or diversification in technology. "Food-collecting," on the other hand, pertains to Upper Paleolithic and post-Paleolithic cultures and to a much greater degree of diversification and specialization.

[2] The domestication of animals was never of primary economic importance in pre-Columbian America. The nearest approach would be the mixed horticulture animal-husbandry economy of the late prehistoric tribes of northern Chile and northwestern Argentina.

38).[3] On the American scene the camp is found in virtually all parts of both continents and at all time periods.

Semipermanent villages are those which give evidence of substantial occupation over an appreciable span of time. Traces of perishable dwellings may be found; accumulated refuse is often deep; and cemeteries may be in or near the site. They were the living places of communities which appear to have occupied such locations sporadically or to have shifted their site locus after every few years. Inhabitants must have numbered several hundred. The semipermanent village marks a shift from a wandering toward a sedentary life. In some American areas it is correlated with a change from food-collecting to intensive collecting or to cultivation.[4]

The *permanent village* settlement is the locus of a community which has occupied the same spot steadily over a long period of time. It reflects this sedentism in permanence of architectural features as well as refuse concentration. Its occupants numbered from a few hundred to over a thousand. The social and political organization of these communities may have rested upon kinship, but there are instances from ethnohistory of chiefs and social class distinctions. In the New World the permanent village, in both archaeological and ethnohistoric times, is most often associated with established cultivation. The American data pertain to two subtypes of permanent villages. The first has a unitary or "undifferentiated" settlement plan. In this, the village is a compact unit of dwellings which may contain within itself a special temple or politicoreligious structure. The second—a dispersed or "differentiated" subtype—has a nucleus or politico-religious center which is surrounded at varying distances by small hamlet or homestead satellites.[5]

The *town-and-temple* settlement pattern and community type is an enlargement and elaboration of the permanent village pattern. The total community is now likely to be a dispersed and differentiated one. The politicoreligious, or "ceremonial," center is the principal focus. It may, or may not, be within the confines of a compact town; but in either case it is usually surrounded by satellite villages and hamlets. The population of the total community probably ran to a few thousand. Their social and political organization was probably hierarchial,

[3] My use of settlement or community types parallels in part that of the seminar paper by Beardsley *et al.* (1956), but I have used archaeological site terms rather than community terms. The "camp" site is presumably the archaeological correlate of what Beardsley *et al.* have defined as "free-wandering" and "restricted-wandering" bands.

[4] My "semipermanent village" is the approximate equivalent of the Beardsley *et al.* (*op. cit.*, pp. 138–41) "central-based wandering" and "semi-permanent sedentary" community.

[5] *Ibid.*, see pp. 141–43, the "simple nuclear centered" community.

with governmental authority supplanting kinship. These towns were the centers of full-time artisans and the residences of rulers, and activities in connection with the temples were directed by an organized priesthood.[6] The town-and-temple settlement is associated with established or intensive cultivation and is best known from Middle America and Peru, although the type is also found in other parts of South and Central America, the West Indies, and in the southeastern and southwestern United States.

The *city,* as distinct from the town-and-temple, cannot be defined upon the basis of settlement features alone, although sheer size and population numbers are among the criteria. The native American city is known only from parts of Middle America and Peru. It was the nerve center of a civilization. It attracted to its precincts the artisans, artists, and specialists. Its temples housed pantheons of deities. It was a focus of trade, of the collection of taxes, and frequently of the military power of the state. It is characterized by monumental architecture, by great arts, and in some places, as in the territory of the lowland Maya, by evidences of true mathematical and astronomic sciences and writing. Its actual settlement pattern may be compact and truly urban, or it may be differentiated into a politicoreligious center surrounded at a distance by villages and hamlets. Its traditions may be relatively homogeneous, or it may have a more cosmopolitan cast. Among the native New World cities, some appear to have maintained a rather limited regional scope of power and influence, while others were undoubtedly the capitals of large empires.[7]

Historical Outlines of Basic New World Life-Patterns

As noted in the introduction, the "basic New World life-patterns" are conceived of as historical entities of great geographic-chronologic scope, each distinguished by the persistences of certain subsistence and community types and cultural forms. The exact number of such life-patterns in the New World is, of course, debatable and somewhat arbitrary. A few major ones emerge from the welter of archaeological facts, and we shall discuss these and refer also to others with less ample data.

pleistocene gatherers (?)

There are suggestions, although the evidence is by no means conclusive, that the earliest inhabitants of the New World were peoples fol-

[6] *Ibid.,* see pp. 143–45, the "advanced nuclear centered" community.

[7] The "city," as I have defined it here, would include some of the communities termed by Beardsley *et al.* (*op. cit.,* pp. 143–46) as "advanced nuclear centered."

lowing a very simple gathering type of subsistence. At Tule Springs, Nevada, there is an artifact assemblage of crude chipped scrapers and chopping tools found in association with extinct camel, bison, and horse bones and with a radiocarbon date of about 22000 B.C. (Harrington, 1955; Simpson, 1955; Wormington, 1957, pp. 197–98). The discoveries in Friesenhahn Cave, Texas, of scrapers and choppers date from well back into the Wisconsin glacial stage, if not earlier (Sellards, 1952; Krieger, 1953; Wormington, 1957, pp. 218–19). Somewhat later, but almost certainly of terminal Pleistocene age, are the rather nondescript scraper-chopper-like stone tools from the lowest levels of Fishbone Cave, Nevada (Orr, 1956; Wormington, 1957, pp. 192–93), and Danger Cave, Utah (Jennings, 1957).[8] None of these artifact groups clearly represents a big-game hunting economy. Their extremely unspecialized nature is vaguely suggestive of Old World Lower Paleolithic cultures (Willey and Phillips, 1958, pp. 82–86). Their presence in the Americas may result from man's very early migrations to these continents, of which we have now only the most imperfect record. On the other hand, it is possible that these finds are merely partial or incomplete assemblages of the Pleistocene big-game hunting pattern.

PLEISTOCENE BIG-GAME HUNTERS

The earliest life-pattern which can be clearly formulated for the New World is that of the Pleistocene big-game hunters (Willey and Phillips, 1958, pp. 86–103). Its origins are uncertain. In the pursuit of large Ice Age mammals and in the employment of pressure-flaked projectile points it has general Old World Upper Paleolithic parallels, but the forms of the American points and other artifacts differ from those of the Old World. Its age is best defined from the period just antedating and concurrent with the final substage of the Wisconsin glaciation. Radiocarbon dates and geological estimates indicate this to be the span from about 12000 to about 7000 B.C., although there are a few dates which go back earlier (Willey and Phillips, 1958, pp. 87 and 91; Crook and Harris, 1958). The Sandia and Clovis projectile points and associated flint scrapers and knives, together with mammoth remains and the bones of other extinct fauna, are the best-known earlier representatives of the big-game hunting pattern.[9] The long, fluted, and fishtailed Clovis type point is widely distributed through North Amer-

---

These would, in effect, be "orthogenetic" cities. The Beardsley *et al.* "supra-nuclear integrated" community would correspond to the "heterogenetic" city.

[8] One projectile point from the earliest Danger Cave level is of long lanceolate form (see Jennings, p. 109; see also Wormington, 1957, pp. 193–95).

[9] For Sandia, see Hibben, 1941, and Wormington, 1957, pp. 85–91; for Clovis, see Howard, 1935; Cotter, 1937, 1938; Wormington, 1957 pp. 47–53.

ica, particularly east of the Rocky Mountains. This fluted point form is peculiarly American; and, although no prototypes have yet been discovered in either the New or the Old World, it is possible to trace its later developments into the Folsom type points. The Folsom complex dates from 9000 to 7000 B.C. and is found in the Colorado and New Mexico High Plains country.[10] The Folsom fluted points display technical improvements over the earlier and somewhat less specialized Clovis forms. The Folsom point is widest just back of the piercing end, and the flutes or channels on each side are longer and occupy a greater surface area of the implement. Such features probably insured a more effective puncturing of the hide of the animal (usually a kind of bison now extinct) and more profuse bleeding. Thus a reasonable argument can be made out for a progressive adaptive efficiency of the Folsom point over the Clovis point.

EARLY POST-PLEISTOCENE HUNTERS

In the subsequent post-Pleistocene (*ca.* 7000–5000 B.C.) the trend in the manufacture of points is to drop out the fluting feature altogether. Such types as the Eden, Scottsbluff, Plainview, and Angostura (Wormington, 1957, pp. 103 ff.), while retaining something of the lanceolate Clovis-Folsom outline, lack the fluting. This trend may mark a lessening in hunting efficiency, although it is possible that with changes to modern fauna the fluted point became less effective.

The Pleistocene big-game hunting pattern apparently spread from North to Middle and to South America. In these areas, in Pleistocene or early post-Pleistocene contexts, lanceolate, well-chipped, but unfluted points have been found associated with mammoth and other extinct animals (Wormington, 1957, pp. 199–205; Willey and Phillips, 1958, pp. 99–103).

As noted, the camp type settlement is associated with the Pleistocene big-game hunting pattern. In fact, many of the sites appear to have been little more than "kills" or "butchering stations." All inferences point to small wandering populations. Very little has ever been found other than the equipment used for the hunt and for the preparation of the meat and hides of the animals. Crude paint palettes, stone beads, and little incised bone disks, taken from a Folsom context, are among the few exceptions to this.

COLLECTORS OF THE NORTH AMERICAN DESERT

The "North American Desert," as the environmental homeland of a food-collecting life-pattern of long persistence, may be defined as the vast arid and semiarid basin, range, and plateau country which in

[10] For Folsom see Roberts, 1935; Wormington, 1957, pp. 23–43.

North America lies between the Rocky Mountains and the mountain systems of the Pacific Coast and stretches from southern Canada deep into Mexico (Jennings and Norbeck, 1955; Jennings *et al.*, 1956, pp. 69–72; Jennings, 1957, pp. 276–87). There are indications that it was in this territory that the moist conditions of the Pleistocene first began to give way to those of the warmer, drier Altithermal climatic period. The onset of the Altithermal is placed at about 5000 B.C., but in the North American Desert basin, as might be anticipated, a collecting life-pattern seems to have been established even earlier than this. Its origins may lie, in part, in the possible earlier Pleistocene gathering pattern just discussed; they may be found, also, in a change-over from the big-game hunting pattern to the hunting of lesser animals and seed-collecting that was brought about by the disappearance of the grasslands and the Pleistocene fauna; and the probability also exists of new population movements or diffusions from the north and, more remotely, from the Old World. Levels representative of a Desert collecting pattern in Danger Cave, Utah, are dated as early as 7500 B.C. (Jennings, 1957, pp. 60–67; Wormington, 1957, pp. 193–95). These contain projectile points of both corner-notched and stemless forms that are reminiscent of other early Desert pattern types. Slab milling stones and twined baskets are also associated with these levels. Among other early dates for what appear to be the beginnings of the Desert collecting pattern are those for the Humboldt phase levels of the Leonard site, in Nevada (*ca.* 9000–5000 B.C.) (Heizer, 1951; Wormington, 1957, pp. 190–92), and for Fort Rock Cave, in Oregon (*ca.* 7000 B.C.) (Cressman, 1951; Wormington, 1957, p. 184). The artifacts at Fort Rock include notched and unstemmed points, scrapers, drills, grinding stones, bone tools, twined basketry, and bark sandals.

Throughout the Desert basin there are numerous locations which show the continuity of these complexes with their varieties of stemmed, notched, and unstemmed projectile points and milling and handstones. The upper levels at Leonard Rockshelter are placed at between 3000 and 2000 B.C. (Heizer, 1951). The Danger Cave stratigraphy has a continuation of the Desert collecting pattern, from its early date until 1800 B.C. (Jennings, 1957). The Cochise continuum of cultures in southern Arizona and southern New Mexico also demonstrates the long existence of the Desert collecting pattern from before 5000 B.C. up to almost the beginning of the Christian Era (Sayles, 1945; Sayles and Anteos, 1941, 1955; Wormington, 1957, pp. 169–73). In the Cochise sequence there is a definite trend through time toward more numerous, larger, and more carefully deepened and shaped milling stones and mortars. This trend is almost certainly expressive of an increased "settling-in" and adjustment to the collecting of seed foods. It is correlated with greater stability of living sites at the end of the Cochise sequence, where there is archaeological testimony to semipermanent villages of houses with prepared floors and storage pits. It is undoubtedly of great significance that remains of a very primitive maize have been found in the Cochise sequence, dating back to 2000 B.C., or before. Such maize appears to have been in its initial stages of cultivation and to have served as a supplement to wild plant foods (Mangelsdorf and Smith, 1949; Dick, 1952; Mangelsdorf, 1958; Willey and Phillips, 1958, pp. 128 ff.).

COLLECTORS OF THE NORTH AMERICAN WOODLANDS

The collecting life-pattern of the Eastern Woodlands of North America is referred to by archeologists as the "Archaic" pattern (Willey and Phillips, 1958, pp. 111 ff.; Griffin, 1952). This mode of existence—the collecting of wild vegetable foods, of shellfish, and the taking of fish and small game—replaced the dependence upon the large Pleistocene fauna that was once present in the East. This replacement was probably a gradual one which kept pace with the disappearance of late pluvial conditions and the big-game animals. Such change is seen in the artifact stratigraphy of Illinois and Missouri caves (Fowler and Winter, 1956; Logan, 1952; Chapman, 1952). The shift-over from big-game hunting to collecting probably began as long ago as 8000 B.C., and by 4000 B.C. it was complete throughout the Eastern Woodlands. The projectile points and tools of the Eastern collecting pattern are large, wide, stemmed, or tanged forms and a variety of mortars, handstones, and pestles.

Between 4000 and 1000 B.C. a number of modifications occur in the Eastern Woodland collecting pattern. These are best seen regionally, and in some cases they appear to be responses to natural settings. It also seems likely that new ideas and migrations of people, of ultimate Asiatic origins, were entering the East at this time. These modifications constitute a definite trend toward adjustment or "living-into" the various niches of the Eastern area. Along seacoasts and river, as in Georgia and Tennessee,[11] shellfishing stations offered opportunities for regular seasonal residence of a semipermanent village kind. With this increased sedentism appear numerous polished stone implements and ornaments, including celts, atlatl weights, vessels, gorgets, and beads. Around the Great Lakes, copper nuggets were fashioned into artifacts by hammering techniques. Finally, toward the close of the period, pottery was made and used. It is with these changes that the

[11] See Willey and Phillips, 1958, p. 115, and extensive footnote bibliography to this subject.

collecting pattern of the East becomes, at least in many localities, intensive collecting.

The intensive collecting pattern of the Eastern Woodlands, or a level of "Primary Forest Efficiency," as it has been called (Caldwell, 1958), climaxes during the first millennium B.C. in certain regions of the Mississippi and Ohio valleys. This is seen in the Poverty Point and Adena cultures with the construction of great ceremonial earthwork centers. The Poverty Point site, in Louisiana (Ford and Webb, 1956), consists of two huge mounds and a concentric series of earth embankments a half-mile in diameter. The Adena centers of Ohio, West Virginia, and Kentucky are marked by a big earth tumuli constructed over interior burial chambers (Webb and Snow, 1945; Webb and Baby, 1957; also Willey and Phillips, 1958, pp. 156–58). The total settlement patterns associated with these monuments are unknown, but it is certain that the societies responsible for such "public works" must have included large numbers of individuals and have necessitated the co-ordination of the efforts of these people, even though their actual living sites may have been of no more than a semipermanent nature. The subsistence basis of these societies is questionable. Some plant cultivation may have been known in the East by this time, but the probable local cultigens seem insufficient as dietary staples, and the presence of maize has not yet been demonstrated. A bit later, between 500 B.C. and A.D. 500, with the advent of the Hopewellian cultures (Caldwell, 1958; Willey and Phillips, 1958, pp. 158–60), maize does come into the picture, and it probably was of real subsistence value as a supplement to the intensive collecting pattern. But it is undoubtedly significant that the Hopewellian cultures continue in the earlier tradition of burial-mound earthwork construction and in the traditions of art and ceremonial paraphernalia which draw their symbolism from religious practices and magic associated with hunting.

OTHER HUNTING AND COLLECTING PATTERNS

The American Plains.—Later, big-game hunting patterns on the Plains of North America developed from the older Pleistocene way of life. This transition came about gradually, and, at first, the technological changes were relatively slight. A modern type of buffalo was pursued by these Plains huntsmen, and they combined this activity with some food-collecting. In later prehistoric times horticulture was brought to certain regions of the Plains and became an important subsistence factor, but buffalo hunting continued. In historic times, after the European introduction of the horse, the buffalo-hunting pattern enjoyed a renaissance (Strong, 1935, pp. 294 ff.; Wissler, 1914). A similar history can be reconstructed for the Argentine pampas and

Patagonia, where the guanaco was the game animal (Cooper, 1946). In both the North and the South American Plains the change-over from Pleistocene to modern climatic conditions appears to have taken place at about 5000 B.C. From this date forward, there are general trends toward smaller projectile points, and these are accompanied by stone food-grinding implements, objects of aesthetic, ceremonial, or non-utilitarian usage, and, later, pottery (Strong, 1935; Mulloy, 1954). These trends are associated with sites which could be classed as semipermanent settlement types. Thus it seems probable that these late big-game hunting cultures of the North and South American Plains, although depending primarily upon the chase for their food resources, were becoming somewhat more sedentary as the result of other subsistence activities; this increased sedentism is, in turn, related to a development of aspects of culture which are not directly a part of the food quest and the adaptation to the natural environment.

The Arctic.—Quite distinct in both natural environmental setting and in history from the later big-game hunting of the Plains is the Arctic hunting pattern. In its earliest phases the Arctic pattern seems to have been oriented toward land hunting. The Denbigh Complex of Alaska is the example. It is possible that there are connections between Denbigh and the early post-Pleistocene big-game hunters farther south. This is suggested by some of the Denbigh parallel-flaked projectile points. But there are other technological traditions in Denbigh flintwork, such as the burins and the micro-core-and-blade forms, which indicate diffusions or migrations from northeastern Asia, perhaps as recently as 2000 or 3000 B.C.[12] Following this horizon, the course of Arctic prehistory is the story of the development of Eskimo culture. This development was not an American Arctic phenomenon alone, as subsequent population movements or strong influences may be traced from North America back to Asia; but, viewing the Asiatic-American Arctic sphere as a whole as the home of this evolving Eskimo culture, there is little doubt that the main trends were those of a gradual supplementation of land hunting by sea-mammal hunting and by fishing and by an increasing adaptation to an environment of ice and snow. Eskimo sites dating back to the beginning of the Christian Era are fully of semipermanent settlement type. Artifacts of all kinds, in ground stone and ivory, abound; and, from the first millennium A.D. forward, Eskimo culture had achieved that unique and amazing ecological adjustment for which it is famous. It is worthy of note that in such customs and value expressions as the burial of the

[12] Giddings, 1951, 1954, 1955. Giddings is of the opinion that, although the Denbigh complex may have lasted this late, its beginnings must be put in the neighborhood of 6500 B.C. See Wormington, 1956, p. 212, in this connection.

dead and ivory-carving art there is no steady mounting elaboration. The climactic developments in these come, rather, with the Old Bering Sea and Ipiutak phases, during the first millennium A.D. Subsequent phases show a lessening in the complexity and elaboration of these themes. The supposition is that these refinements of the cult of the dead and of ornamental art had original Asiatic sources outside the severe Arctic zone and that, with the adaptive evolution of Eskimo culture, they were, to some extent, inhibited or sloughed off.[13]

The North Pacific Coast.—Along the Pacific Coast of North America it is likely that a food-collecting way of life had its beginnings with migrants from the western interior desert areas who brought to the coast the relatively simple subsistence technology of the Great Basin and Plateau countries (Osborne, 1958). As they settled in the Pacific river valleys and along the coast, their adjustment parallels, in many ways, that of the Eastern Woodland collectors. This adjustment is marked, archeologically, by the introduction and increasing use of ground and polished stone implements and ceremonial forms and by greater site size and apparent stability of occupation. By 2000 B.C., if not earlier, the Pacific Coastal pattern was of an intensive collecting kind, accompanied by a semipermanent village settlement type (Wallace, 1954; Beardsley, 1948). In late prehistoric and early historic times some of the California acorn-gathering or fishing societies were living in what must be considered permanent villages of several thousand inhabitants (Heizer, 1958). Farther north, in Oregon, Washington, and British Columbia, there are comparable sequences showing an increased "settling-in" adaptation to the environment.[14] These societies achieved an intensive collecting pattern well back in prehistoric times, and the salmon-fishing Northwest Coast tribes of the historic period are still in this tradition of abundant subsistence.

South America.—In South America there are many and diverse cultures, known both archeologically and ethnographically, which may be classed as collecting or intensive collecting in their subsistence. Among these are the old coastal fishing societies of the Peruvian and Chilian coasts. Here, on coastal bays, are shell refuse deposits which date back to 2500 B.C., if not earlier, and show a history of increasingly efficient adjustment. In the Peruvian sites a period of incipient cultivation marks the later phases (Bird, 1948), and in Chile there are modifications of polished stone, bone, and miscellaneous gear used in fishing and sea-mammal hunting (Bird, 1943). In both instances the

[13] For general treatments of Eskimo archeology, and differences of opinion in reconstruction see Collins, 1953*a*, *b*, 1954; Larsen and Rainey, 1948.

[14] See summarization of this, with bibliographic references, in Willey and Phillips, 1958, pp. 135–37.

shell midden sites appear to be either semipermanent or permanent village locations. On the Brazilian east coast there are similar shell midden sites in which there are some clues for an increase through time in material goods, particularly polished stone utensils and ceremonial objects (Orssich, 1956). Fishers and collectors also lived along the Caribbean coast in Venezuela and in the West Indies. In Venezuela these occupations date from about 2000 B.C. and reveal a thousand-year story of gradual increase in range of tool types and such "luxury" items as bone, stone, and shell beads and pendants (Cruxent and Rouse, 1958). It is likely that these Caribbean tribes of northeastern South America were also practicing some incipient cultivation with root crops. Far to the south, along the Paraná River and on the Pampean and Patagonian plains, there are archeological and ethnohistoric data pertaining to a variety of hunting, fishing, and collecting tribes (Howard and Willey, 1948). As noted, some of these people, particularly those of the pampas and Patagonia, were following a hunting tradition only slightly modified by food-collecting. Others, particularly along the rivers, maintained greater stability of residence and were more devoted to collecting.

CULTIVATORS OF NUCLEAR AMERICA

Nuclear American plant cultivation centered in the territories including and extending from central Mexico to southern Peru. Its beginnings were slow and gradual, and in its incipient forms it was grafted upon, or an adjunct to, pre-existing collecting patterns. This is the case in Tamaulipas, Mexico, on the northern border of Nuclear America, where hints of plant domestication go back to 7000 B.C. and where, by 5000 B.C., it is definite. Between that date and 3000 B.C. squash and beans are identified from dry cave deposits, and by 2500 B.C. small cobs of primitive maize have made their appearance (MacNeish, 1950, 1958). The cultural contexts of these Tamaulipas caves make it evident that these domesticates played but a relatively small part in what may otherwise be considered to be a North American Desert collecting pattern of food-getting. It is noteworthy, however, that in this Tamaulipas sequence of incipient cultivation there is a steady increase in the types and volume of cultigens and also in the varieties and numbers of such seed-grinding implements as mortars, manos, and mullers. With the advent of the final incipient cultivation phase, at about 1600 B.C., improved corn, beans, and Lima beans are all present, and it has been estimated that the cave populations of this period derived about 30 per cent of their sustenance from these crops (MacNeish, 1958). In Peru incipient cultivation also appears to be a kind of epiphenomenon superimposed upon the shellfish and wild-plant collecting patterns of the ancient coastal populations. In this setting, squash, beans (*Canavalia*), and miscellaneous roots and tubers were domesticated as early as 2500 B.C., if not long before (Bird, 1948).

The basic origins and courses of diffusion of the Nuclear American incipient cultivation pattern are difficult to trace. No imperishable and diagnostic tool types can testify indisputably to its presence. Such specialized and definite implements as the maize-grinding metate and mano are more likely to be associated with a fully established agriculture, and the relatively unspecialized seed-grinding implements are as likely to have been used for wild as for domesticated foods. Consequently, unusual conditions of preservation, such as dry caves or completely arid desert sands, arc necessary for an accurate determination of cultivation incipience. It may be for this reason that the arid uplands of Middle America and the rainless Peruvian coast now appear as the earliest centers of the Nuclear American incipient cultivation pattern. For it seems probable that experimentation with plant domestication proceeded in many parts of Nuclear America and that its antiquity throughout the zone goes back to 3000 B.C. or earlier. It is unlikely that any one area or region was the *fons et origo* of all the important food plants. Maize (*Zea mays*), which was to become the basic staple of the New World, probably was domesticated first in Guatemala or southern Mexico and spread from there, although the possibilities of more or less separate and independent cultivation of wild pod corn at various localities in the Nuclear zone cannot be discarded.[15]

The societal and cultural context of Nuclear American incipient cultivation ranges from the camp site, with its relatively limited inventory of basketry and ground and chipped stone tools, to semipermanent and permanent villages with much wider assemblages of textile and stone items plus, in some instances, pottery. The coastal locations appear to have been more propitious for settled life. Pottery, of a relatively simple sort, has been found in a coastal shell midden in Panama, dating at 2100 B.C.;[16] and, in Ecuador, coastal shell refuse stations, going back to 2400 B.C., have also yielded pottery (Evans and Meggers, 1958). On the Peruvian coast the incipient cultivators lived in stone-and-mud-masonry pit-house villages which probably numbered up to more than one hundred persons, and between 1800 and 1000 B.C. they began to manufacture pottery of a kind probably related to the Panamanian and Ecuadorean varieties (Bird, 1948, 1951).

The line marking the shift from what we have called "incipient

[15] Personal communication, P. C. Mangelsdorf, 1958.

[16] Results of Yale Radiocarbon Laboratory, personal communication, E. S. Deevey, 1958.

cultivation" to established cultivation is drawn at that point where communities derive their major support from farming. In the Middle American area this change is marked by the appearance of the first permanent villages.[17] Some of these villages, like those in the Tamaulipas region which follow the incipient cultivation societies, are quite small (MacNeish, 1958). Others, including the early pottery and farming phases of the Valley of Mexico, cover several acres (Vaillant, 1930). Still other village units were probably composed of several scattered hamlets which were nucleated around small ceremonial centers marked by temple mounds.[18] In both Ecuador (Evans and Meggers, 1957, 1958) and Peru (Willey, 1953, pp. 371 ff.) there was a movement away from the immediate coast back into the valleys at this time, a settlement shift suggestive of the rising importance of cultivation over a fishing and shellfish-collecting subsistence. The Peruvian sites of such early farming phases as the Cupisnique were small hamlets dotted over the river valley floor and along the valley edges, apparently focused upon a ceremonial center.

The beginnings of the established cultivation pattern in Nuclear America are dated at about 1500 B.C. in Middle America and at 1000 B.C. in Peru (Willey, 1955, 1958*a*). The next thousand years or so saw an intensification of this pattern, with increasing population densities and a development and elaboration of technology and art. Public building was on the increase in the form of pyramids constructed of adobe, earth, or stone. These pyramids marked the ceremonial or politicoreligious centers and either were surrounded by town type settlements, as in upland Mexico and Guatemala and on the Peruvian coast, or served as nuclei to dispersed hamlets, as in the Maya and Veracruz lowlands. The impressive Middle American art styles and the systems of hieroglyphic writing and calendrics, which were to flower brilliantly only slightly later, had their beginnings in this Formative Period. In Peru, this was also the time of the first great arts, of experimentation in metallurgy, and of the initiation of complex irrigation works (see Bennett and Bird, 1949, pp. 137–53; also Willey, 1958*b*).

Intensive cultivation arose from established cultivation in Peru and in certain portions of Middle America. On the Peruvian scene it is clearly defined by the large-scale irrigation and garden-plot networks

[17] Pottery antecedent to permanent farming-based villages has not yet been revealed in Middle America. I would postulate that it will be found, probably along the Pacific strand in shell midden situations comparable to those of Panama or Ecuador; however, at the present writing, the Panamanian and Ecuadorean pottery appears to be the earliest in Nuclear America and, possibly, the New World.

[18] It seems probable that the earliest Formative Period communities in the Maya lowlands were so "nucleated," although the evidence on this is not clear.

of the coast and the irrigation and terracing of the highland basins. Such works are dated as early as the beginnings of the Classic period, at about A.D. 1.[19] In Middle America intensive cultivation was more limited in geographic scope, and the evidence for it is more difficult to define archeologically. In the Valley of Mexico irrigation and chinampa ("floating-garden") farming can be dated with assurance back to the Toltec civilization at around A.D. 900. Before that the case for intensive cultivation is less certain, although it is possible that irrigation was practiced as early as the first phases of the Teotihuacán civilization at the beginning of the Christian Era (Millon, 1954, 1957; Palerm, 1955). In both Peru and Middle America intensive cultivation is associated with towns and temples and, in some instances, with cities.

The history of Nuclear American cultivation belongs not only to the Nuclear zone but to the outlands bordering it to the north and south. The diffusions or migrations carrying the ideas of plant domestication to these areas must have begun on the incipient cultivation level. We have already noted that primitive maize and other plants were being cultivated in the southwestern United States area in the third millennium B.C. by peoples who were essentially Desert collectors. It is also possible that incipient cultivation spread into northern Chile or northwestern Argentina at a relatively early date, although there is no proof of this. Established cultivation patterns, however, are not introduced to, or developed upon, the Nuclear American peripheries until much later. It was not until the last few centuries B.C. that permanent villages based primarily upon farming were established in the North American Southwest (Jennings *et al.*, 1956, pp. 73 ff.; Willey and Phillips, 1958, pp. 151 ff.), and a comparable date probably is applicable to northern Chile-Argentina.[20] In both the Southwest and in northern Chile-Argentina, elements of intensive cultivation appear sometime after about A.D. 700. The canal systems of the southern Arizona desert Hohokam culture and the agricultural terracing of the southern Andes are examples in point. There seem to be associations between these intensive cultivation patterns and town and temple communities; however, it is also certain that in the southwestern United States such large late prehistoric towns as Pueblo Bonito, in northern New Mexico, came into being without these specialized techniques of complex irrigation or terracing systems.

[19] See Willey, 1953, for dating of irrigation works in Peru; and Willey, 1958*a*, for revised chronological estimates.

[20] Willey (1958*a*) places it slightly later; A. R. Gonzalez, in a seminar presentation given in December, 1958, is inclined to date village agriculture in northwestern Argentina as early as 500 B.C.

OTHER PATTERNS OF CULTIVATION

There are two other native American cultivation patterns whose histories, if not completely separate from that of the Nuclear American pattern, have sufficient independence to deserve comment. Both are found in lowland wooded areas of adequate or abundant rainfall. Unfortunately, in neither case does the archaeological record give more than a hint of these patterns. They have their respective hearts in the North American Eastern Woodlands and its Mississippian River system and in the South American Tropical Forest of the Amazon and Orinoco Rivers.

The Mississippi Drainage of Eastern North America.—It seems likely that the sunflower (*Helianthus*), goosefoot (*Chenopodium*), and pumpkin (*Cucurbita pepo*) were all domesticated here by the beginning of the first millennium B.C. (Goslin, 1957). This may have been in response to remote stimulus diffusion from the Nuclear American cultivation pattern, the plants used being the best that the local environment had to offer; or it may have been an entirely independent development.[21] As previously noted, this Eastern Woodland incipient cultivation was a part of a dominantly collecting or intensive collecting life-way of which the Adena and Hopewell cultures represented a climax (Caldwell, 1958). It is known that, by 500 B.C., such cultures were in contact with the Nuclear American cultivation pattern for maize. By A.D. 500, established cultivation of a Nuclear American variety, derived from Middle America, dominated the Mississippi Valley. Thus the earlier, non-maize, distinctively Eastern Woodland cultivation never proceeded beyond the level of incipience until it became a part of the expanding Nuclear American pattern.

The South American Tropical Forest.—This pattern played a much more important role in native American agriculture. The basic crops were root starches, primarily manioc (*Manihot esculenta*), both bitter and sweet, as well as the sweet potato (*Ipomoea batatas*). Their antiquity as domesticates in Venezuela and Brazil can only be speculated upon, but almost certainly it antedates 1000 B.C. (Sauer, 1952). Whether or not there is any archeological evidence that may be identified, even tentatively, as Tropical Forest incipient cultivation is problematic. Possibly the early phases on Marajó Island, at the mouth of the Amazon, had such a subsistence along with collecting, fishing, and hunting. The sites of these phases appear to be small semipermanent villages marked by little but refuse and simple pottery.[22] Possibly,

[21] MacNeish believes that the pumpkin was domesticated in northeastern Mexico and spread to eastern North America at an early date, while the sunflower, domesticated in the Southwest, spread to the Mississippi Valley by Adena times.

[22] See Meggers and Evans, 1957, for such phases as Ananatuba and Mangueiras.

too, some of the early coastal shell mound sites of Venezuela may have had a partial dependence upon incipient cultivation (Cruxent and Rouse, 1958), but by 1000 B.C. a permanent village location, with well-made, painted pottery, is known from the lower Orinoco River (*ibid.*). Pottery griddles, found at this site, imply the use of manioc flour for cakes. Such a site seems definite evidence of established cultivation, but in the Tropical Forest rather than the Nuclear American tradition.

A blending and interpenetration of the Tropical Forest and Nuclear American cultivation patterns seems to have begun fairly early. In northern Colombia the lower levels of the Momíl site show pottery manioc griddles but no definite clues pointing to the use of maize. The immediately succeeding levels, however, have maize-grinding implements (Reichel-Dolmatoff and Reichel-Dolmatoff, 1956). These early Momíl phases probably date back to 1000 B.C. or even before (Willey, 1958*a*). The spread of maize farther to the east probably took place sometime after this. In both Venezuela and parts of the Amazon system, town and temple type communities were known in pre-Columbian times (Steward, 1948).

New World Culture: History and Evolution

SUBSISTENCE EFFICIENCY

Nearly all the New World hunting or collecting life-patterns demonstrate, or strongly suggest, a trend through time of increasing adaptive efficiency in the realm of subsistence technology.[23] The North American Desert pattern, which is known in the full gamut of its development, reveals the growing importance of seed foods in its middle and later stages. This is seen in the increasing numbers and specialization of food-grinding implements. The North American Woodland pattern development parallels this to some extent, but with a strong emphasis upon specialization in weapons and implements used for forest hunting and stream and coastal fishing. On the Pacific Coast of North America there is a similar sequential record of improved adjustment to local environments in which the collecting of vegetable foods or fishing were the specialties. In the South American collecting patterns these same tendencies also exist, and in the far north the Arctic hunting pattern is the example par excellence of the development of a technology to cope with environmental conditions. Only the very

[23] In a recent paper D. W. Lathrap (1957) has argued that a subsistence technology carried by a human society is analogous to a biological organism which exploits a compatible environment to its farthest geographic limits and develops within the confines of that environment to its highest degree of adaptive efficiency. I am indebted to Lathrap for his stimulating essay and for numerous discussions on this subject.

earliest patterns—those of the Pleistocene gatherers and the Pleistocene hunters—fail to reveal fully this configuration through time of adaptive subsistence efficiency. For the first, the data are too few to make an appraisal. For the second—the big-game hunters—we see a high degree of specialization to an environment at the outset; and it seems likely that the earlier stages of the pattern, leading up to such complexes as the Sandia or Clovis, are as yet undiscovered. Even with only a part of the chronological range of the pattern revealed, however, there is a hint of the specialization trend in the changes that set the later Folsom complex off from the earlier Clovis. Thereafter, the pattern dissolves with the shrinkage and eventual disappearance of the Pleistocene environments.

The American collecting patterns were not pushed back to marginal positions or terminated by natural environmental change but by the propagation of plant cultivation. The history of this cultivation is a story of several millennia of casual, experimental, or incipient cultivation. During this time plant cultivation did not really constitute an independent subsistence pattern but was an adjunct to food-collecting. Incipient cultivation, although progressing slowly, culminated in the Nuclear American and Tropical Forest established cultivation patterns. Offering a subsistence based largely or wholly upon food production rather than collecting, it spread to all those American areas where it was able to find a receptive natural environment and where it could compete successfully with existing subsistence techniques. In parts of America the more productive and more highly specialized intensive cultivation replaced established cultivation.

SOCIAL INTEGRATION

Our surveys of New World life-patterns indicate correlations between the increasing efficiency of subsistence adaptations and the increasing size and stability of settlement or community types. Both the North American Desert and the Woodland collectors lived in camp type settlements in their earlier stages. In their later phases the Cochise populations of the Southwest established semipermanent villages, and the tribes of the eastern United States began to "settle in" along rivers and the coast in semipermanent, or possibly permanent, villages. Elsewhere, as on the Pacific Coast of North America or the South American coasts, the intensification of food-collecting also made it possible for larger social aggregates to live together for longer periods of time.

In the Nuclear American and South American Tropical Forest cultivation patterns the stage of agricultural incipience probably marked some increase in community size and stability, although the evidence here is either lacking or difficult to separate from that of the later stages of food-collecting. Undoubtedly, local conditions and available food plants were decisive variables. It was with the established cultivation level, however, that the permanent village came into existence in these patterns. As previously observed, this does not mean that settled village life was impossible without plant cultivation; we have noted its probable existence in certain circumstances of intensive collecting; but it does signify that when the food economy of a society was predominantly agricultural, that society became "anchored" to a relatively small geographical locus. There is considerable range in size and type of site associated with established cultivation. Although the permanent village was the most usual and widespread form in the Nuclear American and Tropical Forest zones as well as in outlying American areas, there are indications that town-and-temple and even city type communities were supported by this kind of farming.

That the correlations between subsistence efficiency and increase in community size and organization are to some degree causally interrelated seems self-evident. Of the full nature of this interrelationship, it is difficult to be precise or explicit except in particular instances. And when one examines these individual cases, it is clear that the explanation cannot be presented as a mere matter of mathematical ratios: volume of food equals numbers of people equals type and organization of community. The ideas and institutions of culture are always found interposed between the natural, technological, and demographic factors.

CULTURAL FORMS AND INSTITUTIONS

This brings us to our most difficult step in the attempt to follow out the trends and courses of New World cultural evolution. What configurations of change through time may we generalize about those aspects of culture which are not immediately and intimately related to the subsistence technology–environmental settings or to the sizes and groupings of settlements and societies? And how might these configurations correlate, or fail to correlate, with trends of change in the technological and social orders?

To begin at a low and tangible level, it is observed that manufactures of all kinds increase throughout the histories of our American collecting and cultivation life-patterns. This is most noticeable in the North American Desert and Woodland patterns, where, in the early levels, there is relatively little archaeological residue as compared to later phases. In the later phases not only are there more objects, but many of these have no apparent direct relationship to the food quest. Polished stone items of a ceremonial or ornamental nature are examples. Similar trends are also noted in South American collecting patterns. Within the sequences of the cultivation patterns this same increase in material goods, particularly non-utilitarian goods, is characteristic of the earlier periods; later on, it is less evident. There is, then, in cultural items that might be classed as elaborations of life—ornaments, emblems, religious paraphernalia—an increase that more or less parallels the rise of successful food-collecting and plant cultivation and the appearance of semipermanent and permanent villages. This is not to argue that aesthetic or religious experience was lacking in earlier times or on earlier levels, but the time or desire to give frequent material expression to such emotions must have been lacking before the development of some degree of sedentism.[24]

Architecture, particularly communal or public architecture, is another tangible that seems to describe some significant configurations in its occurrences in the American life-patterns. It is absent from both the Pleistocene hunting and the later hunting traditions. It is given some expression with the intensive levels of the collecting patterns. On the North American Pacific Coast it is known from historic times, and it is probable that communal buildings of impressive size were put up here during prehistoric periods, although these, being of wood, would have left little or no archeological traces. In the eastern United States the intensive collectors of the Woodlands built large earth monuments for burial and religious purposes. However, it is with the Nuclear American cultures that public building was greatly elaborated. In both Middle America and Peru the trends show a steady increase in the numbers and size of ceremonial mounds during the Formative Period of the first millennium B.C. The societies responsible for their construction were established cultivators, and this period of the building of great temple mounds is the time of the shift from village to town and temple life. With the Classic Period of the first millennium A.D., ceremonial mounds were still being constructed, but there were now new trends expressed in greater attention to the actual temple buildings surmounting the mounds or in palace constructions. Still later, in the Postclassic Period of the last centuries prior to the Spanish Conquest, there are other architectural changes reflective of shifts in cultural interests and values. In Middle America, fortifications appear in many regions. In Peru there is a marked dropoff in ceremonial mounds and a greater interest in military construction and planned community architecture. The configuration of the architectural trend,

[24] It may be argued that the European cave art of the late Paleolithic is a contradiction to this. I think it more likely, however, that the specialized hunting conditions of these cultures, in their environments of that time, offered a substantial degree of sedentism to these early hunters and artists.

to reduce it to its simplest generalization, is that interest in permanent public or ceremonial building is a correlate of settled life, that it does not appear until after sedentism is reasonably well established, and that then it undergoes changes suggesting a drift away from the unique importance of the temple, to the palace compound, and from this to military construction and urban planning.

Art is perhaps the archaeologist's one best clue to the non-material, non-utilitarian heart of things. At least we shall consider it so, for it is the only means at hand so to interpret the preliterate cultures. In the New World life-patterns there is little evidence of it associated with the Pleistocene big-game hunters, although it may have existed in some perishable form. In the North American collecting patterns, art has its beginnings with the small ornamental and ceremonial objects of sculptured stone or bone. The Hopewellian climax of the Woodland collectors represents a florescence of the art of this particular pattern. In Nuclear America there are varying styles of art in small handicraft objects and ceramics in the earlier phases of established cultivation, and relatively early in the Formative Period in both Middle America and Peru monumental art styles, such as the Olmec and Chavín, make their appearance. In Middle America the Olmec and related early art styles are associated with evidences of writing and calendrics; and both Olmec and Chavín art are found in the contexts of the earliest important temple or public buildings. These styles and their settings imply a considerable sophistication for the cultures and societies which created them as expressions of their religions and value systems. Although in both Middle America and Peru a succession of other art styles follow these and can, in varying degrees, be traced as having developed from them, there appear to be no evolutionary trends that could be conceived of as "increase," "improvement," or "refinement" in these aesthetic developments. Nor can we say that certain types of economic and social milieus are conducive to artistic achievement over and beyond the level of established cultivation and the emerging town and temple community. Monumental Olmec art appears in Middle America in the tropical lowlands of the Gulf Coast, where a true urban development seems to have been precluded by the necessities of scattered village or hamlet-based forest farming. The later Maya art of the Classic Period, generally held to be the apogee of Middle American and New World aesthetic achievement, had its formation and growth in a similar ecological and social setting. The same, of course, holds for Maya hieroglyphics and astronomy. Conversely, Classic Period Teotihuacán of the upland Valley of Mexico, apparently a city in the true urban sense, did not enjoy a comparable artistic or intellectual flowering, although its political power may have been greater than that of the contemporaneous Maya. Thus, from an examination of the American data, the expression of cultural values in art seems to demand a base of sedentary life with some economic surpluses and some moderate community size. Beyond this, it is not clear that increase in volume or particular organization of these factors resulted in any measurable elaborations or intensifications of aesthetic or intellectual institutions.

THE NEW WORLD AS A WHOLE

Up to now we have considered the question of the evolution of culture in the New World within the confines of the historical units designated as life-patterns. In each of these patterns we have observed certain trends or configurations of occurrence in social and cultural traits. There are similarities in these trends, from pattern to pattern, and there are also distinct differences. These differences, together with the variable chronological and distributional aspects of the life-patterns, suggest that it is possible to conceive of New World prehistoric events as a unitary, if highly complex, history and evolution.

The earliest of the definite American life-patterns—that of the Pleistocene big-game hunters—marks a stage in New World culture history and culture development referred to as the *Lithic,* and *Paleo-Indian,* or the *Paleo-American.*[25] It was a highly specialized way of life, and the pattern was broken with the changing of the natural environment at the end of the Pleistocene. Some of the peoples and societies involved may have perished, but it is likely that others sought new modes of subsistence adaptation. Possibly minor food-getting techniques, of little importance in an economy geared to the mammoth hunt, became the bases of a new orientation. Also, there was probable recourse to borrowing either from peoples newly arrived from Asia or from those descendant in the possible ancient traditions of Pleistocene food-gathering. In some such way the collecting patterns must have come into being.

The food-collecting way of life, although represented in several historically distinct life-patterns, comprises a stage in the development and history of New World subsistence technology. This has been called the *Archaic stage.*[26] The food-collectors were, at first, less well adapted to their environments than the big-game hunters; but, in losing the specialization of their ancestors, relatives, or neighbors, they

[25] Willey and Phillips (1958) use the term "Lithic"; Suhm, Krieger, and Jelks (1954) suggest "Paleo-American"; "Paleo-Indian" has had a long and fairly extensive usage.

[26] Willey and Phillips (1958) derived this term from its specific usages in eastern North America and in California but have extended it beyond specific historical limitations as a New World "stage."

had been launched upon a course that would eventually lead to a much greater subsistence efficiency and economic security than the chase of the animal herds. After a few thousand years the results of this new specialization, or variety of subsistence specializations, is apparent in the semipermanent villages, the manufacture of "luxury" items, and the beginnings of art and ceremonial constructions.

Carried somewhat like a parasite in the host of the food-collecting patterns was the minor element of plant domestication. Of considerably antiquity, it appears to have remained little more than dormant for millennia. Gradually this incipient cultivation assumed more importance, especially in certain geographical regions where the native wild flora and fauna were not overly abundant. Finally, plant cultivation emerged as the agricultural way of life. This threshold of village communities sustained by farming marks the *New World Formative stage.*[27]

Like the early food-collectors, the first established village agriculturists had a less successful economic adaptation than many contemporary intensive collectors, such as those of the North American Eastern Woodlands or Pacific Coast. But the potentialities were before them, and it was not long before the agriculturally based societies of Nuclear America surpassed the wealthiest of the food-collectors in surpluses and population numbers. With the establishment of the secure cultivation threshold, there were no more major changes in native American subsistence patterns. Intensive cultivation, achieved in some places, was a specialization and a refinement, not a new departure. Thus it is at somewhere near this point that the course of New World culture history and culture evolution can no longer be traced in terms of subsistence efficiency.

Community size, concentration, and complexity of organization do, however, continue to show a configuration of increase after the establishment of cultivation as the primary means of subsistence. Although, to some extent and in some places, this may be attributed to intensive cultivation and to population rise, it is certain that cultural choice entered increasingly into the equation. As towns and cities arose in Nuclear America, during what have been called the *Classic* and *Postclassic stages* of culture development,[28] it is significant that absolute numbers of people did not change so much as did the settlement patterns by which they were organized. The town and the city were not simply the means of containing population masses; they were an or-

[27] The term "Formative," used as a "stage" designation, was derived by Willey and Phillips (1958) from its specific period applications in Middle America and Peru.

[28] The terms "Classic" and "Postclassic," used as "stage" designations, are also derived from their applications as periods to Peruvian and Middle American sequences (Willey and Phillips, 1958).

ganic part of societies organized and integrated over large territories or states.

We have remarked that both monumental art styles and temples occur relatively early in the rise of Middle American and Peruvian civilizations, perhaps less than a thousand years after the first establishment of village life based on cultivation and at the beginnings of what might be considered the town-and-temple community. There followed, then, a series of architectural and artistic developments whose trends cannot be generalized in the sense of "increase," "improvement," or "refinement" but rather as reflective of changes in social and political types and in moral values. Such trends as are suggested are those marking a shift from sacred to secular emphases, from relative peace to times of war, and from the smaller city or regional state to the larger territorial empire. It is from an appraisal of such trends that the distinction between a Classic and a Postclassic stage has been based (Willey and Phillips, 1958, pp. 182–99).

We come, then, to sum up the case for the evolution of culture in the native New World. I would conclude that whatever the contacts with the peoples and cultures of the Old World (and these have been but briefly alluded to in this paper), New World culture did, indeed, evolve in an essentially independent manner. This evolution can be traced in subsistence modes which, within several major historical patterns, describe a configuration of increasing efficiency and environmental adaptation. This same general trend of subsistence efficiency for production can also be projected across these various historical patterns so that it describes an over-all configuration of increase despite the fluctuations which mark the junctures of the patterns. Paralleling subsistence increase is the enlargement of the social unit and its geographic stability. Sedentism and food surpluses are also seen as the conditions necessary for the creation of material wealth and for the memorializing of religious and aesthetic emotions in art and architecture. Great skill and sophistication were attained in New World art, including monumental art, well in advance of the rise of the power of the city or the state. The evolution of the city and the state in native America appears to be linked with the growing power of military and secular forces and with attempts to propagate single religious, political, and social views—unified moral orders—to ever widening spheres of influence.

References

Beardsley, R. K. 1948. "Culture Sequences in Central California Archaeology," *Amer. Antiquity*, XIV, 1–28.

Beardsley, R. K., *et al.* 1956. "Functional and Evolutionary Implications of Community Patterning. In *Seminars in Archaeology,* ed., ROBERT WAUCHOPE. ("Memoirs of the Society for American Archaeology," No. 11.)

BENNETT, W. C., and BIRD, J. B. 1949. *Andean Culture History.* ("Handbook Series," No. 15.) New York: American Museum of Natural History.

BIRD, J. B. 1943. *Excavations in Northern Chile.* ("Anthropological Papers of the American Museum of Natural History," Vol. XXXVIII, Part 4.) New York.

———. 1948. "Preceramic Cultures in Chicama and Virú." In *A Reappraisal of Peruvian Archaeology,* ed. W. C. BENNETT. ("Memoirs of the Society for American Archaeology," No. 4.)

———. 1951. "South American Radiocarbon Dates." In *Radiocarbon Dating,* ed. FREDERICK JOHNSON. ("Memoirs of the Society for American Archaeology," No. 8.)

BRAIDWOOD, R. J. 1958. "Prelude to Civilization." Manuscript prepared for the symposium on "The Expansion of Society," at the Oriental Institute, University of Chicago, December, 1958.

CALDWELL, J. R. 1958. *Trend and Tradition in the Prehistory of the Eastern United States.* ("Scientific Papers of the Illinois State Museum," Vol. X, and "Memoirs of the American Anthropological Association, No. 88.)

COLLINS, H. B. 1953*a*. "Recent Developments in the Dorset Culture Area." In *Asia and North America: Transpacific Contacts,* ed. M. W. SMITH. ("Memoirs of the Society for American Archaeology," No. 9.)

———. 1953*b*. "Radio Carbon Dating in the Arctic," *Amer. Antiquity,* XVIII, 197–203.

———. 1954. "Arctic Area," *Program of the History of America, Indigenous Period,* Vol. I, No. 2. Mexico, D.F.: Comisión de Historia, Instituto Panamericano de Geografiá e Historia.

COOPER, J. M. 1946. "The Patagonian and Pampean Hunters," *Handbook of South American Indians,* I, 127–68. (Bull. 143, Bureau of American Ethnology, Smithsonian Institution.)

COTTER, J. L. 1937. "The Occurrence of Flint and Extinct Animals in Pluvial Deposits near Clovis, New Mexico. IV. Report on the Excavations at the Gravel Pit in 1936," *Proc. Amer. Acad. Nat. Sci.,* LXXXIX, 2–16.

———. 1938. "The Occurrence of Flints and Extinct Animals in Pluvian Deposits near Clovis, New Mexico. VI. Report on Field Season of 1937," *Proc. of the Philadelphia Acad. Nat. Sci.,* XC, 113–17.

CRESSMAN, L. S. 1951. "Western Prehistory in the Light of Carbon 14 Dating," *Southwestern Jour. Anthropol.,* VII, No. 3, 289–313.

CROOK, W. W., JR., and HARRIS, R. K. 1958. "A Pleistocene Campsite near Lewisville, Texas," *Amer. Antiquity,* XXIII, 233–46.

CRUXENT, J. M., and ROUSE, IRVING. 1958. "An Archaeological Chronology of Venezuela." Manuscript to be published as a Social Science Monograph, Pan American Union.

DICK, H. W. 1952. "Evidences of Early Man in Bat Cave and on the Plains of San Augustin, New Mexico." In *Indian Tribes of Aboriginal America: Proceedings, 29th International Congress of Americanists,* ed. Sol Tax, III, 158–63. Chicago: University of Chicago Press.

Dobzhansky, T. 1958. "Evolution at Work," *Science,* CXXVII, 1091–97.

Evans, Clifford, and Meggers, B. J. 1957. "Formative Period Cultures in the Guayas Basin, Coastal Ecuador," *Amer. Antiquity,* XXII, 235–46.

———. 1958. "Valdivia—an Early Formative Culture on the Coast of Ecuador," *Archaeology,* XI, 175–82.

Ford, J. A., and Webb, C. H. 1956. *Poverty Point: A Late Archaic Site in Louisiana.* ("Anthropological Papers of the American Musuem of Natural History," Vol. XLVI, Part I.)

Giddings, J. L. 1951. "The Denbigh Flint Complex," *Amer. Antiquity,* XVI, 193–203.

———. 1954. "Early Man in the Arctic," *Scient. American,* CXC, No. 6, 82–89.

———. 1955. "The Denbigh Flint Complex Is Not Yet Dated," *Amer. Antiquity,* XX, 375–76.

Goslin, R. M. 1957. "Food of the Adena People." In *The Adena People,* No. 2, by W. S. Webb and R. S. Baby. Columbus: Ohio State University Press.

Griffin, J. B. 1952. "Culture Periods in Eastern United States Archaeology." In *Archaeology of Eastern United States,* ed. J. B. Griffin, pp. 352–64. Chicago: University of Chicago Press.

Harrington, M. R. 1955. "A New Tule Springs Expedition," *Masterkey,* XXIX, No. 4, 112–13. Southwest Museum, Los Angeles.

Heizer, R. F. 1951. "Preliminary Report on the Leonard Rockshelter Site, Pershing County, Nevada," *Amer. Antiquity,* XVII, No. 1, 23–25.

———. 1958. "Prehistoric Central California: A Problem in Historical-Developmental Classification," *Papers on California Archaeology,* pp. 63–69. ("Reports of the University of California Archaeological Survey," No. 41, pp. 19–26.) Berkeley: Department of Anthropology, University of California.

Hibben, F. C. 1941. *Evidences of Early Occupation of Sandia Cave, New Mexico, and Other Sites in the Sandia-Manzano Region.* ("Smithsonian Miscellaneous Collections," Vol. XCIX, No. 23.)

Howard, E. B. 1935. "Occurrence of Flints and Extinct Animals in Pluvial Deposits near Clovis, New Mexico. 1. Introduction," *Proceedings of the Philadelphia Acad. Nat. Sci.,* LXXXVII, 299–303.

Howard, G. D., and Willey, G. R. 1948. *Lowland Argentine Archaeology.* ("Yale University Publications in Anthropology," No. 37.)

Jennings, J. D. 1957. *Danger Cave.* ("Memoirs of the Society for American Archaeology," No. 14.)

Jennings, J. D., and Norbeck, E. 1955. "Great Basin Prehistory: A Review," *Amer. Antiquity,* XXI, 1–11.

Jennings, J. D., *et al.* 1956. "The American Southwest: A Problem in Cultural Isolation." In *Seminars in Archaeology: 1955,* ed., Robert

WAUCHOPE. ("Memoirs of the Society for American Archaeology," No. 11.)

KLUCKHOHN, CLYDE. 1958. "The Moral Order in the Expanding Society." Manuscript prepared for the symposium on "The Expansion of Society," at the Oriental Institute, University of Chicago, December, 1958.

KRIEGER, A. D. 1953. "New World Culture History: Anglo-America." In *Anthropology Today,* ed. A. L. KROEBER, pp. 238–64. Chicago: University of Chicago Press.

LARSEN, HELGE, and RAINEY, FROELICH. 1948. *Ipiutak and the Arctic Whale Hunting Culture.* ("Anthropological Papers of the American Museum of Natural History," Vol. XLII.)

LATHRAP, D. W. 1957. "Radiation: The Application of Cultural Development of a Model from Biological Evolution." Manuscript, Seminar Paper, Harvard University.

MACNEISH, R. S. 1950. "A Synopsis of the Archaeological Sequence in the Sierra de Tamaulipas," *Rev. Mex. estud. antropol.,* XI, 79–96. Mexico, D.F.

———. 1958. "Preliminary Archeological Investigations in the Sierra de Tamaulipas, Mexico." Manuscript to be published by American Philosophical Society, Philadelphia.

MANGELSDORF, P. C. 1958. "Ancestor of Corn," *Science,* CXXVIII, 1313–20.

MANGELSDORF, P. C., and SMITH, C. E. 1949. "New Archaeological Evidence on Evolution in Maize," *Bot. Mus. Harvard Univ. Leaflets,* XIII, No. 8, 213–47.

MEGGERS, B. J., and EVANS, CLIFFORD. 1957. *Archaeological Investigations at the Mouth of the Amazon.* (Bull. 167, Bureau of American Ethnology, Smithsonian Institution.)

MILLON, R. F. 1954. "Irrigation at Teotihuacán," *Amer. Antiquity,* XX, 177–80.

———. 1957. "Irrigation Systems in the Valley of Teotihuacán," *ibid.,* XXIII, 160–67.

MULLOY, WILLIAM. 1954. "The McKean Site in Northeastern Wyoming," *Southwestern Jour. Anthropol.,* X, 432–60.

ORR, P. C. 1956. *Pleistocene Man in Fishbone Cave, Pershing County, Nevada.* (Bull. 2, Nevada State Museum, Department of Archaeology.)

ORSSICH, ADAM, and E. S. 1956. "Stratigraphic Excavations in the Sambaquí of Araujo II," *Amer. Antiquity,* XXI, 357–69.

OSBORNE, DOUGLAS. 1958. "Western American Prehistory—an Hypothesis," *Amer. Antiquity,* XXIV, 47–53.

PALERM, ANGEL. 1955. "The Agricultural Bases of Urban Civilization in Mesoamerica. In *Irrigation Civilizations: A Comparative Study,* pp. 28–42. "Social Science Monographs," No. 1, Pan American Union.)

REDFIELD, ROBERT. 1953. *The Primitive World and Its Transformations.* Syracuse, N.Y.: Cornell University Press.

REICHEL-DOLMATOFF, G. and A. 1956. "Momíl, Excavaciones en el Sinú," *Rev. Colombiana Antropol.,* Vol. V. Bogotá.

ROBERTS, F. H. H. 1935. *A Folsom Complex: Preliminary Report on Investigations at the Lindenmeier Site in Northern Colorado.* ("Smithsonian Miscellaneous Collections," Vol. XCIV.

SAUER, C. O. 1952. *Agricultural Origins and Dispersals.* New York: American Geographical Society.

SAYLES, E. B. 1945. *The San Simon Branch, Excavations at Cave Creek and in the San Simon Valley.* I. *Material Culture.* ("Medallion Papers," No. 34.) Globe, Ariz.: Gila Pueblo.

SAYLES, E. B., and ANTEVS, ERNST. 1941. *The Cochise Culture.* ("Medallion Papers," No. 24.) Globe, Ariz.: Gila Pueblo.

———. 1955. Report given at the 1955 Great Basin Archaeological Conference, *Amer. Antiquity,* XX, 311.

SELLARDS, E. H. 1952. *Early Man in America.* Austin: University of Texas Press.

SIMPSON, R. D. 1955. "Hunting Elephants in Nevada," *Masterkey,* XXIX, No. 4, 114–16. Southwest Museum, Los Angeles.

STEWARD, J. H. 1948. "The Circum-Caribbean Tribes: An Introduction." In *Handbook of South American Indians,* IV, 1–42. (Bureau of American Ethnology, Smithsonian Institution, Bull. 143.)

STRONG, W. D. 1935. *An Introduction to Nebraska Archaeology.* ("Smithsonian Miscellaneous Collections," Vol. XXIX, No. 10.)

SUHM, D. A., KRIEGER, A. D., and JELKS, E. B. 1954. *An Introductory Handbook of Texas Archaeology* (Bulletin of the Texas Archaeological Society, Vol. XXV.)

VAILLANT, G. C. 1930. *Excavations at Zacatenco.* ("Anthropological Papers of the American Museum of Natural History," Vol. XXIII, Part I.)

WALLACE, W. J. 1954. "The Little Sycamore Site and the Early Milling Stone Cultures of Southern California," *Amer. Antiquity,* XX, 112–23.

WEBB, W. S., and BABY, R. S. 1957. *The Adena People,* No. 2. Columbus: Ohio State University Press.

WEBB, W. S., and SNOW, C. E. 1945. *The Adena People.* ("University of Kentucky Reports in Anthropology and Archaeology," Vol. VII, No. 1.)

WILLEY, G. R. 1953. *Prehistoric Settlement Patterns in the Virú Valley, Peru.* (Bull. 155, Bureau of American Ethnology, Smithsonian Institution.)

———. 1955. "The Prehistoric Civilizations of Nuclear America," *Amer. Anthropologist,* LVII, 571–93.

———. 1958*a*. "Estimated Correlations and Dating of South and Central American Culture Sequences," *Amer. Antiquity,* XXIII, 353–78.

———. 1958*b*. "The 'Intermediate' Area of Nuclear America: Its Prehistoric Relationships to Middle America and Peru." Manuscript presented to the 33d International Congress of Americanists, San José, Costa Rica.

WILLEY, G. R., and PHILLIPS, PHILIP. 1958. *Method and Theory in American Archaeology.* Chicago: University of Chicago Press.

WISSLER, CLARK. 1914. "The Influence of the Horse in the Development of Plains Culture," *Amer. Anthropologist,* XVI, No. 1, 1–25.

WORMINGTON, H. M. 1957. *Ancient Man in North America.* ("Popular Series," No. 4, Denver Museum of Natural History.) 4th ed.

ROBERT J. BRAIDWOOD

LEVELS IN PREHISTORY: A MODEL FOR THE CONSIDERATION OF THE EVIDENCE

This short essay,* with its delineative device, is based on an intuitive assessment of a still highly incomplete body of archeological evidence. The background for the assessment—perhaps necessary for most non-anthropological readers—is suggested in two recently published papers (Braidwood, 1959 and 1960). The whole maneuver is clearly intended to be a heuristic one. It hints at dissatisfaction with the current classifications and nomenclature of cultural levels in prehistoric times, and with certain unilinear and duolinear schemes, "trees," or bar diagrams often used in their delineation (see also Childe, 1956). It attempts to include the peculiarly human factor of cultural diffusion, and to consider the factor of environmental differences as cultures evolved.

In the interpretation of the prehistoric archeological evidence, those bands of the once complete spectra of all cultural activities which now show through most clearly for us are the bands representing subsistence and settlement types. While there are doubtlessly varying (not yet clearly understood) subtle interrelationships between "subsistence" and "settlement," the scheme simply lumps the two together as "subsistence-settlement types." These are *not* to be equated in the reader's mind with any particular typological tradition in the preparation of chipped stone tools or with other artifactual traditions, although our own understanding of the histories of the various tool-preparation traditions naturally affected the delineation. Nevertheless, the consideration of homotaxis was as much in our mind as was actual similarity in tool-preparation tradition.

It will be readily apparent, however, in the outline that follows that the different types are indeed not precisely defined. The prehistoric

* The title of this paper has been abbreviated from the author's "Levels of Subsistence-Settlement Types, Multilinear Evolution, Environment and Diffusion, in Human Prehistory—a Model for the Consideration of the Evidence"—EDITOR'S NOTE.

ROBERT J. BRAIDWOOD is Professor of Anthropology at the University of Chicago. Specializing in Old World prehistory dealing with the origins of village life, he has conducted archeological field work in Iraq, Iran, and Syria.

archeological record is seldom complete enough, as it is yet available, to allow such definition. This is particularly true of the earlier eras and sub-eras of both of the stages we treat. An orderly assessment of the degree of efficiency in extracting food (as well as for the raw materials for clothing and shelter) is rarely possible for long-extinct human groups. Nor, for the same earlier eras and sub-eras, is our evidence for human settlements at all clear. The wording of the outline is meant to be connotative rather than explicit. As regards settlement, we have adapted the terminology of a useful paper by Beardsley, Meggers, *et al.* (1956) to our own convenience.

As the thresholds of urban civilizations are reached in the ascending archeological record, evidence for the interpretation of other bands in the full cultural spectra increasingly show through. Hence we frankly abandon this classification and delineative device at the threshold of urban civilizations, even though the outline form of our classification does suggest that it might be continued. We believe our levels of subsistence-settlement types in themselves may well not be the most useful basis for the consideration of the cultural evolution of civilizations, considering the expanding interpretative possibilities which civilizational remains offer in the archeological evidence.

In the delineation, it will be obvious that we have selected only a few environments, and that these are treated in a very generalized way. We are, of course, conscious that many more environments existed, that all these should ideally be considered specifically and, in fact, considered with respect to their changing biomes and habitats throughout Pleistocene time. The evolving behavioral potentialities of the human (or—for the earliest levels—semi-human?) unit itself are also, for the moment, consciously ignored.

The concentric circles in the delineation represent the factor of diffusion (this part of the device frankly suggests the pebble dropped in a pool of still water), but the time intervals between the several levels of circles are set purely for delineative convenience. Obviously, the spacing of the levels of the circles should represent every moment of the long time range which the device shows. We believe the diminishing sizes of the circles, as time proceeds, is meaningful. We take this to be a graphic representation of Dorothy Garrod's very astute observation that "the speeding up of change and development which begins to show . . . is reflected . . . not only in the greater number of industries having enough individual character to be classified as distinct . . . but in their restrictions in space, since [cultural] evolution now starts to outstrip diffusion" (Garrod, 1953, p. 14).

The increasingly intensive adaptations of the subsistence-settlement types to more restricted environments, and the "living into" these environments, form the conceptual basis for our classification-shift from an era of *gathering* to an era of *collecting*.[1] We are quite cognizant of the fact that this shift first begins to appear in the archeological record at about the time of the appearance of anatomically modern man in the record of human paleontology.

At the point of transition from the earlier era of food-gathering (*I–A* of our outline and delineation) to that of food-collecting (*I–B*), our delineation shows two lines of level *I–B-1*, save in the column for temperate North America where this was necessary in view of the paucity of present knowledge. This doubling of the lines is an admitted awkwardness of the delineative device. It is *not* our intention to have the device suggest "trees"; there was only one level of *I–B-1* as an ideal type. In order to allow the smaller concentric circles representing diffusion to overlap at this level, we were forced to double the vertical lines. It is certainly true that there were many more different variations within the general level of the *I–B-1* subsistence-settlement type than there had been in level *I–A-3*, but the device must *not* be read to mean that there were different kinds of *I–B-1*. The fact is that, in the sense of variations within *I–B-1*, two lines would be far from enough.

It should be emphasized that diffusion does not end as the concentric circles grow smaller; the circles still overlap, and an ideally complete model would show increasingly more of them overlapping as regional variations increased. There is discontinuity for the Far Eastern column on our device only because we lack archeological evidence there. It may also be noted in passing, if an ideally complete model be visualized, that especially as the concentric circles grew smaller, they would often appear to have been more responsive to compression out of symmetry, by mountain barriers, oceans, and deserts, than would appear to have held in earlier times.

The use of heavy arrows at the top of the device suggests the diffusion of the wheat, barley, sheep, goat, cattle, etc. pattern out from southwestern Asia, and of the maize, beans, squash, etc. pattern out from the American Southwest and Mesoamerica. It need not follow, from our change in delineative technique from concentric circles to

[1] Cf. Childe (1956), p. 86, where the implications of Menghin's "protolithic" and "miolithic" terminology is preferred to that of the prevailing "paleolithic" and "mesolithic." Before Childe's book was available to us, Braidwood and Reed (1957, p. 20) had written the same point. If my memory serves me, the idea came to all of us during the course of a luncheon with Gordon Childe in the Zoological Society's restaurant in Regent Park, London, in July, 1955. Gordon Willey's paper in the present volume utilizes the idea as set down in Braidwood, 1960. The distinction between the two words, "gathering" and "collecting," is based on the notion that "gathering" seems to connote a somewhat more unspecialized and random activity, while "collecting" connotes a more selective and intensified activity.

arrows, that we are assured that there were new kinds of cultural mechanics involved in the diffusion of food-production. This remains a question for further consideration. Certainly, however, this diffusion must have proceeded at a vastly accelerated rate relative to most earlier diffusions.

In the classification which follows, *no* attempt is made to conceptualize the innumerable blends of one level with another which certainly happened, at least in the later reaches of the time range concerned. In this sense, each of the levels suggested is certainly to be conceptualized as an ideal type. In our column for southwestern Asia, we very consciously allow the first sub-era of food-production (*II–A-1*) to arise directly from the first sub-era of food-collection (*I–B-1*). We believe the evidence suggests that such was the case, within the zone of the natural habitat for the potential domesticates. In pertinent parts of southwestern Asia, in fact, the first sub-era of food-production (*II–A-1*) was apparently restricted to this natural habitat zone and was also apparently of relatively short duration, soon being completely superseded by the second sub-era (*II–A-2*). In the New World, on the other hand, it is not clear that either the restriction or the relatively short duration were so characteristic for level *II–A-1*).

SUGGESTED LEVELS OF SUBSISTENCE-SETTLEMENT TYPES TO THE THRESHOLDS OF URBAN CIVILIZATIONS

I. *Food-gathering stage*

A. Food-gathering era

1. Sub-era of naturally determined mammalian subsistence and free-wandering (cf. general anthropoid) level; tools *fashioned* but not yet *standardized*.

   This is the general level of the Australopithecinae and of the very crude and typologically variable pebble-tools from the Australopithecus layer at Sterkfontein (Dart, 1957). Extension of this level to the Far East? (Cf. Meganthropus?)

2. Sub-era within which food-gathering and free-wandering began to be significantly culturally determined; tools of earliest *standardized* traditions (i.e., later pebble-tools; the earlier core-bifaces, flakes, chopper-chopping tools). Very broad distribution for a given tool type (e.g., Casablanca to Capetown to Madras for the core-bifaces).

   This is the general level within which a few such tool types as the Acheulean, Levalloisian, Anyathian, etc. developed, with consistent standardization within each

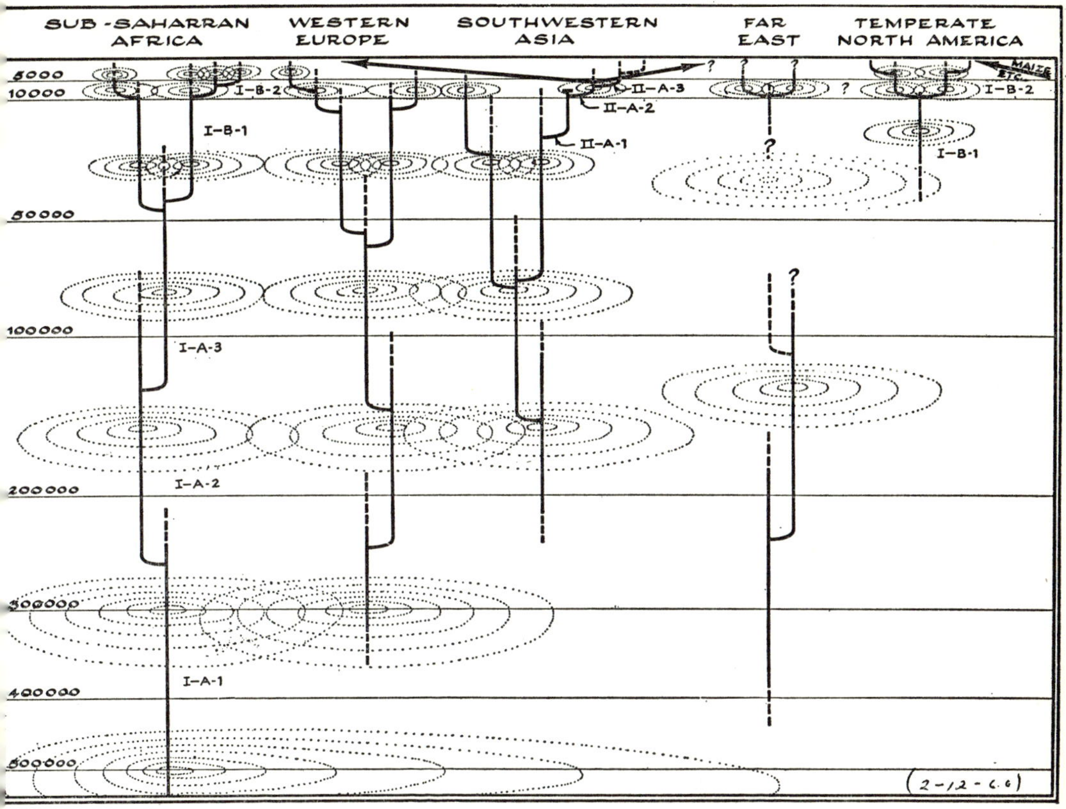

FIG. 1. A model for the consideration of the evidence for levels of subsistence-settlement types, multilinear evolution, environment and diffusion, in human prehistory.

tool type in hundreds of thousands of examples. Open air "living" or "kill" sites, such as Isimila in Tanganyika (Howell, Cole, and Kleindienst, 1960), for which "hunting" would be "too dignified" a word to describe the subsistence activities.

3. Sub-era of elemental restricted wandering, hunting. Caves commonly occupied, where they existed. Tool standardization now extends to some variety of standardized tool forms within one industry. Some regional restriction in the distribution of any one given industry.

   This is the general level suggested by the great variety of "Mousterian," "Levalloiso-Mousterian," "Acheuleo-Levalloisian," etc. industries. Note: do intentional burials and such things as the "bear cult" give the first artifactual hints of the moral order? Was a basic sexual division of labor, with men hunting and women collecting plant and insect foods, already existent?

B. Food-collecting era

1. Sub-era of selective hunting and seasonal collecting patterns for restricted-wandering types of groups. Considerable typological variety and "tools to make tools." Rather marked regional restriction of any given industry, although a generalized tool-preparation tradition (i.e., the blade-tool tradition) may still be wide-spread. Adequate artifactual evidence that a moral order existed (i.e., Franco-Cantabrian art).

   This is the era of the "kill sites" of the horse stampeders at Solutrè and of the mammoth killers at Dolni Vistoniče. Hut communities in central Europe. There may have been variants of this basic food-collecting level which persisted into the ethnological present, but it is conventionally conceptualized as an adaptation to various Pleistocene environments. This may not be strictly correct, and it poses questions about Pleistocene environments below the now temperate zones.

2. Sub-era of intensified hunting and collecting, season-bound activities; some proclivity towards plant manipulation, the taking of fish, fowl, molluscs, fleeter post-glacial mammals. Restricted-wandering to center-based wandering groups.

   The "mesolithic" is conventionally conceptualized as a cultural readaptation to post-glacial environments. This conception is probably not valid for all parts of the world (cf. Braidwood (1960) and Willey's paper elsewhere in this volume). Contrast between the "desert culture" and the "archaic" is suggestive of the variations and elaborations which obtained at this level. Extends to the ethnological present.

3. Sub-era of highly specialized food-collecting adapted to certain very specialized environments, which allow semi-permanent to permanent sedentary types of groups.

   An obvious example would be that of the "salmon reapers" of the Pacific Northwest. Extends to the ethnological present.

II. *Food-producing stage*

A. Primary era, based on individual or family scale efforts, non-mechanized.

1. Sub-era of incipient cultivation (and animal domestication in some regions) involving experimental manipulation of the potential domesticates within a subsistence milieu of types *I–B-1* or *-2*, and (at least initially) within an environment of natural habitat for the potential domesticates; restricted wandering to semi-permanent settlement groups.

   This level is hardly "pure" in the sense of an ideal type, as it is inconceivable that its food-supply can have been entirely the result of cultivation. It is an exasperatingly difficult sub-era to identify archeologically, as the artifacts are mainly of types also pertaining to the previous levels—there must have been a considerable amount of "making do" with older tools. Thus there is frankly an *ex post facto* element in the conception of this level. As it is conceptualized (in the Old World, at least) it was soon displaced and did not reach to the ethnological present as an ideal type.

2. Sub-era of the primary village-farming community (or its functional equivalent, settlement-wise), in which a marked proportion of the dietary intake is of produced food. Semi-permanent to predominantly permanent settlement types. It is this level (and not the foregoing *II–A-1*) which certainly in the Old World, at least, had such great diffusionary potential. It took many varieties, as well as blends, and one dependent variety appears to have been the beginning of pastoral nomadism.[2]

   This level is the "neolithic" in Childe's sense of "a self-sufficing food-producing economy" (although not *all*

[2] This fact came out clearly in a joint seminar held with Lawrence Krader in 1957. Cf. Krader, 1955.

food was produced, some being still collected). The artifacts now reflect the new technical order (and—perhaps?—a corresponding swing in the moral order, but no basic change). Extends to the ethnological present.

3. Sub-era of the expanded village-farming community (often also characterized as the sub-era of towns and temples or of "incipient urbanization"); permanent settlements dependent on a subsistence pattern of predominantly produced food. In southwest Asia, the plow and draft animals. Craft specialization now definitely assured (e.g., metallurgy), and the swing towards the technical order of urban type is under way. There are probably also beginnings of change in the moral order.[3]

   Extends to the ethnological present, but probably only in its developed sense, as part of a much broader *oikoumené* which included full urbanization.

As the reader refers from the above classification to the delineative device may we again remind him (1) that our levels of subsistence-settlement types depend only on our own intuitive assessment of the incomplete archaeological evidence; (2) that the individual columns are not intended to be translated directly into the typologies of tool-preparation traditions; (3) that while some of our levels—across the individual columns—are in the main contemporary, other levels (or parts of them) may be only homotaxial; and (4) that the complete sequence of levels may not be represented in every column. The levels are conceptualized as ideal types, and the representation of blends of one level with another is not attempted in the delineation.

Summary

It will be obvious that I have been influenced by the writings of Robert Redfield and Julian Steward. I had much benefit from a long conversation with Gordon Willey in December, 1958, and also from reading the draft of his paper for the present volume. I certainly also benefited from participation in the Wenner-Gren Foundation seminar (autumn, 1958), organized by Joseph R. Caldwell and Creighton Gabel at the Chicago Natural History Museum. Finally, it is doubtful whether any set of ideas born within the Department of Anthropology at the University of Chicago can ever be individually identified as to author. The fingers of most of our colleagues, of our good graduate

[3] Cf. Thorkild Jacobsen in H. Frankfort *et al.*, 1951. See also suggestive consideration of trends away from the folklike in Robert M. Adams, 1956.

students, and of such recent visitors as Desmond Clark, Bruce Howe, and Jean Perrot are doubtlessly also in this pie. The idea for the delineative device dawned on me during an animated discussion with Thorkild Jacobsen and Milton Singer at the end of the January 30, 1959, session of Singer's Anthropology 241, but the hands of Robert Adams, Clark Howell, and Arthur Jelinek have been on it since then.

References

ADAMS, ROBERT M. 1956. "Some Hypotheses on the Development of Early Civilizations," *Amer. Antiquity,* XXI, 227–32.

BEARDSLEY, RICHARD K., MEGGERS, BETTY J. *et al.* 1956. "Functional and Evolutionary Implications of Community Patterning." In *Seminars in Archaeology: 1955* (Memoir 11, Society for American Archaeology).

BRAIDWOOD, ROBERT J. 1959. "Archeology and Evolutionary Theory," pp. 76–89 in *Evolution and Anthropology: A Centennial Appraisal.* (The Anthropological Society of Washington.)

———. 1960. "Prelude to Civilization," in the proceedings of the Oriental Institute symposium, *The Expansion of Society in the Ancient Near East and its Cultural Implications,* December, 1958. In press, University of Chicago Press.

BRAIDWOOD, ROBERT J. and CHARLES A. REED. 1957. "The Achievement and Early Consequences of Food-production: a Consideration of the Archeological and Natural-historical Evidence." *Cold Spring Harbor Sym. on Quant. Biol.,* XXII, 19–31.

CHILDE, V. GORDON. 1956. *Piecing Together the Past.* New York: Frederick A. Praeger.

DART, RAYMOND. 1957. In *Nature,* CLXXX, 521–24.

GARROD, DOROTHY A. E. 1953. "The Relations Between South-West Asia and Europe in the Later Palaeolithic Age," *Jour. World Hist.,* I, 13–38.

HOWELL, F. CLARK, COLE, G. H., and KLEINDIENST, M. R. 1960. "Isimila. An Acheulean Occupation Site in the Iringa Highlands, South Highlands Province, Tanganyika," *Proc. 4th Pan-African Congress on Prehistory, Leopoldville, Belgian Congo, 1959,* edited by G. MORTELMANS. In press.

JACOBSEN, THORKILD. 1951. In *Before Philosophy,* H. FRANKFORT *et al.* London: Penguin Books.

KRADER, LAWRENCE. 1955. "Ecology of Central Asian Pastoralism," *Southwest Jour. of Anthrop.,* XI, 301–326.

REDFIELD, ROBERT. 1953. *The Primitive World and its Transformations.* Ithaca: Cornell University Press.

STEWARD, JULIAN. 1955. *The Theory of Culture Change.* Urbana: The University of Illinois Press.

ROBERT M. ADAMS

THE EVOLUTIONARY PROCESS IN EARLY CIVILIZATIONS

It is not difficult to construe the origin and spread of food production as a biologically adaptive process, contributing directly and fundamentally to an increasingly intensive and secure utilization of an expanding life-zone. But, with the growth and superimposition of later patterns in human history, the pertinence of the strictly biological evolutionary model is less assured. The adaptive significance of some of the most far-reaching sociocultural changes of historical periods may frequently have been an unimportant consideration in their initial success, and it certainly has manifested itself only gradually, in derived and questionable form or through a host of mediating institutions (Adams, 1959). Not surprisingly, then, the evolution of human society and culture has sometimes been viewed only as very broad accumulative trends that retain a biologically adaptive character but that have little or no relevance to particular historical sequences (Childe, 1944; White, 1959).

Numerous contributions to centenary symposia honoring Darwin show, however, that cultural evolutionism remains a viable doctrine even where the familiar biological processes can be recognized only by increasingly remote analogy. Given the empirical proclivities of most anthropologists, historians, and archeologists, the reason for the continuing—currently growing—concern with evolution cannot be found in the attractiveness of schemata recognizing only long-term progressive tendencies above the field of concrete historical interplay. Instead, it signifies that cultural evolutionism embodies an empirical orientation that is based on command, rather than on rejection, of descriptive cultural detail. With the increasing complexity of society and culture, the direct interaction of man as a biological organism with his environment is less and less important as a problem. We deal

ROBERT M. ADAMS is Assistant Professor of Anthropology at the University of Chicago and Research Associate and Assistant to the Director at the Oriental Institute, a division of the University. Over the past four years he has conducted fieldwork in Iraq and in Mexico, aimed at studying the rise of early civilizations in their natural settings.

in the main with regularities of a different order: changing patterns of organization, attitude, and technical equipment that are culturally interposed in this relationship, as discussed by Professor Willey in his contribution to this Centennial Celebration. Thus the primary entities of cultural evolution have few direct biological counterparts, although I believe that they support an understanding of regularities of sociocultural change no less specific than is encouraged for the paleontological record by the classic doctrines of evolutionary biology.

This paper seeks to enumerate some significant features of an evolutionary approach as it may be applied to the more detailed understanding of early civilizations like those in Mesopotamia, Egypt, China, Mesoamerica, and Peru. It draws illustrations especially from work on Mesopotamia and Mesoamerica, since the origin and development of these areas can be relatively the most satisfactorily accounted for with evidence available at present. All the early civilizations might be simply and uniformly characterized as having seen the emergence of extensive social stratification, numerous full-time craft specialists, and the political-military authority of a primitive state. And although the illustrations used here are selectively drawn, all the early civilizations share three major qualifications for contributing to the evolutionary study of human societies more generally: (1) they are well-documented, complex entities, with deep historical roots, permitting the fuller reconstruction that is possible (vis-à-vis all precivilized periods) with complementary archeological and historical lines of inquiry into interrelated institutions and activities; (2) they are essentially autochthonous in the major lines of their development, underlining the significance of any observed similarities; and (3) they exhibit many developmental parallelisms (cf. Steward *et al.,* 1955). However, the objective of this paper is not to define the place of early civilizations in the broad evolutionary record of all human society. Instead, it considers the requirements of an evolutionary mode of analysis that can usefully be applied to early civilization as a particular organizational level or stage out of the succession of emergent grades into which man's social and cultural history clearly is divisible.

Successive Stages as Major Structural Changes

A widely recognized procedure of—and prerequisite for—evolutionary studies is the systematic identification of developmental stages or substages into which the continuous historical-archeological record of particular areas can be divided, as Dr. Steward has told us. Accepting as the broadest possible objective of an evolutionary approach the presentation of sociocultural change and variety as "an orderly . . .

rational and intelligible process" (Childe, 1951, pp. 175, 179), it is clear that stage designations need to reflect, as far as possible, an orderly progression and to be free from idiosyncratic or stylistic features which defeat cross-cultural comparison. Unfortunately, agreement on taxonomic indicators has not been possible even among those workers in the field who more or less explicitly accept a "structural" orientation in stage terminology.

A few recent examples may serve to illustrate the range of differences in contemporary stage classifications. V. Gordon Childe, following, in general, the original views of L. H. Morgan, has distinguished Savagery from Barbarism by the introduction of food production at the beginning of the latter, and Barbarism from Civilization mainly by the presence in the latter of writing as "a convenient and easily recognizable index of a quite revolutionary change in the scale of the community's size, economy, and social organization" (Childe, 1951, pp. 22, 24). Elsewhere he has provided a fuller discussion of the characteristics of civilization as a stage, including urbanism, monumental public works, the territorial organization of society in the form of a state, the creation of a centrally controlled social surplus out of tribute or taxation, the beginnings of exact and predictive sciences, class stratification, full-time craft specialists, expanded foreign trade, and the reappearance of naturalistic art—all in addition to the art of writing (Childe, 1950).

Julian H. Steward outlines a briefer and more inclusive classificatory scheme, having conceptualized cultural development as a matter not only of increasing cultural complexity but also of the emergence of successively higher "levels of sociocultural integration." While recognizing the possibility of intervening levels, he distinguishes the nuclear family, the folk society, and the state as primary. These, he observes, "are qualitatively distinctive organizational systems, which represent successive stages in any developmental continuum and constitute special kinds of cultural components within higher sociocultural systems" (Stewart, 1955, p. 54). Still another suggested ordering of the data is provided by Morton H. Fried, who identifies social differentiation as the basic evolutionary determinant. Distinguishing between "pristine" and "secondary" state formation, he suggests that the emergence of at least the former (which alone is of interest in connection with the early civilizations as defined here) can be viewed as a historically inevitable transition from egalitarian organization to rank society, to stratification society, and then to state society (Fried, 1959). My own view, perhaps overgeneralizing from the "pristine" examples of Mesopotamia and Mesoamerica, defines the emergence of civilization mainly in terms of the contrast between independent temple-centered communities and the stratified, militaristic, urban state (Adams, 1956). While there is general agreement on the organization of society in terms of a state at the end of the sequence, at least among these schemata, there is less agreement on the order of importance of associated sociocultural features and correspondingly still less on the characterization of a preceding stage or stages.

Differences in theoretical position have undoubtedly contributed to differences in stage taxonomy, but I suggest that the variation just described is better understood as positioned along an axis recording relative degrees of commitment to "theory" and "fact." Childe, an archeologist, was content never to define civilization save as the congeries of traits for which he might find tangible evidence. "The sole advantage of technological criteria over political or ethical criteria is that they are more likely to have been recognizable in the archeological record" (Childe, 1951, p. 9). In consequence, of course, Childe's characterization of civilization as a stage tends to overstress the importance and constancy of technological criteria and to overstate the role of technological developments like the introduction of metallurgy as independent causative factors. My own view, also primarily archeological in origin, has the already expressed weakness that it may be too closely tied to idiosyncratic features of those examples of early civilization whose origins and emergence, for historically accidental reasons, are best understood. Thus I am frankly unprepared to cope with the apparent absence of cities in civilized Pharaonic Egypt (Wilson, 1954) and with the (happily inconclusive!) evidence that during the Old Kingdom there the priesthood slowly attained, rather than was slowly forced by the expanding state to restrict, an independent political and economic role (Kees, 1933, pp. 252 ff.).

On the other hand, there are difficulties of a different order with the stages reconstructed by social-anthropological theorists like Fried and Steward: they provide little help to the specialist in ordering his data. Fried interposes distinctions, particularly that between egalitarian and rank society, which may be rooted in contemporary ethnographic observation but can hardly be traced satisfactorily in the archeological record of precivilized societies; hence, while heuristically valuable, they are not directly useful to the archeological empiricist. Steward's concept of the succession of folk and state levels of integration is a seminal one, but it tends to overlook or blur a basically contradictory aspect of the various specific ways in which the succession took place. In at least our two best "pristine" examples, the strongest, most successful impulses to state formation seem to have come not from the politically most advanced occupants of urban centers but from marginal groups, recently arrived in the nuclear areas, whose relatively strong folk and tribal ties were an important source of strength under politically fluid conditions. Both the reliance placed by Sargon on the cohesiveness of his Akkadian kinsmen (King, 1910, pp. 239–40; Frankfort, 1951, p. 74) and the prominence achieved by successive incoming "barbarian" groups of Toltecs, Chichimecs, and Aztecs in central Mexico (Jiménez, 1958, pp. 6 ff.; Palerm and Wolf, 1957, pp. 3–5) are cases in point.

It is within the capacity of this paper neither to reconcile these and other differences in stage terminology nor to reify the more comprehensive approaches of Steward and others, from whom in any case I have drawn heavily. But it does appear that different classifications respond to different thcorctical purposes and degrees of responsibility to primary historical-archeological data, and this lack of agreement emphasizes the fact that stage formulations by themselves do not comprise an evolutionary study. They provide only a framework for considering the concrete problems of transition between quite hypothetical states of equilibrium and a taxonomy which may facilitate cross-cultural comparison.

Determinate Processes as Connecting Links in Stage Succession

In other words, stage formulations and a classificatory approach in general merely establish the conditions for the subsequent, more penetrating study of historical transitions. Their value is primarily heuristic: useful schemata help to identify and emphasize key institutional changes and to introduce their functional interrelations and varying rates of sociocultural change as important research problems. Since stages necessarily fail to reflect agreement, their usefulness can be measured only against the content of real historical sequences; hence stage formulations have already reached a point of diminishing returns as a scholarly activity. A more frequent objective of future studies probably will be the identification of determinate processes of change involving (or pertaining significantly to) a transformation of one stage, or level of integration or set of institutions, into a successor. Another future research trend will probably involve comparative diachronic study that seeks to account for variation within a single sociocultural type and sequence of stages.

It is perhaps worth noting that this kind of shift from classificatory to processual studies is also in progress in fields to which the classic evolutionary doctrines are more clearly central:

> The most important future studies of North American recent mammals will not be classificatory. Almost everything is still to be learned about the

actual variation in wild populations of mammals and the real geographic and ecological distribution of such variation—a study that has been as much confused as furthered by the traditional descriptions [Simpson, 1959, p. 1353].

However, the abruptness of this shift should not be overstressed. Traditional taxonomy, after all, will remain the starting point for future investigations. Moreover, as I. M. Lerner has recently pointed out, a processual approach has been implicit in the doctrine of evolution from the very beginning. Darwin's own work was directed less toward identifying the systematic categories of evolutionary descent than toward the demonstration of natural selection as an efficient evolutionary process or mechanism (Lerner, 1959, p. 173).

To be convincing, processual studies need to synthesize and explain economically the widest and deepest possible range of historical detail. Without full historical control, generalized accounts of sequential development are little more than a priori hypotheses or programmatic statements. It seems to follow that the opposition sometimes thought to exist between a culture-historical approach and evolutionism is largely artificial. Where detailed analyses in the culture-historical tradition have been concerned with the development of institutions reflecting or embodying primary structural features and where they have been written so as to make possible an increased understanding of the intervening links in a particular changing situation that are amenable to generalizing or comparative analysis, they belong firmly within the spectrum of evolutionary studies.

A. L. Smith's careful analysis of the slow transformation of a group of small temple-shrines into a greatly expanded "palace"-complex at Uaxactun in the Maya lowlands (Smith, 1955, esp. Figs. 2–5), for example, is an important source of insight into the changing forms of "theocratic" control assumed for that area and period as well as for its parallels elsewhere. Again, Sir Leonard Woolley's grave-by-grave account of the cemeteries of Early Dynastic Ur in southern Mesopotamia (Woolley, 1934) makes possible at least a limited quantitative statement about the onset of social differentiation as measured in grave-goods and on the rate of expansion of the metallurgical crafts. It would be a pity if essential contributions like these to evolutionary studies were all undertaken by specialists in culture-history, while the self-proclaimed evolutionists were all armchair "generalists." Fortunately, this is less and less the case in recent years, although it has been a damagingly accurate characterization at times in the past. Today, as Leacock notes, a growing trend is in prospect "toward the synthesis of the historical empiricism associated with the name of

Boas, and the theoretical evolutionism associated primarily with Morgan" (Leacock, 1958, p. 193).

Historical Explanation as the Analysis of Complex Causal Processes

Perhaps the most unequivocal position on the determination of cause-and-effect relationships as the primary objective of current evolutionary studies has been taken by Steward in a number of recent papers:

> The methodology of evolution contains two vitally important assumptions. First, it postulates that genuine parallels of form and function develop in historically independent sequences or cultural traditions. Second, it explains these parallels by the independent operation of identical causality in each case. The methodology is therefore avowedly scientific and generalizing rather than historical and particularizing. It is less concerned with unique and divergent (or convergent) patterns and features of culture—although it does not necessarily deny such divergence—than with parallels and similarities which recur cross-culturally [Steward, 1955, p. 14].
>
> It is only when one compares two historically independent areas that it is possible to stipulate that a particular set of phenomena are causally interrelated; that is, to state that certain conditions presuppose or depend on certain other conditions [*ibid.*, p. 88].
>
> *If the more important institutions of culture can be isolated from their unique settings so as to be typed, classified, and related to recurring antecedents or functional correlates, it follows that it is possible to consider the institutions in question as the basic or constant ones, whereas the features that lend uniqueness are the secondary or variable ones* [*ibid.*, p. 184; italics in original].

For several reasons I cannot accept this position. One that has already been alluded to is that Steward's equation of evolutionary methodology with generalizing and of historical methodology with particularizing creates a false dichotomy. Perhaps large-scale cause-and-effect relationships are easier to assert, but their validation, at least, requires a meticulous working-out of their sequential and functional particulars. Nor is it clear that causal interrelationships can be established *only* by comparison of historically independent sequences. This view depends, as Steward recognizes, on the asserted existence of "genuine" formal and functional parallels; yet how are genuine and spurious parallels to be distinguished without a detailed, empirical analysis that does not confine itself to identical features in two presumably parallel cultural traditions but also seeks to explain those features which are unique or variable? A more reasonable position on this question would seem to involve the recognition that cause-and-effect relationships are not absolutes but instead are working hypotheses, probably to be refined or altered as understanding develops further. Causal processes that are of evolutionary interest, then, are those that most economically order and "explain" all the available data from a particular area that is pertinent to the major structural changes which its historical sequence records. And cross-cultural comparison serves to highlight differences in explanatory syntheses prepared for different areas, not only in order to formulate broad generalizations applying, for example, to all early civilizations, but also to assist in the re-examination and refinement of the more detailed causal processes assumed to operate on a particular historical scene.

A frequent defect in the very general statements of historical cause-and-effect relationships that are derived from an exclusively comparative method is that they tend to rest on unilinear sequences of abstractions having little relation to real historical entities or processes. In connection with the rise of early civilizations, two of these misleadingly self-generating abstractions have played an especially prominent role. One is that competition for land as a result of "population pressure"—apparently brought about by an assumed tendency for population to increase to or beyond the limit that can be sustained by the available resources—is an important cause for the beginnings of large-scale militarism (Steward, in Steward *et al.,* 1955, pp. 68, 71).

I am inclined to doubt that this process ever occurred anywhere prior to the special conditions created by modern medicine and sanitation practices, but at any rate it can be stated more categorically that there is no evidence to associate a high and rising population (relative, of course, to what is known of the available subsistence resources) with the onset of militarism in any of the early civilizations. Instead, what evidence there is, deriving mainly from surveys of ancient settlement patterns, very clearly suggests that "population pressure" was not a significant factor at all. In Mesopotamia the appearance of militarism probably coincided with a rise in population, but only to a level representing a small fraction of that attained in later antiquity with essentially similar subsistence methods (Jacobsen and Adams, 1958; Adams, 1958). On the Peruvian north coast, fortifications provide evidence of institutionalized warfare before a level of complexity comparable to that in Mesopotamia was achieved, but also at a time when population and settlement density were very much less than they became later (Willey, 1953).

The second highly dubious abstraction—that the administrative requirements of large-scale irrigation management brought into being the early civilized states and largely shaped the political forms they took (Steward [in Steward *et al.*, 1955, p. 71, and *passim*] here follows tentatively one feature of Karl A. Wittfogel's more comprehensive theory of the relation of "hydraulic societies" to "Oriental despotism" [see Wittfogel, 1957])—rests on the assumed temporal priority of large-scale irrigation over the appearance of the early states. This assumption, too, is not in accord with the available evidence, as I have attempted recently to show (Adams, 1959).

If the evolutionary approach is not to lead us into the cul-de-sac of self-contained causal theories, it seems essential that we recognize the complexity and interdependence of events leading to the major stage transformations. For all their undoubted institutional similaritics, each of the early civilizations was the unique product of alternately conflicting and mutually reinforcing trends, and it is futile to hope that even their basic features will be explained in all cases as the predictable, predetermined outcome of some sort of general law. In a word, it appears that comparable occurrences of systems of complex, interacting forces are a more useful focus of evolutionary studies than the gross formal similarities which these systems sometimes engendered. And to understand the growth of civilization in these terms, we need once more to turn away from broad generalizations and toward detailed historical analysis.

The stress on historical particulars in the foregoing paragraphs is not intended to imply an abandonment of the study of evolutionary processes in different cultural traditions. In fact, it is precisely on the level of detailed historical trends that the study of parallels may prove most illuminating. With the beginnings of large-scale militarism, for example, a comparable series of changes can be observed in Mesopotamia of the early to mid-third millennium B.C. and central Mexico at around 1000 A.D., using contemporary historical and literary accounts and later redactions of oral traditions. Since the cultural settings do not reflect historical contact, the precise forms of these changes are entirely unlike; yet in functional terms their configurations are intriguingly similar.

In Mesopotamia, Early Dynastic rivalries can be traced in a succession of positions and functions denoting different kinds of authority (Hallo, 1957). In the case of the *en* or "lord," the earliest to appear in the record, "the political side of the office is clearly secondary to the cult function. The *en*'s basic responsibility is toward fertility and abundance" (Jacobsen, 1957, p. 107). The *lugal* or "king," on the other hand, found his source of strength in a large private manor not connected with the temple. With the essentially secular figures rising to claim this title through military prowess was associated "par définition la prétention d'étendre leur domination sur un territoire dépassant les limites de la Cité-État" (Falkenstein, 1954, p. 795). Still a third figure, the *ensik,* normally appears during the Early Dynastic period as the ruler of a single city-state. As the manager of the estate of the main temple and as the chief executor of the god's decisions on major questions like rebuilding the temple, the *ensik* appears to have retained some of the sacerdotal qualities of the *en* while engaging in protracted political and military strife with neighboring *ensiks* or their secular counterparts. If the character and background of this contention are generally obscure, its outcome is not. Ultimately, the unstable patterns of alliance and domination characteristic of the time (cf. Poebel, 1926) came to an end with the triumph of dynastic political authority in the person of a *lugal* controlling a territorial state and successfully putting into effect his imperial pretensions.

The comparable sequence of events in Mexico rests on thinner and more ambiguous documentation that purports to describe the rivalry and succession of two semilegendary figures—Quetzalcoatl-Topiltzin and Tezcatlipoca—during the period of the supremacy of Tula. Leading students of the problem are in sharp disagreement not only on the dating of the succession but also on the significance and timing of the incidents of which it is composed (cf. Kirchoff, 1955; Jiménez, 1958). There is even debate on the possible identity of Quetzalcoatl with still a third figure, Huemac. Nevertheless, there exists enough common ground between the alternative interpretations to suggest that behind the dynastic intrigues and unexplained changes in personal allegiance a subtle but irreversible shift took place in the social (and perhaps ethnic) group and political ideology from which political leadership was drawn, as well as in the character of the dominant religion. The priest-representative of older gods personifying natural forces was ultimately replaced in Tula after a period of more or less open struggle by a figure whose identification with a cult of human sacrifice found favor with newly formed groups of warriors bent on predatory expansion. Our awareness of close parallelisms between this process and the attainment of dominant political and military authority by the Mesopotamian *lugal* is only heightened when we learn that this transformation in the character of community leadership was accompanied in both areas by tendencies toward deifying the temporal officeholder. Moreover, the powers and pretensions of the Aztec kings, as seen by the Spaniards several hundred years later, had apparently continued to grow during that interval after the fall of Tula almost precisely as did those of their Sumero-Akkadian counterparts during the later third millennium B.C.

It would be desirable to carry this analysis to a level of still greater specificity in order to determine at what point (if any) significant parallelisms between the two sequences are submerged beneath historical details applying uniquely to each. But for that purpose it would be necessary to provide not only a full account of known historical features and "events" but also a full evaluation of the historical sources themselves. It would be desirable also to determine how the process by which relatively secular and autonomous political institutions gradually supplanted the sacred and traditional authority of the temple was matched by similar and complementary processes affecting other institutions. But this would require a relatively full spelling-out of sociocultural contexts in the two areas. Both these extensions of the analysis, in other words, require digressions beyond limits that are reasonable in a general paper. However, perhaps even this single, briefly sketched example will serve to support the contention that detailed comparisons are potentially as useful a source of evolutionary insight as are statements of broad trends and stage successions that override and ignore the complexities with which sociocultural change is normally attended.

Prevailing Orientation of Change As Focus for Processual Studies

It has become a commonplace that evolutionary change is markedly disjunctive, in the sense both that it does not proceed at a uniform over-all rate and that its structural effects are sharply focused rather than diffuse. In cultural terms the latter sense is of particular significance. It implies as Steward points out, that stage transformations consist of relatively rapid, qualitative transformations in particular institutional complexes and not of any necessarily equivalent or simultaneous increase in total cultural complexity. In time, of course, the effect of new structural features will be felt more generally; even where older cultural forms are retained, they may be endowed with new functions.

An example particularly relevant to this paper is the formation of bodies of peasantry out of precivilized village folk-societies, a process whose general character has been penetratingly discussed by Robert Redfield (1953, pp. 31 ff.). He sees gradual, but cumulatively important, changes in the political, economic, and moral life of the village as it was more or less forcibly brought into interdependence with the dominant institutions of the expanding city. But it seems clear that this process is essentially one of secondary response to more decisive changes going on in the early urban centers. It is also apparent that the strategic trends whose interplay led initially to the emergence of civilization were largely confined to, or at least disproportionately focused on, the city. Childe's formulation of the appearance of early civilization as an "urban revolution" aptly reflects both the restricted locus of the transformation and its rapid, qualitative character within that locus.

An implicit concern with the disjunctive aspects of change is a distinguishing feature not only in general or theoretical formulations of cultural evolution but also in empirical studies undertaken with a similar approach. Perhaps this is most evident in the prevailing emphasis on the growth of class stratification and associated changes in the forms of land tenure. This is widely regarded as a strategic process in the accumulation of the surpluses upon which the early civilizations were based, even where the socioeconomic conditions of the average producer may have changed little in consequence.

Again to take research on Mesopotamia and Mesoamerica as an example, early students of the economic history of both areas tended to minimize the extent of differentiation in access to productive resources. They insisted that agriculture was carried on largely within a traditional framework dictated at most by status, not class, relationships. In Mesoamerica this was believed to have been structured mainly along corporate kin-group lines (Bandelier, 1878), while in Mesopotamia the shared concern of the community for the cults of its gods was thought to have sustained a virtually all-inclusive "Tempelwirtschaft" under the control of the priesthood (Schneider, 1920; Deimel, 1931).

More recent studies of early Mesopotamian economic texts have tended to stress the size and importance of secular (palace and private) capital accumulations and estates alongside those of the temple, forcing a substantial revision of earlier views partly by the introduction of new evidence but also in part by vigorously advocating the importance of class stratification as a developmental social force (Diakonoff, 1954). In Mesoamerica the recent reinterpretations have placed greater stress on the decay of the traditional kin groups as effective landholding units that was going on just prior to the Spanish Conquest and on the beginnings of a substantial accumulation of land in the hands of a hereditary nobility (Monzón, 1949; Kirchoff, 1954–55). Since these interpretations rest on substantially the same body of evidence as that available to Bandelier eighty years ago, it is obvious that they must derive primarily from theoretical considerations of an evolutionary character.

To generalize from these remarks, an evolutionary approach necessarily seems to involve an emphasis on rapidly developing, *emergent* institutions. In the context of the early civilizations, urbanism, class stratification, kingship, trade and markets, craft specialization, and militaristic expansionism comprise the essentials of this emergent configuration. Even though perhaps these institutions initially involved a relatively small proportion of the population and may have had only a remote or sporadic influence on the life-cycle of an "average" individual, it was primarily their growth that allows us to describe orderly, intelligible, and significant evolutionary stages. Moreover, it was the pattern of growth of these institutions that set the course and limits for subsequent, more widely ramifying development. In a word, the elucidation of evolutionary processes provides an expression of the prevailing orientation and type of change. It does not itself provide a description of the total cultural reality in which disjunctive change is taking place.

Growth of Early Civilizations as an Aspect of Evolutionary Change

Turning from substantive historical sequences, in conclusion we need to consider the position of cultural evolution within the wider field of evolutionary studies. It was maintained at the outset that evolutionary principles could be extended to the study of complex cultural phenomena only by retaining an attitude of strong historical empiricism and by focusing mainly on entities essentially without direct counterparts in the traditional fields of biology. In light of the example furnished by the early civilizations, two qualifications to this statement are now apparent.

The first and most obvious is that the growth of civilization exhibits a broadly adaptive character in spite of its cultural, rather than biological, basis. By having become civilized, a larger, more functionally specialized population, bound together with ties demonstrating a new level of organizational complexity, was enabled to subsist within a given region. Moreover, conditions were created whereby the internal limitations of the region as an ecological niche could be transcended more and more effectively. The coercive state made possible an extension of redistributive institutions which (at least in Mesopotamia) had already appeared earlier, providing an unprecedented degree of control over local and periodic fluctuations in food supply. It also made possible planned, artificial improvements in subsistence technology and in the environmental conditions within the region with respect to subsistence; large-scale irrigation, canal construction, centralized agricultural management, and the support of specialists in ancillary operations like tool production and exchange are all cases in point. And, finally, it made possible the transcending of particular resource limitations of the immediate region through conquest and state-supported trade. If many of these developments were slow to appear, or were even absent in some of the early civilizations until introduced by some external agency, it can hardly be denied that they form a functional configuration when considered as a group. And the essentially independent development of this configuration in several Old and New World centers, followed by its apparently relentless (if periodically interrupted) outward spread, only serves to emphasize its adaptive advantages.

The second qualification is more abstract but no less important. If the entities dealt with by cultural evolution are largely unique in the spectrum of evolutionary studies, its basic outlook and postulates are not. Considered as variant forms taken by a single, but vital, phase in human history, the early civilizations provide as powerful an affirmation as any analogous development in paleontology for Julian Huxley's [1] comment:

> All reality is in a perfectly proper sense evolution, and its essential features are to be sought not in the analysis of static structures or reversible changes but through the study of the irrevocable patterns of evolutionary transformation [Huxley, 1953, p. 88].

Bibliography

ADAMS, R. M. 1956. "Some Hypotheses on the Development of Early Civilizations," *Amer. Antiquity,* XX, 227–32.

———. 1958. "Survey of Ancient Watercourses and Settlements in Central Iraq," *Sumer,* XIV, 101–4.

———. 1959. "Early Civilizations, Subsistence, and Environment." In *Oriental Institute Symposium on the Expansion of Society in the Ancient Orient* (in press).

BANDELIER, A. F. 1878. "On the Distribution and Tenure of Lands, and the Customs with Respect to Inheritance, among the Ancient Mexicans," *Peabody Museum of Harvard University, 11th Ann. Rept.,* II, 385–448.

CHILDE, V. G. 1944. *Progress and Archaeology.* London: Watts & Co.

———. 1950. "The Urban Revolution," *Town Plan. Rev.,* XXI, 3–17.

———. 1951. *Social Evolution.* London: Watts & Co.

DEIMEL, A. 1931. *Sumerische Tempelwirtschaft zur Zeit Urukaginas und seiner Vorgänger.* (Analecta Orientalia, No. 2.) Rome.

DIAKONOFF, I. M. 1954. *Sale of Land in Pre-Sargonic Sumer* (papers presented by the Soviet delegation at the XXIII International Congress of Orientalists, Assyriology Section), pp. 19–29. Moscow: Publishing House of the U.S.S.R. Academy of Sciences.

FALKENSTEIN, A. 1954. "La Cité-temple sumérienne," Cahiers hist. mond., I, 784–814.

[1] I am indebted to Sir Julian Huxley and Elmer R. Service for critical comments.

FRANKFORT, H. 1951. *The Birth of Civilization in the Near East.* London: Williams & Norgate.

FRIED, M. H. 1959. "On the Evolution of Social Stratification and the State." In *Paul Radin Memorial Volume* (in press).

HALLO, W. W. 1957. *Early Mesopotamian Royal Titles: A Philologic and Historical Analysis.* New Haven: American Oriental Society.

HUXLEY, J. S. 1953. Review of *Life of the Past,* by G. G. Simpson, *Scient. American,* CLXXXIX, 88–90.

JACOBSON, T. 1957. "Early Political Development in Mesopotamia," *Zeitschr. Assyriol.,* LII, 91–140.

JACOBSEN, T., and ADAMS, R. M. 1958. "Salt and Silt in Ancient Mesopotamian Agriculture," *Science,* CXXVIII, 1251–58.

JIMÉNEZ, M. W. 1958. *Historia antigua de Mexico.* 3d. cd. Mexico, D. F.: Escuela Nacional de Antropologia e Historia.

KEES, H. 1933. *Ägypten.* ("Handbuch der Altertumswissenschaft, Kulturgeschichte des alten Orients," Vol. I.) Munich.

KING, L. W. 1910. *A History of Sumer and Akkad.* New York: Stokes.

KIRCHOFF, P. 1954–55. "Land Tenure in Ancient Mexico," Rev. mexi. estud. antropol., XIV, 351–62.

———. 1955. "Quetzalcóatl, Huemac, y el fin de Tula," Cuad. Amer., XIV, 163–96.

LEACOCK, E. 1958. "Social Stratification and Evolutionary Theory: Introduction," *Ethnohistory,* V, 193–99.

LERNER, I. M. 1959. "The Concept of Natural Selection: A Centennial View." In *Commemoration of the Centennial of the Publication of the "Origin of Species" by Charles Darwin: Proceedings of the American Philosophical Society,* CIII, No. 2, 173–82.

MONZÓN, A. 1949. *El Calpulli en la organización social de los Tenochca.* ("Pub. del Instituto de Historia," No. 14.) Mexico, D. F.: Univ. Nacional Autón. de Mexico.

PALERM, A., and WOLF, E. R. 1957. "Ecological Potential and Cultural Development in Mesoamerica." In *Studies in Human Ecology,* pp. 1–37. ("Pan American Union, Social Sciences Monographs," No. 3.)

POEBEL, A. 1926. "Der Konflikt zwischen Lagaš und Umma zur Zeit Eannatums I und Entemenas." In *Paul Haupt Anniversary Volume,* pp. 226–67. Baltimore: Johns Hopkins Press.

REDFIELD, R. 1953. *The Primitive World and Its Transformations.* Ithaca, N.Y.: Cornell University Press.

SCHNEIDER, A. 1920. *Die Anfänge der Kulturwirtschaft: Die sumerische Tempelstadt.* Essen: Bädeker.

SIMPSON, G. G. 1959. Review of *The Mammals of North America,* by E. R. Hall and K. R. Kelson, *Science,* CXXIX, 1353–54.

SMITH, A. L. 1955. *Uaxactun, Guatemala: Excavations of 1931–37.* ("Publications of the Carnegie Institution of Washington," No. 588.)

STEWARD, J. H. 1955. *Theory of Culture Change.* Urbana: University of Illinois Press.

STEWARD, J. H., *et al.* 1955. *Irrigation Civilizations: A Comparative Study.* ("Pan American Union, Social Sciences Monographs," No. 1.)

WHITE, L. A. 1959. *The Evolution of Culture,* Vol. I. New York: McGraw-Hill Book Co.

WILLEY, G. R. 1953. *Prehistoric Settlement Patterns in the Virú Valley, Peru.* (Bureau of American Ethnology, Bull. 155.)

WILSON, J. A. 1954. "Cities in Ancient Egypt," *Economic Development and Cultural Change,* III, 74.

WITTFOGEL, K. A. 1957. *Oriental Despotism: A Comparative Study of Total Power.* New Haven: Yale University Press.

WOOLLEY, C. L. 1934. *The Royal Cemetery, Ur Excavations,* Vol. IV. London and Philadelphia.

JULIAN H. STEWARD

EVOLUTIONARY PRINCIPLES AND SOCIAL TYPES

THE MEANING OF CHANGE

One can add little to the kudos so abundantly bestowed on Darwin for his role in establishing the idea that change or historicity is inherent in all phenomena. For the biological sciences, Darwin not only convinced the world of change but provided evolutionary principles which describe the essential orderliness of biological transformations. The physical scientists, too, have formulated change in the inorganic realm, both microscopically and macroscopically, in such a way as to reveal orderly process (see papers by Gaffron, Evans, and Shapley in our companion volume, "The Evolution of Life"). It is, in fact, inconceivable that change in biological and physical phenomena should be random, quixotic, chaotic and without regularities which are described as "causes." Moreover, hypotheses, descriptions, or formulations of recognized order are presumed to apply to all places and times within the universe.

That cultures have changed is also unquestioned, but the social sciences and humanities have achieved little agreement regarding what constitutes orderly principles in cultural change. It is sometimes held that cultural change is not amenable to formulations; or that a series of special histories must constitute our only categories; or, again, that there are principles underlying all cultural change. If no principles whatever are recognizable in cultural change, it is probably inappropriate to insist that there has been cultural evolution. On the other hand, contention that there has been cultural evolution certainly imposes the obligation to identify the orderly alterations of structures or systems, the principles that have operated at all times and places. This poses extremely difficult problems of cultural taxonomy and of determination of process.

JULIAN H. STEWARD is Research Professor of Anthropology and member of the Center for Advanced Study at the University of Illinois, a member of the National Academy of Sciences, and recipient of the Viking Fund Medal for 1952. His specialization includes aboriginal America and acculturation of native populations.

Some Views of Cultural Evolution

There have probably been more papers written on cultural evolution during this centenary than during the last half century, but they exhibit a very wide variety of individual interests. While some are more or less directed to the problem of orderliness in culture change, they constitute little more than a barely discernible trend. The essential meaning of cultural evolution has by no means been clarified, and some of the views expressed in the present symposium are so disparate that they are hardly even in conflict.

One view, which is expressed by Huxley (in "Evolution of Life") if I understand him correctly, and by others, is that cultural evolution expresses the distinctive creativity of the human mind. A somewhat different emphasis is Kroeber's interest in the major streams of world cultural history. A deliberate effort to cut across these streams and, in fact, to disregard them as major categories, is found in the hypotheses of worldwide developmental stages postulated by Leslie A. White (1959a) and to a certain extent by the nineteenth century unilinear evolutionists. A rather different means of seeking order is the more recent tendency to begin with the particulars of change in culture areas or special streams of history and then formulate tentative and limited cross-cultural hypotheses of what seems significant in structure and process.

These views are not necessarily in direct conflict. The creativity of the human mind is beyond doubt. If culture is conceived as the continuity of a social pattern or value system, it unquestionably has flowed in major channels. At the same time, it is perfectly clear that the agricultural revolution of several millennia ago and the later industrial revolution profoundly altered structures within all the cultural historical traditions. There is also no question that transformations have occurred repeatedly, on a smaller scale, within each of the many streams of history. The obvious need at the moment is to reconcile these diverse interests in a common search for orderliness in culture change.

Emphasis upon the capacity for reason, creativity, and communication naturally highlights the distinctively human qualities resulting from the biological evolution of the human brain, hands, and bipedal locomotion. This evolution created the preconditions of cultural evolution and thus suggests that cultural evolution is an extension of biological evolution. But this does not imply that cultural evolution follows the principles of biological evolution. Cultural behavior is really phenotypical behavior, as contrasted with the rigid patterns of genotypical or instinctual behavior, but it is infinitely more adaptive in man than in any other animal. More specifically, the ability to use speech and other symbols, to reason, and to manipulate tools underly the development of the vast variety of known human behavior patterns.

Acknowledgement that man is a rational and creative thinker—or at least can be rational upon occasion—in no way tells us what he will think, and it can easily involve us in sterile discussion of free will. We have to assume that the application of reason is reasonable in the sense of following some orderly pattern, and that creativity, as manifest in ethos, value systems, styles, religions, and philosophy, has some comprehensible relationship to other activities and is not wanton, random, causeless, or without relevant antecedents and cultural correlates.

The more fundamental difficulty in discussing cultural evolution is the absence of any generally accepted system of classification. Kroeber stresses repeatedly elsewhere in this volume that anthropology's cultural taxonomy is pre-Linnean; it lacks systematization and classificatory criteria for cultural categories, except those which are evident in distinctive areas, histories, or traditions. It is as if culture had no order other than the accident of each history. Kroeber's own classification reflects his lifelong interest in history, which gives primary emphasis to the many streams of history that develop through time, criss-cross, diffuse, and interact. The main streams are more often characterized by their intellectual achievements, specific inventions, philosophies, and value systems than by structural components. Their identification and characterization result more from deep insights, sometimes from what is more nearly intuitive feelings, as in the recognition of styles, than from emphasis upon formal criteria. Structural transformations during time are made secondary to the stylistic continuities.

The nineteenth century unilinear evolutionists have been attacked for their philosophical presuppositions, especially for their use of the idea of progress as a basic principle, and for the inadequacy of their facts. I believe, however, that their more serious limitation was that they never came to grips with the question of structures and types of structures. Preoccupied with cultural origins, as many of them were, and inclined to deal topically with religion, the family, technology, law, and other special subjects, primitive culture was treated as a very generalized but not well-characterized category. It is germane to this point that anthropologists left the great cultural transformations or growth of civilizations in Egypt, Mesopotamia, the Indus Valley, and China to the orientalists and sinologists, and that, although they accepted Middle America and the Andes as a legitimate part of their subject matter, they have understood the structural changes in these areas only recently.

The question of the developmental typology of whole cultures never became a major issue in nineteenth century anthropology, as pointed out by Bordes in his paper elsewhere in this volume, except that a small number of scholars offered world schemes of cultural evolution. For example, L. H. Morgan (1877) postulated classificatory diagnostics for each of seven stages from savagery to civilization and thereby became vulnerable on many scores. Most of the nineteenth century anthropologists treated the whole primitive world as a single category from which facts were drawn at random to illustrate their points, while the question of social change was left to Engels, Marx and their followers. The great impact that V. Gordon Childe (1951) has had upon understandings of culture-change arises less from theoretical claims than from the fact that he bridged the gap from the primitive to the civilized in the Middle East and was directly concerned with the profound structural changes involved. But Childe knew little about other areas.

Certainly the clearest taxonomy has come from Leslie A. White (1949; 1959 *a, b*), who proclaims complete allegiance in principle to the nineteenth century writers. White, however, deals not only with the origins of primitive culture but with the great transformation —"revolution," he calls it—that occurred among all societies which were fundamentally affected by plant and animal domestication. He has no place in his scheme for Kroeber's streams of history or for culture areas or local traditions. White's two main structures are: kin-based tribal societies, which controlled little energy; and the internally-differentiated, class-structured, territorial states, which controlled high energy. If we understand him correctly, he postulates that a third major structural change occurred after the industrial revolution and believes that a fourth is being initiated by the use of nuclear energy.

It can hardly be argued that White's stages or categories do not represent major structural transformations, although it may be noted parenthetically that they do not constitute a tree of culture comparable to the tree of life. Presumably, each culture was transformed independently whenever the impact of domestication was sufficiently great; or rather, primitive culture as a single entity was transformed.

There is question, however, concerning certain implications of White's categories and evolutionary principles. Since anthropologists have traditionally dealt with preliterate, primitive societies, few accept White's contention that variations within the primitive world that resulted from local cultural-ecological adaptations, area traditions, diffusion, and other factors are unimportant. Moreover, White's characterization of primitive society as essentially based on classificatory kinship systems, *i.e.*, extended kinship, disregards the principles of sexual and age divisions. Again, his discussion of agrarian states or civilized societies after the agricultural revolution draws freely from such diversified societies as those of the Andes, Middle America, Egypt, Mesopotamia, Islam, the ancient Hebrews, and the Roman Empire, as if all these were essentially similar (*cf.* Steward (1960) review of White, 1959*a*). What is similar taxonomically among these societies represents a high degree of generalization. The role of such factors as religion, irrigation, warfare and commerce in their development can only be understood through much fuller archeological information and detailed comparisons.

Much of the discussion of evolution has hinged upon the question of what order of generalization is required to justify using the term. White seems to argue that evolutionary changes must be universal. I have maintained that processes may operate cross-culturally only a few times to produce similar structures and suggested that the many lines of development could be described as "multilinear evolution." But, if evolution requires universal principles, I willingly forego the term and continue, along with many others, to interest myself in limited generalizations before undertaking to make the grand scheme.

Limited and detailed comparison, I believe, characterize the present trend. It grows out of the influence of Franz Boas, who introduced unrelenting empiricism and field research into cultural studies. In its extreme of "cultural relativism," Boas' influence led to the denial of evolutionary categories and causal relationships. It also led to re-examination of earlier evolutionary claims. Perhaps today, proceeding from the particular to the general, we can arrive at evolutionary principles that fill the bill better than those offered in the past.

The Implications of Cultural Relativism and the Culture Area

In American research, following Boas' influence several decades ago, the culture area became the basic taxonomic category. Defined in terms of distinctive element-content and unique integration, the so-called "pattern" of each area was sometimes conceived more stylistically than structurally, as in Ruth Benedict's *Patterns of Culture* (1934). While no one, of course, claimed that these area patterns were god-given, they were treated as if they were part of an original creation. Little interest was taken in their origins, except to trace the diffusion of their elements. When the concept of cultural personality entered social science, the normative aspects of patterns became emphasized, and interest centered on how the distinctive cultural content and the psychological patterning of individual behavior constituted mutually-reinforcing factors that tended to resist change of culture patterns. Explanations of origins and transformations were avoided.

Some scholars became definitely antihistorical. However, others who retained an interest in history and prehistory operated with area-bound cultural taxonomy, which ascribed developmental periods secondary importance. Style, as manifest in ceramics, weaving, architecture, and other material remains rather than structure, became diagnostic of culture areas, spheres, traditions, and co-traditions. The very use of such terms as "developmental" and "florescent," even for the long and complicated sequences of the Central Andes and Meso-America (for example, by Strong, Bennett, and Willey in Bennett, 1948), implied change that was more like the germination, growth, and flowering of a plant than the transformation of one species into another.

In part, this emphasis upon aesthetic florescence is an Americanist characteristic, derived from the primacy of the culture area and from the exclusion, until recently, of Old World civilizations from anthropological classifications and formulations. Yet we find that historians, with their strong humanistic orientation, still ascribe major importance to style; for example, Coulborn (1959) sees aesthetic style as the enduring characteristic of each of the early civilizations in his discussion of the concepts of water-gods.

The Present Trend

To judge by many recent writings which bear "evolution" in their titles, and by many more which do not, there is increasing interest in finding causes of culture change other than diffusion. It should not be necessary to state that no one denies the massive diffusion of cultural elements and element complexes. The point is that acceptance of diffused traits generates internal changes, which have become the focus of interest. More specifically, attention is increasingly paid to changes in social structure.

In prehistoric archaeology, this trend has brought a shift from a predominating concern with implement typology to settlement pattern in relationship to environment, for example, Braidwood (in this volume), Beardsley *et al.* (1955), Willey (1953), and Steward (1937). The universality of stages postulated by Braidwood and by Beardsley *et al.* is certainly not confirmed, and I question on ethnographic grounds whether there has ever been a "free wandering" stage. Whether this approach will fill the requirements of evolution is less important than the new problems which orient it. While diffusion is amply evident in the distribution of types and stylizations of early implements, the nature of human organizations which use these implements in the food quest is now commanding attention.

Intensive analyses, with attention to structural change and process within the Old World, have been carried out especially by V. Gordon Childe (1934, 1946, 1951) and by Braidwood (1952, 1958, 1959), and within the New World by Willey and Phillips (1953, 1955) and others. These understandings have been generalized in cross-cultural hypotheses concerning the development of irrigation civilizations in a symposium by Donald Collier, Angel Palerm, Robert Adams, Karl Wittfogel, Ralph Beals, and the present writer (Steward, 1955, ed.), and by Adams (1956). Wittfogel (1957) has written far more broadly upon the subject.

Social organization is also the center of interest in Murdock's typological and developmental studies, which are treated essentially statistically (1959). Kinship, marriage, residence, and descent are primary taxonomic criteria, but universal principles as such are not suggested. Goldman, in an excellent review of the theory of cultural evolution (1959) and in two studies of Polynesia (1955, 1957), deals with the critical change in structure from kinship to class. Among Polynesian variations, he sees status rivalry as "basic in all evolutionary sequences leading to civilization" (1959, p. 74.) He has not demonstrated, however, that status rivalry is a universal principle. Moreover, the very fact of status differences in kinship societies generally implies some control of surplus or other perquisites which themselves have to be explained if any causality is to be found.

Goldman in 1941 and the present author in his contribution to the Radin Festschrift—"Carrier Acculturation"—studied change of simple bands or lineages into a strong status system among the Carrier Indians; the fur trade was the key factor in introducing the required surplus. The relationship of the status system to the wealth of the fur trade has also been the subject of many other analyses of Northwest Coast Indians. Along somewhat similar lines, studies of the conflict of the principles of kinship and state institutions in West Africa have received much attention since the writings of Maine in the last century.

Other studies directed at the problem of cultural typology with reference to causality are Oberg's (1957) discussion of South American lowland types and Steward and Faron's (1959) on five South American types. These studies are less concerned with the universality of their findings than with significant structural diagnostics and the relationship of these to land use, environment, and socially integrating factors.

The Problem of Principles

It is noteworthy that most attempts to take a large view of cultural evolution focus upon change in structure, principles, and processes at that crucial point when internal specialization and social classes begin to supersede kinship groups, that is, when productive surplus and means of controlling it become central considerations. This is as true of Goldman's limited evolutionary sequence in Polynesia as of Willey's (Centennial Paper) shift of evolutionary criteria from technological to social features in the High civilizations of America and White's world stages.

White seems most keenly aware of any writer that evolutionary principles should be universal. He points to some principles that are common not only to the cultural realm but to the universe, such as a tendency to segmentation, and to others common to biological organisms and societies, such as utilization of external energy. So far as culturological phenomena are concerned, however, the general principle that improved technology increases command of energy and thus transforms societies and, in turn, their ethos is presumed to apply at all times and places. It might be contended, however, that the utilization of energy really involves many mechanisms that differ greatly according to the specificity of local histories. White does not solve the difficulty that after the agricultural revolution, new processes appeared, resulting in structures based upon the non-kinship principles of internal specialization, class structure and state institutions. The required universality of evolutionary principles seems to break down.

There is a possible way of viewing these phenomena, however, which may enable us better to hew to the line of principle. Primitive societies are, with few exceptions, structured basically not only upon lines of kinship but also of age and sex. Kinship takes many forms; age entails assignment of roles according to maturity and often is formally graded; and sex is manifest not only in reciprocity within the family but in organizations that follow sex lines. These might all be considered variations upon the universal, biological fact that human beings have chronic sexual interest, that the human infant has a prolonged period of helplessness, that the human learning potential and language capacity make complicated socialization possible, and that the nuclear family is universal because of sexual, procreational, and socializational needs. Modified and often elaborated as these principles are, they are rarely superseded or eliminated in the primitive world. Purely culturological principles had not taken complete command. It is, perhaps, as if we were studying the primordial, chemical-laden swamp where self-replicating atoms were taking form but the principles of biological evolution as known today had not yet come clearly into being.

My point is not to minimize culture. But Hallowell's observation (1959) has cogency—that, since primate society has fairly definite form in pre-cultural times, we cannot look upon the emergence of culturally determined societies without reference to their biological basis. In the primitive world the cultural arrangements have not suppressed the biological themes. It is only in societies which have specialists and hereditary classes that wholly non-biological structural principles begin to emerge.

The primitive variations upon sex, age, and kinship principles are determined largely by two additional factors: tools and environment. Everywhere the nuclear family, whether polygynous, polyandrous, or monogamous, is the basic expression of the biological facts of life, but whether such families constitute independent societies, or are grouped in patrilineages, matrilineages, or other larger structures, depends largely upon how their technology is used in a particular environment, that is, upon the process of cultural ecology. Important as this process is in determining the organization of independent primitive societies, where it is somewhat analogous to biological ecology, it becomes a minor factor in complex states.

Ecological adaptations are highly inferential where we have only a scant archaeological record. Evidence as to the nature of society consists of settlement pattern and environment as much as tool inventory. Howell and Washburn earlier in this volume conclude from cave middens that the australopithecines changed from scavengers to hunters, even though archaeology discloses only stone hand tools. Hunting, however, normally either requires spears or bows to kill medium and large mammals at a distance, fairly elaborate traps and snares, or else such co-operation as surrounds and drives. Unless the culture utilized stone- or bone-pointed spears or arrows, or, as Leakey suggests, bolas, we would have no archaeological record of the techniques involved in very early periods.

The causal implications of technology, cultural ecology, and other purposes for which human beings organize are much clearer in ethnographic cases. The Great Basin Shoshoni (Steward, 1938) and the southern Chilean Alacaluf and Chono (Steward and Faron, 1959) were split into independent families because their subsistence was based upon sparse and scattered foods that were collected—seeds and small mammals in the Great Basin, shellfish in Chile—and because such food collection is competitive rather than co-operative. Although both groups had bows and in other environments could well have been large game hunters and although the Shoshoni held occasional co-operative rabbit and antelope hunts, the family was the basic social unit in each case. Technology per se and environment in its totality had very secondary importance. The adaptation to a collecting subsistence caused the family to assume virtually all cultural functions: sexual co-operation in the provision of subsistence and shelter, socialization of children in the mode of life, and care in sickness. Additional cultural functions, such as Chono initiation ceremonies, brought clusters of families together briefly at intervals of several years.

The similarity or typological category of these independent family units represents a certain order of abstraction, though not nearly so great an abstraction as that which places all societies prior to the agricultural revolution in a single category. Structurally, these societies are more similar to one another than they are to patrilineal hunting bands or sedentary farm communities. The similarities are still greater when the functional rather than formal aspects of the culture content are emphasized. Houses, despite formal differences, were temporary and improvised, clothing was made of locally available materials, supernaturalism, except for the Chilean deities connected with initiation ceremonies, concerned individual and family welfare, and so on. These provide the synchronic basis for the type. Diachronically, also, identity of process is implied. Whatever the cultural antecedents of the Shoshoni, Alacaluf and Chono, the cultural ecological adaptation was the same in each case.

With reference to the streams of culture history, these people would be classified differently. The Chilean societies might be viewed as part of an archaic American cultural stratum, or, with reference to their stone tools, as of an "early lithic" type, or again in view of their relationship to higher cultures of South America they would be "marginal." The Shoshoni would also be "marginal" with reference to the Southwestern climax of the Anasazi area from which they probably drew much of their basketry, rabbit skin blankets, rabbit nets, and other technological features. But in each case, the population density and distribution and the organization of the people for the meager repertory of cultural activities, which were principally concerned with survival, developed internally from evolutionary processes rather than externally from diffused patterns.

This family-type society is not a universal stage or survival of such a stage, but those who claim that all human societies consist of territorial grouping of clusters of families are in error. Such independent family units must have existed in the past under suitable conditions. A very different type, the patrilineal hunting band, is found in Tierra del Fuego, southern California, Australia, Tasmania, and among the Bushmen, and Congo Negritoes. This type consists of an extended patrilineal family, that is, a half dozen or so nuclear families, which are related in the male line and which own territory and bring wives from outside the band. This essential structural similarity is related to hunting of fairly sparse but non-migratory game; it represents maximum efficiency in terms of co-operative hunting in a terrain which hunters can know and traverse, and it involves an optimum number of co-operating families.

Other types of hunters and gatherers are the matrilineal bands of the Sirionó and Guayakí forest nomads of Bolivia and Paraguay (Steward and Faron, 1959), the Eskimo family clusters of arctic hunters, which is seemingly a unique type, the multifamily bands of Canadian caribou-hunters, and various more settled communities of acorn-gatherers of California and fishermen of the Northwest Coast.

These types are not definable in terms of culture-element content nor mere technology. They are not, as White (1959) implies, all based upon extended and classificatory kinship systems. The Shoshoni have a descriptive system, and the other societies have many kinds of systems. They can be categorized only partly by Murdock's (1959) criteria of kinship, descent, marriage, and residence.

Viewed in its particulars, any primitive society is an organization of human beings which is necessarily adapted to the requirements of survival in its habitat. But the organization involves sexual unions and child-rearing, which are biological constants. Upon this biological basis develop reciprocal sexual divisions of labor and sometimes formal structures which follow sexual lines, as in men's tribal societies. Co-operation beyond the nuclear family tends usually, although not always, to involve relatives in one line or the other and hence provide a basis for extended, formal kinship-structuring. The long human span of life tends to assign the individual different roles as a child, adult, and elder, and often it underlies formal age-grading.

In the absence of internal specialization and social classes, primitive societies are organized upon the basis of these three principles. In the varied streams of cultural development, especially where more abundant resources permit denser populations and more complex societies, these principles frequently find elaboration and emphasis that amounts to structural transformation, but rarely a break-through to new principles. Thus, certain East African cattle breeders are distinguished by their age-grade societies and clan development; the Hopi Indians, by a complex system of matrilineal clans, kiva groups, a kachina society, and associational groups; certain Ge tribes of eastern Brazil, by several pairs of moieties and in some instances age-grades; and many Amazon tropical forest farm villages, by patrilineal clans and a men's house or men's society.

The Evolution of Agrarian States

The transformation from primitive to civilized communities entails some conflict between the egalitarian principles of the former and the differentiation of status and role of the latter. This conflict has long been noted, more in sociological than in anthropological literature, and it is reflected in the dichotomy expressed by such pairs as *gemeinschaft* and *gesellschaft, societas* and *civitas,* and folk and urban. The conflict does not mean that societies cease to be structured along lines of sex, kinship, age, and associations. Instead, the earlier structures are modified and adapted to the functions of the newer and larger structure. The family and household surrender certain functions to the community, the community becomes integrated within the state, and so on, as a series of internal sublevels within higher levels of cultural integration. Such levels are merely constructs for analyzing particular societies and histories. They do not represent cross-cultural abstractions or evolutionary stages, although they may be employed for this purpose.

The most profound transformation was that which followed the agricultural revolution. Some of the very general effects of this revolution are clear. There was an agricultural surplus which supported non-food-producers, dense and stable populations, class stratification, and political, religious, or military institutions that controlled the state. Beyond these simple generalizations, the problem of cultural classification is greatly complicated by the complexity of structures and apparent multiplicity of processes. At the same time, it is facilitated because the historic record, in part a written record, discloses diachronic processes which are far less inferential than processes assumed in the case of primitive societies. It becomes possible to define change partly in terms of culminations of one or more major processes.

The process of cultural ecological adaptation becomes increasingly subsidiary to other processes which underly state formation, for local societies cease to be independent structures that are organized primarily for survival in situations of very direct man-to-nature relationships and become sub-societies, or part-societies, whose interaction with the environment is increasingly modified by a special relationship to the larger society and culture. Adaptation to environment continues to be a factor in culture change, but the higher civilizations provide a greater spectrum of adaptative possibilities, such as purely agrarian, commercial, militaristic, or theocratic communities whose place in the total cultural structure is maintained by the larger society. Meanwhile, the larger society evolves in response to new kinds of processes which really do not have homologues among primitive societies. Thus, multi-community structures may attain cohesion as states because of co-operation in irrigation works, militarism, theocratic controls or other factors. Merely to list these processes so succinctly, let alone to ascribe primary importance to one or another, however, is to make much too high a level of generalization.

It is a highly generalized statement to say that in the Andes, Egypt, Mesopotamia, the Indus Valley, and Yangtze Valley irrigation was obviously a key factor in increased farm productivity, and at some point the water systems were so great as to require a state managerial control. We need to know, in each case, what this point was and whether some pre-existing institution, such as a priesthood, took over this function, or whether a new institution evolved. We need to know better how the growth of population related to expansion of farming and water works. We are by no means certain when and under what circumstances militarism became an institution of state aggression, which enforced amalgamation of local groups into a single political unit. The nature and role of commerce in state growth needs clarification. The same is true of other institutions, such as property, slavery, or peonage of different kinds, and urbanization.

The processes just mentioned seem to have operated cross-culturally and produced rather similar structures, but this is so highly generalized a statement that it adds little to knowledge. The processes themselves can be broken down into smaller connecting links. For example, instead of proposing the broad theory that military conquest created states, we would have to examine the emergence of militarism in the apparent theocratic, non-militaristic state of Teotihuacan in Mexico and compare it with militarism in early Andean states. One hypothesis might suggest that militarism came to Mexico as conquest from the outside, whereas it began in Peru with the capture of sacrificial victims, then became an instrument of state control, and only later a means of territorial conquest.

With reference to processes, cultural structures may then be viewed as culminations of predominant processes rather than as static, formal structures. This does not obviate the necessity of some kind of classification, however, for processes can only be recognized through their concrete manifestations at particular moments. Prior to the industrial revolution, there were many kinds of states in which agriculture, commerce, militarism, religion and other factors played different roles. None was static, but all tended strongly to consist of hereditary classes and thus to have structures which were fairly enduring until conquest or revolution overthrew them or until they were affected by the industrial revolution.

We return at this point to the role of man's creative capacity in culture development. The food revolution that was the basis of these agrarian states resulted from the application of certain scientific knowledge, while state administration, conquest, and religious and aesthetic developments were all creative expressions of sorts. But it is doubtful whether anyone in the course of the rise of agrarian states was aware of their outcome, let alone consciously directed their growth beyond the pattern in which he was involved.

The age of exploration, science, invention, and the industrial revolution applied reason on a more massive scale. The technological effect of this is an incredible acceleration of change in means of production. The social effect is the continued and accelerating individual mobility, which increasingly destroys the possibility of hereditary status and classes. There is no need to discuss the mechanisms of education, mass communications, and the like involved. The central fact is that new skills are demanded almost every year for all statuses and roles throughout society. This was no more foreseen by political philosophers of the west than by Marxian philosophers who envisaged a dictatorship of the working class. Russian communism has changed radically during its brief forty years owing to an increase of individual mobility which has probably accelerated faster than in the western democracies.

While this acceleration of internal rearrangements has, of course, been recognized, its magnitude could not have been foreseen, and it is difficult even now to project its consequences so as to alter them through intelligent planning. And, of course, planning involves social or moral standards that are not susceptible to scientific validation. The most we can say perhaps is that with new sources of energy and automation foreseeable, society will face the problem of leisure and what to do with it.

Some Conclusions

This paper is largely an admission of the general uncertainty now surrounding the concept of cultural evolution. Excellent critical analyses of the concept can be found in White's writings, in Lesser (1952), in the papers by Kluckhohn, Murdock, Hallowell, Haag, Braidwood, and Greenberg in the Centennial Symposium of the Anthropological Society of Washington (1959) and in Goldman (1959), but the methodological difficulties of cultural evolution still lack clarification.

In the physical and biological universes, evolution implies change which can be formulated in principles that operate at all times and places, although the particular principles of biological evolution differ from those of the physical realm. Expectably, or at least by analogy, then, cultural evolution should contain its own distinctive principles, which also underly all cultural change.

By this criterion, no one has yet demonstrated cultural evolution. Some of White's principles are common to physical, biological and cultural phenomena. Others differ when primitive, kinship-based societies are transformed into civilized societies. The "principle" of energy levels per se has little meaning; the question is how energy is used. The individual mobility that followed the industrial revolution was the process or connecting link between energy and society; it destroyed the earlier hereditary classes of agrarian states. If an age of nuclear energy brings further transformations, it will be through the agency of automation and other arrangements contingent upon quantities of energy rather than the nature of the source.

At the other extreme are the interests in histories, culture areas, or traditions, which become "evolutionary" only by being so designated.

In between is an increasingly large number of papers, some that deal with the particulars of what is called "evolution" within single culture areas, and others that compare change between areas. The latter necessarily generalize in some degree in order to recognize structures that are similar formally or functionally and to ascertain relevant processes or causes of change.

What can we say, then, of cultural evolution? Are transformations within a single area or tradition evolutionary? Can we say that limited cross-cultural generalizations that formulate different but recurring lines of change are evolutionary? Or must we conclude that cultural evolution, unlike physical and biological evolution, generates new principles as it evolves? This last proposition is implied in the emphasis upon the creativity of the human mind, except that reference to principles surely requires recognition of ascertainable order in culture change. It challenges us to show what new principles of order emerge at successive stages.

Perhaps the wholesale proclamation of allegiance to cultural evolution in 1959 is principally to do honor to Darwin. Nevertheless it has constituted an important theoretical stocktaking, even though the new evolutionists will undoubtedly continue to do what they have been doing during the last two or three decades. This recent research, however, constitutes an important trend, consisting of interest in transformations, however large or small, and a search for causes. Emphasis is now upon society, which is conceived more narrowly as kinship systems in the case of primitive peoples and more holistically in the case of civilized peoples. As hypotheses are constructed and modified, there is reason to hope that some sort of solid taxonomic basis will be found. As hypotheses are validated and broaden, perhaps universal principles will emerge. These may not differ greatly from those of the nineteenth century writers, but they will have a more solid empirical basis.

Bibliography

ADAMS, ROBERT M. 1956. "Some Hypotheses on the Development of Early Civilizations," *Amer. Antiquity,* XXI, No. 3.

BEARDSLEY, RICHARD K. *et al.* 1956. "Functional and Evolutionary Implications of Community Patterning" in *Seminars in Archaeology: 1955.* (Memoir No. 11, Society for American Archaeology.)

BENEDICT, RUTH. 1934. *Patterns of Culture.* Boston: Houghton Mufflin.

BENNETT, WENDELL C. 1948. "The Peruvian Co-tradition," pp. 1–7 in *A Reappraisal of Peruvian Archaeology,* WENDELL C. BENNETT, ed. (Memoir No. 4, Society for American Archaeology.)

BRAIDWOOD, ROBERT J. 1952. *The Near East and its Foundations for Civilization.* (Chicago Natural History Museum Popular Series in Anthropology, No. 37.)

———. 1958. "Near Eastern Prehistory," *Science,* CXXVII, 1419–30.

———. 1959. "Archaeology and Evolutionary Theory," pp. 76–89, in *Evolution and Anthropology: A Centennial Appraisal.* Anthropological Society of Washington.

CALDWELL, JOSEPH R. 1958. "Trend and Tradition in the Prehistory of the Eastern United States," *Amer. Anthropologist,* LX, No. 6, Part 2. (Memoir No. 88, Scientific Papers of the Illinois State Museum, Vol. X.)

CHILDE, V. GORDON. 1934. *New Light on the Most Ancient East.* London: Routledge and Kegan Paul.

———. 1946. *What Happened in History.* London: Penguin Books.

———. 1951. *Social Evolution.* London and New York: H. Schuman.

COOPER, JOHN M. 1942. "Areal and Temporal Aspects of Aboriginal South American Culture," *Primitive Man,* XV, 1–38.

COULBORN, RUSHTON. 1959. *The Origin of Civilized Societies.* Princeton: Princeton University Press.

GERARD, R. W., CLYDE KLUCKHOHN, and ANATOL RAPAPORT. 1956. "Biological and Cultural Evolution," *Behavioral Science,* I, 6–34.

GOLDMAN, IRVING. 1955. "Status Rivalry and Cultural Evolution in Polynesia," *Amer. Anthropologist,* LVII, 680–97.

———. 1941. "The Alkatcho Carrier: Historical Background of Crest Prerogatives," *Amer. Anthropologist,* XLIII, pp. 396–418.

———. 1957. "Cultural Evolution in Polynesia," *Journ. Polynesian Soc.,* LXVI, 156–64.

———. 1959. "Evolution and Anthropology," *Victorian Studies* (September, pp. 55–75.)

GOODENOUGH, W. H. 1957. "Oceania and the Problem of Controls in the Study of Cultural and Human Evolution," *Journ. Polynesian Soc.*, LXVI, 146–53.

HAAG, WILLIAM A. 1959. "The Status of Evolutionary Theory in American Archeology," pp. 90–105 in *Evolution and Anthropology: A Centennial Appraisal.* Anthropological Society of Washington.

HALLOWELL, A. IRVING. 1959. "Behavioral Evolution and the Emergence of the Self," pp. 36–60 in *Evolution and Anthropology: A Centennial Appraisal,* Anthropological Society of Washington.

HARRIS, MARVIN. 1959. "The Economy Has No Surplus?" *Amer. Anthropologist,* LXVII, 185–99.

HUXLEY, JULIAN S. 1958. "Cultural Process and Evolution" in *Behavior and Evolution,* ANNE ROE and GEORGE GAYLORD SIMPSON, eds. New Haven: Yale University Press.

KLUCKHOHN, CLYDE. 1959. "The Role of Evolutionary Thought in Anthropology," pp. 144–57 in *Evolution and Anthropology: A Centennial Appraisal.* Anthropological Society of Washington.

KROEBER, A. L. 1939. "Cultural and Natural Areas of North America." *University of California Publications in American Archaeology and Ethnology,* Vol. 38 242 pp.

———. 1945. "The Ancient Oikoumene as an Historic Cultural Aggregate." (Huxley Memorial Lecture for 1945) London: Royal Anthropological Institute of Great Britain and Ireland. Reprinted in 1952 in his *The Nature of Culture,* Chicago: The University of Chicago Press.

———. 1946. "History and Evolution," *Southwest. Jour. Anthrop.,* II, 1–15.

LESSER, ALEXANDER. 1952. "Evolution in Social Anthropology," *Southwest Journ. Anthrop.,* VIII, 134–46.

MORGAN, LEWIS H. 1877. *Ancient Society, or Researches in the Lines of Human Progress from Savagery, through Barbarism to Civilization.* New York: Henry Holt & Co.

MURDOCK, GEORGE P. 1959. "Evolution in Social Organization," pp. 126–43 in *Evolution and Anthropology: A Centennial Appraisal.* Anthropological Society of Washington.

MURPHY, ROBERT F., and JULIAN H. STEWARD. 1956. "Tappers and Trappers: Parallel Process in Acculturation," *Econ. Develop. and Culture Change,* IV, 335–55.

OBERG, KALERVO. 1955. "Types of Social Structure among the Lowland Tribes of Central and South America," *Amer. Anthropologist,* LVII, 472–88.

PHILLIPS, PHILIP and GORDON R. WILLEY. 1953. "Method and Theory in American Archaeology: An Operational Basis for Culture-Historical Integration," *Amer. Anthropologist,* LV, 615–34.

SINHA, S. 1955. "Evolutionism Reconsidered," *Man in India,* XXXV, 1–18.

SOUTH, STANLEY. 1955. "Evolutionary Theory in Archeology," *Southern Indian Studies,* VII, 10–32.

STEWARD, JULIAN H. 1937. "Ecological Aspects of Southwestern Society," *Anthrop.,* XXXII, 87–104.

———. 1938. *Basin-Plateau Sociopolitical Groups.* (Bureau of American Ethnology, Bulletin 120.)

———. 1956. *Anthropological View of Contemporary Culture Change, "Kyoto American Studies Seminar Publications,"* No. 2: "Anthropology." Kyoto, Japan.

———. 1956. "Cultural Evolution," *Scien. Amer.,* CXCIV, 69–80.

———. 1958. "Problems of Cultural Evolution," *Evolution,* XII, 206–10.

———. 1960. Review of Leslie White's *The Evolution of Culture* in *Amer. Anthropologist.* In press.

STEWARD, JULIAN H., ed. 1955. *Irrigation Civilizations: A Comparative Study.* Pan American Union, Social Sci. Monographs, No. 1, Washington, D. C.

STEWARD, JULIAN H. (with ROBERT F. MURPHY). See Murphy and Steward, 1956.

STEWARD, JULIAN H. and LOUIS C. FARON. 1959. *Native Peoples of South America.* New York: McGraw-Hill Book Co.

WHITE, LESLIE A. 1949. *The Science of Culture.* New York: Farrar, Strauss.

———. 1959*a*. *The Evolution of Culture.* New York: McGraw-Hill Book Co.

———. 1959*b*. "The Concept of Evolution in Cultural Anthropology," pp. 106–25 in *Evolution and Anthropology: A Centennial Appraisal,* Anthropological Society of Washington.

WILLEY, GORDON R. 1953. *Settlement Patterns in the Virú Valley.* (Bureau of American Ethnology, Bulletin 155.)

WILLEY, GORDON R. and PHILIP PHILLIPS. 1955. "Method and Theory in American Archaeology II: Historical-Developmental Interpretation," *Amer. Anthropologist,* LVII, 723–819.

WISSLER, CLARK. 1922. *The American Indian.* New York and London: Oxford University Press.

———. 1923. *Man and Culture.* New York: Thomas Y. Crowell.

WITTFOGEL, KARL. 1957. *Oriental Despotism: A Comparative Study of Total Power.* New Haven: Yale University Press.

H. W. MAGOUN

EVOLUTIONARY CONCEPTS OF BRAIN FUNCTION FOLLOWING DARWIN AND SPENCER

The year 1859 was marked by the publication of a controversial volume, *The Origin of Species by Natural Selection, or the Preservation of Favored Races in the Struggle for Life;* its influence in many fields is being celebrated in this, its centennial year. The author, Charles Darwin, was the son and grandson of physicians, and he himself began the study of medicine at Edinburgh, where he found the anatomical lectures of the third Professor Alexander Munro "intolerably dull." Shifting to Cambridge and the study of theology, Darwin graduated finally in the sciences and embarked as naturalist on the voyage of a surveying vessel, *H.M.S. Beagle,* circumnavigating the globe.

The observations and experiences of his journey (Darwin, 1839) gradually led Darwin to formulate the theory of evolution by natural selection—in his Victorian setting, a concept that was revolutionary in the extreme. The *Scala naturae,* in which living beings were arranged in a spectrum of increasing complexity, was familiar to both earlier naturalists and to biologists of the eighteenth century, but its order was generally conceived as the immutable product of divine creation. Darwin's conception proposed instead that natural selection, working within the range of normal variations, led finally to survival of the fittest, and so, in a naturalistic way, accounted for both evolution and the adaptation of existing forms to their environment.

Darwin's later writings, *The Descent of Man* (1871) and *"The Expression of the Emotions in Man and Animals* (1872), called more direct attention to the phylogenetic development of the brain, mind, and behavior. The extensive elaboration of concepts of the organization and function of the brain in evolutionary terms was primarily a post-Darwinian development, however, and its ètiological relationship to

H. W. MAGOUN is Professor of Anatomy at the School of Medicine, University of California at Los Angeles, and affiliated with the Veterans Administration Hospital at Long Beach, California. He is best known for his work in experimental neurology.

Darwin's ideas does not seem previously to have been extensively examined. One cannot but be curious concerning the derivation of the views of Hughlings Jackson in neurology, of Pavlov in physiology, of Freud in psychiatry, and of Edinger in anatomy, each of which accounted for the phylogenetic elaboration of the central nervous system in terms of a series of superimposed levels, added successively as the evolutionary scale was ascended.

In each of these conceptual systems (Fig. 1), the management of primitive, innate, stereotyped behavior, having to do with the preservation of the individual and the race, was attributed to older, subcortical, neuraxial portions of the central nervous system, which formed Jackson's lowest level and subserved the Pavlovian unconditioned reflex and the Freudian id.

Next, the more mutable, adaptive, learned behavior of Pavlov's conditioned reflex, together with the capacity of the Freudian ego for perception and the initiation of movement were ascribed to higher neural structures, including the sensorimotor cortex of Jackson's mid-

ENGLISH NEUROLOGY Hughlings Jackson	RUSSIAN NEUROPHYSIOLOGY Ivan P. Pavlov	COMPARATIVE NEUROANATOMY Edinger, Kappers, Herrick	PSYCHOANALYTIC PSYCHIATRY Sigmund Freud	SYNTHESIS
HIGHEST LEVEL	SECOND SIGNAL SYSTEM	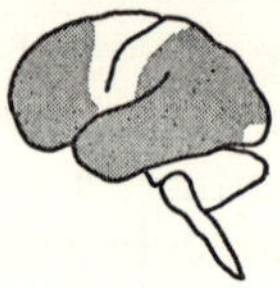	SUPER EGO	ABSTRACTION DISCRIMINATION SYMBOLIZATION COMMUNICATION
MIDDLE LEVEL	CONDITIONED REFLEX	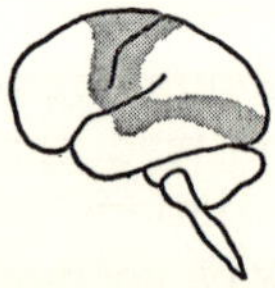	EGO	ACQUIRED ADAPTIVE BEHAVIOR
LOWEST LEVEL	UNCONDITIONED REFLEXES	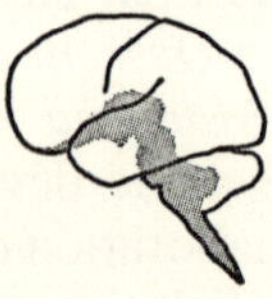	ID	INNATE STEREOTYPED PERFORMANCE

FIG. 1.—Chart comparing the evolutionary concepts of the organization and function of the brain which developed after Darwin and Spencer. (Modified from a chart by Stanley Cobb, 1949. "Human Nature and the Understanding of Disease." In FAXON, N. W., *The Hospital in Contemporary Life*. Cambridge: Harvard University Press.

dle level, which developed above or upon the older subjacent parts.

Finally, in the brain of man, hypertrophy of the associational cortices of the frontal and parieto-occipito-temporal lobes, forming Jackson's highest level, was correlated with the capacities of Freud's superego and, in the dominant hemisphere, with the capabilities of Pavlov's second signal system for symbolization and communication by means of spoken and written language.

Further testimony for the evolution of neurological function in these terms was provided by Jackson's view of dissolution, or regression by reversal of the phylogenetic process, when clinical impairment proceeded from highest, through middle, to lowest levels during neurological disease in man. Jackson specified that the resulting deficit was usually accompanied by some release of lower activity, normally subjugated to higher control. This latter feature was elaborated also in the Freudian system, in which conflicting interests of the different levels were emphasized as a source of psychic disturbance.

In much the same way that increased complexity and specialization appeared as the supposed ladder of nature was ascended by the earlier classificationists, more and more elaborate functions came to view as one climbed cephalically up the successive levels of the central nervous system. In this progressive encephalization, the brain came to resemble the earth itself, not simply in its globular form, but in consisting as well of a series of strata laid down like those of geology, one upon the other, in evolutionary time. Each neural accretion was associated with a characteristic increment of function. Following Jackson, a dissolutionary school of neurophysiology developed in which encephalization was reversed by operative transection and evolution traced backward by observing residual capacities diminish in the increasingly truncated, decorticate, decerebrate, and spinal preparations.

Probably because such views are still so contemporary, little attention has been given to exploring the role of Darwin and the interest in evolution excited by his work in establishing these concepts of neural organization and function.

HUGHLINGS JACKSON

While the views of Hughlings Jackson (1958) might be presumed to be the most directly Darwinian, they were, on the contrary, derived chiefly from Thomas Laycock, with whom Jackson began his career, and from Herbert Spencer, whom he admired greatly. Both Laycock and Spencer had applied evolutionary principles to concepts of the organization and function of the brain independently of and preceding Darwin. In his *Mind and Brain,* first published in 1859, Laycock wrote:

As we ascend the scale, the differentiation of tissue takes place and instincts of plants or animals appear. As we ascend still higher in animal life, the instincts gradually lose their unknowing character and the mental faculties emerge with their appropriate organic basis in the encephalon. Finally, with the highest evolution, we find man evincing in art and science the results of the operation of mental powers which in the lower animals are purely instinctive and in the lowest organisms simply vital processes.

There can also be found in Laycock an expression of the conflicting interests of the different levels, with the higher holding the lower in check, to be elaborated later by Jackson and by Freud:

This entire group of corporeal appetites and animal instincts is characterized by the quality of necessity. In lower organisms they are performed blindly. In man and higher vertebrates, in whom there is a development of cognitive faculties, states of consciousness, termed motives, will coincide with a knowing restraint exercised over them. But even with the highest and strongest of human motives, it is often found difficult to curb them effectually. Those classed under the head of primordial instincts or corporeal appetites, most necessary to the well-being and maintenance of the organism and the species, are the farthest removed from the will and consciousness.

His clinical observations in neurology led Laycock to consider disease as "retrocession," in which changes taking place were the inverse of evolutionary. He proposed a law of "*dis*volution" in certain kinds of brain disease, when there was a decay of the mental powers and return to an earlier, infantile status. Concepts of evolutionary levels of function, conflicting in their interests and exhibiting dissolution in neurological disease, can thus be detected in a germinal stage in Laycock's views.

Jackson's psychological concepts were strongly influenced by Herbert Spencer, from whom Darwin borrowed the term "survival of the fittest." After having been an evolutionist for some time, in 1851 Spencer formulated the basic principles that were to be elaborated in most of his later work. He had been asked to write a notice of a new edition of Carpenter's *Principles of Physiology* and "in the course," he noted (1904), "of such perusal as was needed to give an account of its contents," came across the theory of von Baer—that the development of all plants and animals was from homogeneity to heterogeneity. This concept of progressive differentiation, added to that of Lamarckian adaptation, became his distinctive evolutionary view.

In the first edition of his *Principles of Psychology*, published in 1855 and thus four years before the *Origin of Species*, Spencer pointed out that his arguments "imply a tacit adhesion to the development hypothesis," as the theory of evolution was then called, "that Life in its multitudinous and infinitely varied embodiments has risen out of the lowest and simplest beginnings, by steps as gradual as those which evolve a homogeneous, microscopic germ into a complex organism, by progressive unbroken evolution, and through the instrumentality of what we call natural causes. Save for those who still adhere to the Hebrew myth or to the doctrine of special creation derived from it, there is no alternative but this hypothesis or no hypothesis."

Applying his "development hypothesis" to psychology, Spencer reasoned:

If the doctrine of Evolution is true, the inevitable implication is that Mind can be understood only by observing how Mind is evolved. If creatures of the most elevated kinds have reached those highly integrated, very definite and extremely heterogeneous organizations they possess, through modification upon modification accumulated during an immeasurable past, if the developed nervous systems of such creatures have gained their complex structure and functions little by little; then, necessarily, the involved forms of consciousness, which are the correlates of these complex structures and functions, must also have arisen by degrees.

In the study of mind, "in its ascending gradations through the various types of sentient beings," Spencer conceived of

a nascent Mind, possessed by low types in which nerve centers are not yet clearly differentiated from one another, . . . [consisting of a] confused sentiency formed of recurrent pulses of feeling having but little variety or combination. At a stage above this, while yet the organs of the higher senses are rudimentary, Mind is present probably under the form of a few sensations which, like those yielded by our own viscera, are simple, vague and incoherent. From this upwards, mental evolution exhibits a differentiation of these simple feelings into the more numerous kinds which the special senses yield; an ever increasing integration of such more varied feelings, an ever increasing distinctness of structure in such aggregates; that is to say, there goes on subjectively a change from an indefinite, incoherent homogeneity to a definite, coherent heterogeneity.

Returning to the problem "of how such higher coordinations are evolved out of lower ones and how the structure of the nervous system becomes progressively complicated," Spencer proposed the interpolation of new plexuses of fibers and cells between those originally existing. In diagrammatic sketches, apparently of an invertebrate ganglion, Spencer distinguished (Fig. 2) "a nervous center to which afferent fibers bring all order of peripheral feelings, and from which efferent fibers carry to muscle the stimuli producing their appropriately combined contractions." If a part of the coordinating plexus (*A*) "takes on a relatively greater development in answer to new adjustments which environing conditions furnish, we may expect one part of this region

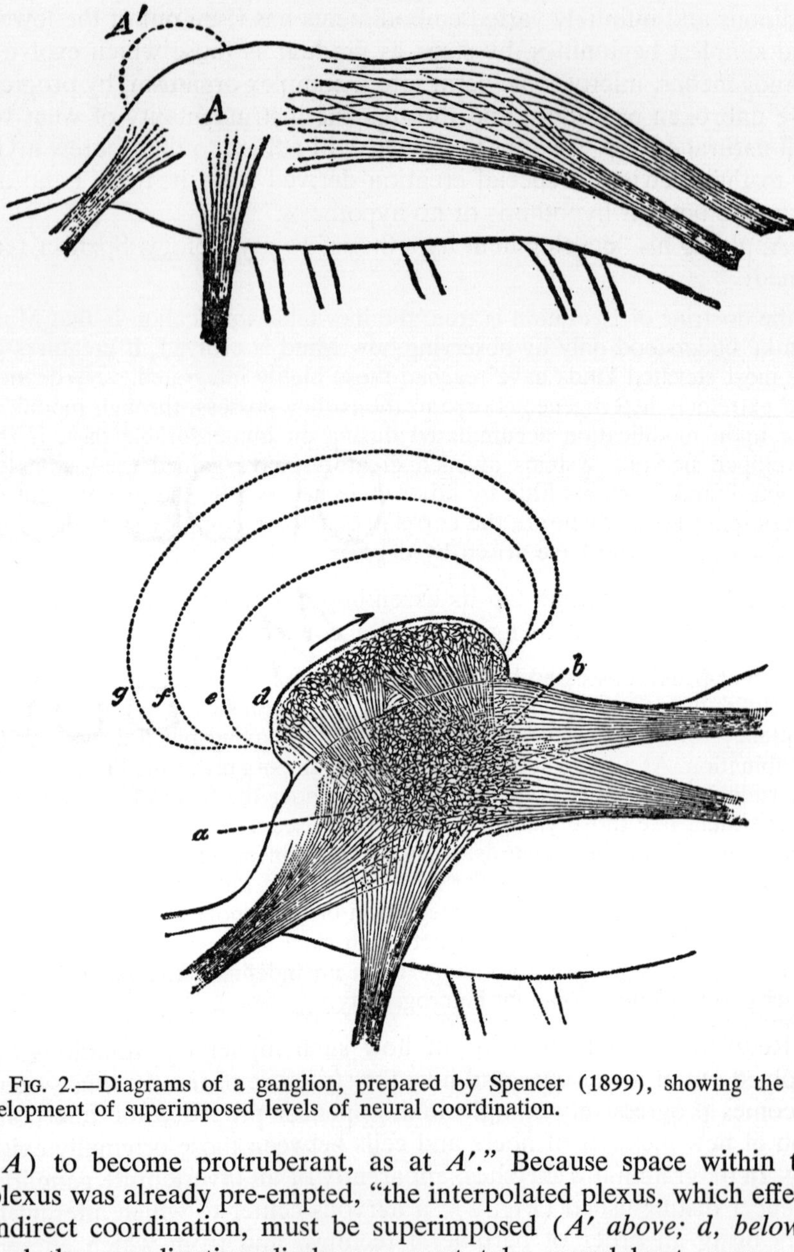

FIG. 2.—Diagrams of a ganglion, prepared by Spencer (1899), showing the development of superimposed levels of neural coordination.

(*A*) to become protruberant, as at *A′*." Because space within the plexus was already pre-empted, "the interpolated plexus, which effects indirect coordination, must be superimposed (*A′ above; d, below*), and the coordinating discharges must take roundabout courses as shown by the arrow. Little by little, there is an enlargement of the superior coordinating center by the interpolation of new coordinating plexuses at its periphery (*e, f, g, below*)."

IVAN P. PAVLOV

Though Pavlov's work in the physiology of the central nervous system did not commence until his fifties, its conceptualization was influenced strongly by the ideas of Darwin and of Spencer, encountered in his youth through the writings of Pisarev and Sechenov. In his *Autobiography,* Pavlov wrote: "I was born in the town of Ryazan in 1849 and received my secondary education at the local theological seminary. Influenced by the literature of the sixties, and particularly by Pisarev, our intellectual interests turned to natural science and many, myself included, decided to take the subject at the university" (Pavlov, *Selected Works,* 1955).

Pisarev was a writer and critic whose articles in the *Russkoye Slovo* promoted revolutionary-democratic and materialistic ideas among the intelligentsia of the sixties. There seems little doubt that Pavlov first became captivated by Darwin and the theory of evolution from reading Pisarev's lengthy, systematic, popular exposition of the *Origin of Species* entitled "Progress in the Animal and Vegetable Worlds" (in Pisarev, *Selected Philosophical, Social and Political Essays,* 1958).

The ecstatic attitude toward Darwin, which Pavlov preserved to the end of his days, can easily be identified with Pisarev's lofty expression in writing of Darwin:

> This brilliant thinker, whose knowledge is enormous, took in all the life of nature with such a broad view and penetrated so deeply into all its scattered phenomena that he discovered, not an isolated fact, but a whole series of laws according to which all organic life on our planet is governed and varies; and he told of them so simply, proves them so irrefutably and bases his arguments on such obvious facts, that you, a common human, uninitiated in natural science, are in a state of continual astonishment at not having thought out such conclusions yourself long ago.
>
> For us ordinary and unenlightened people, Darwin's discoveries are precious and important just because they are so fascinating in their simplicity, so easy to understand; they not only enrich us with new knowledge, they give fresh life to all the system of our ideas and widen our mental horizon in all dimensions. In nearly all branches of natural science, Darwin's ideas bring about a complete revolution. Even experimental psychology finds in his discoveries the guiding principle that will link up the numerous observations already made and put investigators on the way to new fruitful discoveries.

A second early influence upon Pavlov was provided by the writings of Sechenov (*Selected Works,* 1935). Later in his career, Pavlov (1928) referred to his beginning study of higher nervous activity with the objective techniques of conditional reflex physiology (Fig. 3):

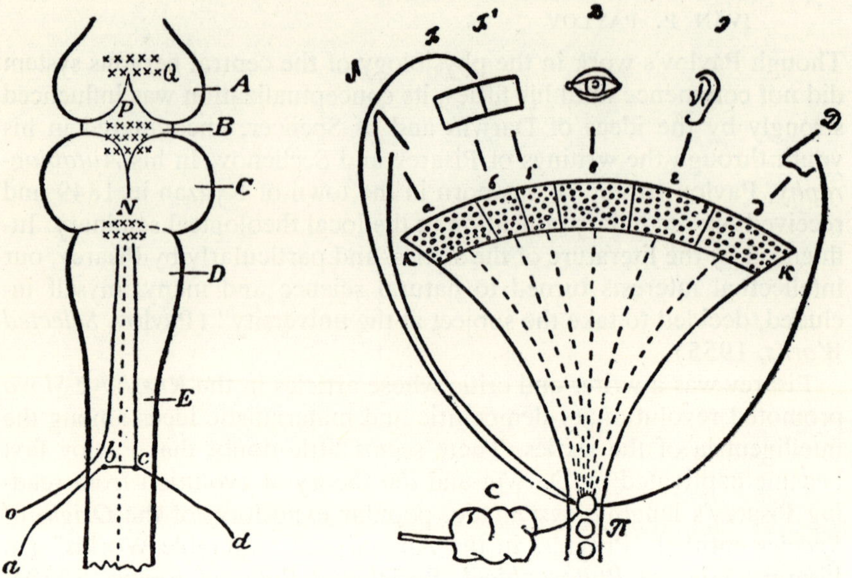

FIG. 3.—Diagram of the central nervous system of the frog (*left*). Stimulation of the sites marked by crosses inhibited spinal reflexes, illustrating the hierarchy of neural levels and the domination of higher over lower. (From Sechenov, 1935.)

Right, a diagram of the mechanism of the Pavlovian conditioned reflex by which the animal makes adaptive adjustments to its environment through new links between the cortical analyzers, and connections from them to older, subcortical, uncondition reflex arcs. (From Pavlov, *Selected Works,* 1955.)

The most important motive for my decision, even though an unconscious one, arose out of the impression made upon me during my youth by the monograph of I. M. Sechenov, the Father of Russian physiology, entitled *Reflexes of the Brain* and published in 1863. The influence of thoughts which are strong by virtue of their novelty and truth, especially when they act during youth, remains deep and permanent, even though concealed. In this book, a brilliant attempt was made, altogether extraordinary for that time, to represent our subjective world from the standpoint of neurophysiology.

In another report on objective study of higher nervous activity, Pavlov (1928) began:

With full justice, Charles Darwin must be counted as the founder and instigator of the contemporary comparative study of the higher vital phenomena of animals; for, as is known to every educated person, through his highly original support of the idea of evolution, he fertilized the whole mentality of mankind, especially in the field of biology. The hypothesis of the origin of man from animals gave a great impetus to the study of the higher phenomena of animal life. The answer to the question as to how this study should be carried out and the study itself have become the task of the period following Darwin.

It is interesting to note that Sechenov, like Jackson a contemporary of both Darwin and Spencer, was more directly influenced by Spencer's writing than by that of Darwin. Though not appearing in time to influence *Reflexes of the Brain* (1863), both Darwin's and Spencer's works were early translated into Russian, the *Origin of Species* in 1864 by Professor S. A. Rashinsky of Moscow University, of whose efforts Pisarev was highly critical, and a year later in shorter exposition by K. A. Timiryazev, the leading Russian Darwinist (Platonov, 1955). Spencer (1904) learned of a Russian translation of his *First Principles* in 1866 and, a decade later, heard with surprise, from Professor Sontchitzici of the University of Kiev, that all of his works had then been translated into Russian, excepting the *Sociology,* which was soon to be added to the list.

In his *Elements of Thought,* published in 1883, Sechenov wrote:

> Darwin's great theory of the evolution of species has placed the idea of evolution on such a firm basis that it is at present accepted by the vast majority of naturalists. This logically necessitates the recognition of the principle of evolution of psychical activities. Spencer's hypothesis may actually be called the application of Darwinism to the sphere of psychical phenomena (Quoted in Sechenov, 1935).

And later:

> Another and no less important success in the study of the mental development of man in general we owe to the famous English scientist, Herbert Spencer. It is only on the ground of Spencer's hypothesis, concerning the sequence of stages of neuropsychical development from age to age, that we can solve the ancient philosophical problem of the development of mature thought from initial infantile forms. To Spencer we owe the establishment, on the basis of very wide analogies, of the general type of mental development in man, as well as the proofs of the fact that the type of evolution of mental processes remains unchanged through all stages of the development of thought. The present essay is based on the theories of Spencer; therefore, our first task will be to expound the main principles of his theory. It even appeared at the same time as Darwin's theory and is practically a part of the general theory of organic evolution.

SIGMUND FREUD

Passing now to Freud, his autobiography (in Jones, 1953–57) refers to the influences leading him to medicine as a career: "At the time the theories of Darwin, which were then of topical interest, strongly attracted me, for they held out hopes of an extraordinary advance in our understanding of the world; and it was hearing Goethe's beautiful essay on *Nature* read aloud at a popular lecture, just before I left school, that decided me to become a medical student."

There are singularly few other allusions to Darwin in Freud's writings, and the factors responsible for his visualization of the psychic apparatus as spatially stratified were, doubtless, unconscious ones. It seems an exaggeration to propose that a continuum can be detected, in any literal sense, in Freud's anatomical, neurological, and psychoanalytical works. Instances of a recurring effort to interpret neural organization and function in evolutionary terms can, however, be noted. In his monograph on *Aphasia,* published in 1891, Freud wrote:

> In assessing the functions of the speech apparatus under pathological conditions, we are adopting as a guiding principle Hughlings Jackson's doctrine that all these modes of reaction represent instances of functional retrogression (disinvolution) of a highly organized apparatus, and therefore correspond to earlier stages of its functional development. This means that under all circumstances an arrangement of associations which, having been acquired later, belongs to a higher level of functioning, will be lost, while an earlier and simpler one will be preserved. From this point of view, a great number of aphasic phenomena can be explained (Freud, 1953).

In a letter to Fleiss in 1896, Freud discussed a revision of his *Project for a Scientific Psychology* and referred to his "latest bit of speculation, the assumption that our psychical mechanism has come about by a process of stratification" (quoted in Freud, 1954). A quarter of a century later, Freud made two attempts to diagram these ideas, with interesting differences in the form of the figures. The first (Fig. 4, *left*), prepared in 1923, resembled an inverted brain, although reference was made to it as an ovum. The second (Fig. 4, *right*), prepared a decade later, was really egg-shaped. In his lecture on *The Anatomy of the Mental Personality,* Freud (1933) elaborated upon the contents of these figures:

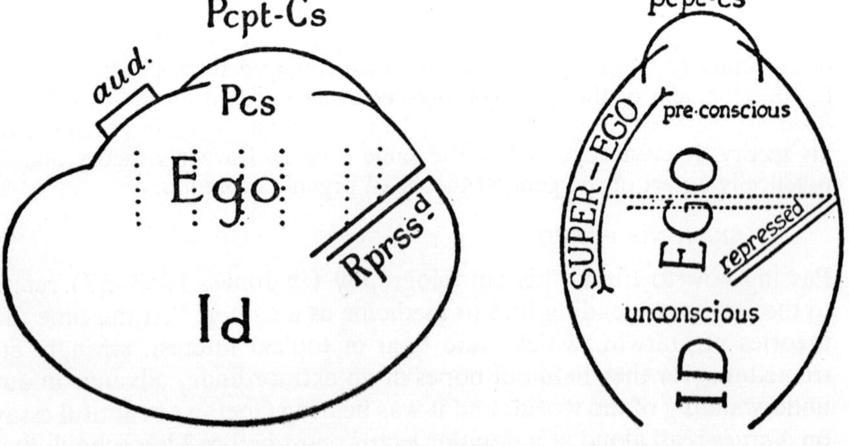

FIG. 4.—Two diagrams by Freud (1950, 1954), presenting the mental apparatus as though spatially stratified.

Superego, ego and id are the three realms, regions or provinces into which we divide the mental apparatus of the individual, and it is their mutual relations with which we shall be concerned.

The *id* is the obscure, inaccessible part of our personality and can only be described as being all that the ego is not. We can come nearer to the id with images, and call it a chaos, a cauldron of seething excitement. We suppose that it is somewhere in direct contact with somatic processes and takes over from them instinctual needs. These instincts fill it with energy, but it has no organization and no unified will, only an impulsion to obtain satisfaction for the instinctual needs in accordance with the pleasure principle. Contradictory impulses exist side by side in it, without neutralizing each other or drawing apart; at most they combine in compromise formations under the overpowering pressure toward discharging their energy. In the id, there is nothing corresponding to the idea of time. Conative impulses which have never got beyond the id, and even impressions which have been pushed down into it by repression, are virtually immortal and are preserved for whole decades, as though they had only recently occurred. They can only be recognized as belonging to the past, deprived of their significance, and robbed of their charge of energy, after they have been made conscious by the work of analysis, and no small part of the therapeutic effect of analytic treatment rests upon this fact. Naturally the id knows no values, no good and evil, no morality. There is nothing in the id which can be compared to negation. Instinctual cathexes seeking discharge—that, in our view, is all that the id contains.

The *ego* is directed onto the external world; it mediates perceptions of it and in it are generated, while it is functioning, the phenomenon of consciousness. The ego has taken over the task of representing the external world for the id. In the fulfillment of this function, it has to observe the external world and preserve a true picture of it in the memory traces left by its perception. The ego also controls the path of access to motility, but it interpolates between desire and action the procrastinating factor of thought, during which it makes use of the residues of experience stored up in memory. In this way, it dethrones the pleasure principle, which exerts undisputed sway over the processes in the id, and substitutes for it the reality principle, which promises greater security and success. The relation to time, too, is contributed to the ego by the perceptual systems; indeed, it can hardly be doubted that the mode in which this system works is the source of the idea of time. What, however, especially marks the ego out in contradistinction to the id is a tendency to synthesize its contents, to bring together and unify its mental processes, which is entirely absent from the id. In popular language, we may say that the ego stands for reason and circumspection, while the id stands for the untamed passions. One might compare the relation of the ego to the id with that between a rider and his horse: the horse provides the locomotive energy, and the rider has the prerogative of determining the goal and of guiding the movements of his powerful mount.

The role which the *superego* undertakes later in life is at first played by an external power, by parental authority. It can be traced back to the influence of parents, teachers and so on, and is based upon an overwhelmingly important biological fact; namely, the lengthy dependence of the human child on his parents. We have allocated to the superego the activities of self-observation, conscience and the holding up of ideals. It is the representative of all moral restrictions, the advocate of the impulse toward perfection. In short, it is as much as we have been able to apprehend psychologically of what people call the "higher things in human life." It becomes the vehicle of tradition and of all the age-long values which have been handed down from generation to generation. The ideologies of the superego perpetuate the past, the traditions of the race and the people, which yield but slowly to the influence of the present to new developments.

In discussing the interrelations of these parts, Freud, like Spencer, appeared to invoke Lamarckian views:

The ego has the task of bringing the influence of the external world to bear upon the id. In the ego, perception plays the part which, in the id, develops upon instinct. The experiences undergone by the ego seem at first to be lost to posterity; but, when they have been repeated often enough and with sufficient intensity in the successive individuals of many generations, they transform themselves so to say into experiences of the id, the impress of which is preserved by inheritance. Thus in the id, which is capable of being inherited, are stored up vestiges of the existences led by countless former egos; and, when the ego forms its superego out of the id, it may perhaps only be reviving images of egos that have passed away and be securing them a resurrection.

Drawing an analogy between the practices of mystics and psychoanalytic therapy, Freud (1954) continued:

It can easily be imagined that certain practices of mystics may succeed in upsetting the normal relations between the different regions of the mind so that, for example, the perceptual system becomes able to grasp relations in the deeper layers of the ego and in the id which would otherwise be inaccessible to it. Whether such a procedure can put one in possession of ultimate truths, from which all good will flow, may be safely doubted. All the same, we must admit that the therapeutic efforts of psychoanalysis have chosen much the same method of approach; for their object is to strengthen the ego, to make it more independent of the superego, to widen its field of vision, and so to extend its organization that it can take over new portions of the id. Where id was, there shall ego be. It is reclamation work, like the draining of the Zuyder Zee.

These concepts that developed in the latter part of the nineteenth century in the fields of psychology, neurology, physiology, and psychiatry began to be associated, from the nineties on, with contributions to comparative neuroanatomy. Once work in this latter field began, a growing number of investigators were recruited; the movement may be clearly traced to two pioneer instigators, however—Ludwig Edinger in Germany and Clarence Herrick in the United States.

LUDWIG EDINGER

The man who inaugurated modern programs of research in comparative neuroanatomy and who in 1904 established a Neurological institute in Frankfurt devoted primarily to study in this field, tells of entering Heidelberg in 1872 as a medical student to study under Gegenbaur:

> This was the man who, with his friend Haeckel, had laid the foundation for evolutionary viewpoints in the study of anatomy. Science at the time stood wholly under the influence of Darwin's evolutionary teachings, and Gegenbaur and Huxley were the creators of a new conception of comparative anatomy. I, who during my gymnasium education had longed for such a leader, looked forward to my new teacher with enthusiasm (only to be disappointed later by his "dry and uninspiring lectures") (Quoted in Krücke and Spatz, 1959, pp. 1–25).

Dr. Tilly Edinger (personal communication) has most kindly provided this additional information:

> I can say that everything my father did scientifically—*der ganze Papa!*—was influenced by Darwin and Darwin's prophet, Haeckel. Papa devoured Darwin's writings when still in school and had so long a row of Darwin's books that I suppose he had everything. When other fathers tend to tell their children stories about I don't know what, he told us—if not about Goethe—about Darwin. Papa also went to a Darwin celebration in England and regarded it as a most wonderful experience; can it have been the 1909 Centenary of Darwin's birth? My sister remembers visiting with Papa, Haeckel, the two discussing she doesn't know what and Papa exclaiming, "I'm out-haeckeling you there!," thus creating a new verb.

In 1874 Edinger shifted to Strassbourg for his clinical training, and also commenced research in Waldeyer's laboratory in a high tower room above the entrance to the hospital. His later popular text on *The Anatomy of the Central Nervous System of Man and of Vertebrates* (1899) bears this inscription: "To his teacher, Professor Wilhelm Waldeyer, this first attempt at a comparative anatomy of the brain is dedicated in reverance and gratitude." Edinger's approach to comparative neuroanatomy began in 1885 with recognition that "perhaps the brain consists of two different parts, one for elementary functions and a second which develops along with the evolution of the animal series." He went on to differentiate the *paleoencephalon,* or subcortical brain, containing the elementary apparatus and showing little evolutionary change, from the cerebral and cerebellar hemispheres forming the

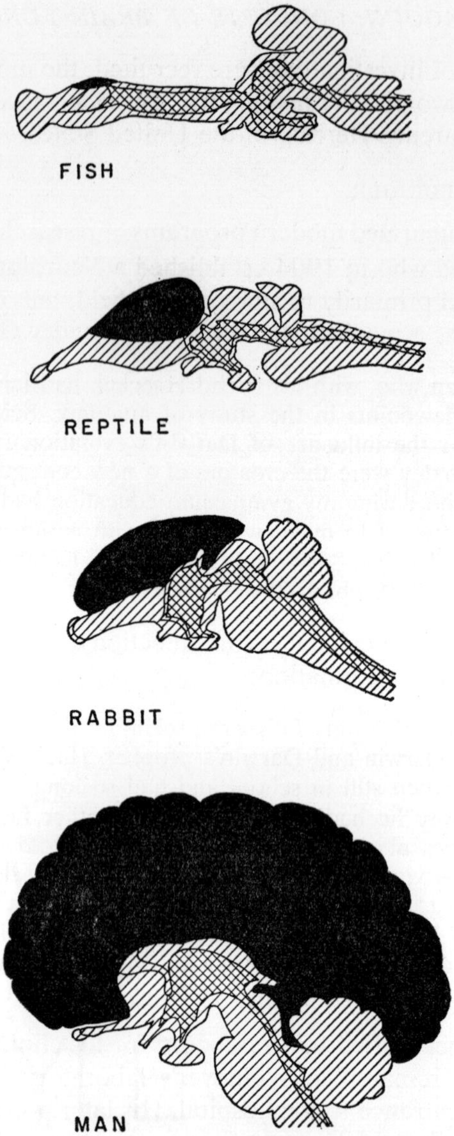

FIG. 5.—Midsagittal views of the brain of the fish, reptile, rabbit, and man, showing the development of the neoencephalon superimposed upon the paleoencephalon. (Adapted from Edinger, 1885.)

neoencephalon, which grew in substance with the evolution of vertebrates and achieved their highest development in man.

Study of the neopallium then became Edinger's major interest, and at first he thought it would be an easy task: "All one had to do would be to study the development of the cerebrum throughout the animal series; however, this work actually occupied me for twenty years, day in and day out, because new questions constantly arose."

With the customary artistic skill and clarity of exposition which characterized all of Edinger's work, the pictorial generalizations of Figure 5 illustrate his views of the evolutionary development of the neoencephalon (*black*) above the paleoencephalon (*gray*). Research in comparative neuroanatomy, which had been opened by Edinger's effort, was extended in 1908 by the establishment in Amsterdam of the Dutch institute for Brain Research under the direction of C. U. Ariens-Kappers, who had spent the period of 1906–1908 with Edinger at Frankfurt. Without any such direct derivation, an independent school of comparative neurology developed also in the United States.

CLARENCE L. HERRICK

The founder in the 1890's of this vigorous American school was Clarence L. Herrick, the elder brother of C. J. Herrick, who writes (1955):

> A generation earlier, the static natural philosophy of the day was shaken from its foundations by Lyell in geology, Darwin in biology and a few other adventurous pioneers who recognized that nature is process, not stable structure. Their search was for laws of change rather than the immutable absolutes of traditional metaphysics. . . . It is evident that my brother very early in his scientific career planned to devote the later years of his life to neurological and psychological research and the vital relationships of these two domains of human experience. His work (as a young man) was concerned chiefly with field natural history, animal behavior and comparative anatomy. In view of this preparation, he naturally chose to approach the program planned for his mature years by way of comparative neurology. When the opportune time to put this plan into execution came, in 1888, he found that he was following the lead of another pioneer, Ludwig Edinger. These two men are generally regarded as the founders of comparative neurology as an organized scientific discipline.

In the six short years of health available to him at Denison University and the University of Cincinnati, Clarence Herrick contributed a remarkable amount of detailed information concerning the fish brain. In 1891 he founded the *Journal of Comparative Neurology,* which has since continued as the major international medium of publication in this field (C. J. Herrick, 1954).

The remarkable breadth of Herrick's views of neural organization and function, as well as of their evolutionary orientation, is indicated (1891) by two editorials, "Problems of Comparative Neurology" and "Neurology and Psychology," in the first volume of his new journal, in which he says,

> The broad generalization of modern biology that function precedes and in a sense creates its organ leads to the belief that in tracing the evolution of the nervous system, we are to a very considerable extent determining the progressive revelation of that which differentiates the animal from the inanimate residue. . . .
>
> There can be no doubt that the theoretical problem of morphology which has excited most interest of late and which seems to have the strongest hold on the imagination of investigators generally is the question as to the origin of the head and its various correlated structures. One reason for this prolonged and persistent effort to solve a problem which is, essentially, simply a theoretical one lies in the fact that the formation of the head is a culmination of the whole series of progressive changes which constitutes organic evolution. Whatever view may be taken of "cephalization" in its technical form, all must agree that the tendency of evolution has been to subordinate more and more structures and functions to the purposes of the head. Moreover, it has been more or less distinctly seen that the solution of the anatomical and physiological problems connected with the head is essential to the completion of any systematic theory respecting the connection between mind and body. It has been felt that if the head with its structures so obviously intended to serve as an avenue of expression for the mind can be explained as a compound of the somewhat modified simple elements occurring in each segment of the body of a lower animal, then the mind itself might prove but the sum of all the functions represented by these several organs, though rendered ever so "psychical" by reason of their complex interaction. Probably few biologists would care to commit themselves to so extreme a view as this, yet the great problem remains: What is the relation between the functions of the nervous elements and the phenomena of mind as such, or, in other words, just what has the nervous system and especially the brain to do with thought.

In discussing "the bearing of the progressive differentiation of cerebral localization upon psychical evolution," Clarence Herrick continued,

> The writer has endeavored to show that consciousness in the limited sense is of comparatively late origin and that a rigid application of the doctrine of natural selection would exclude it from all participation in nervous activities until such time as the struggle for survival had become ameliorated to an extent, making conscious selection possible without involving direct loss or destruction. After such a field for spontaneity had been opened, consciousness would become a valuable and then a necessary adjunct, and the effect of the reaction of conscious beings upon each other

would be to widen the arena for its display and increase the complexity of its activities. . . .

It may be added that if psychology really needed a material sensomotorium commune, or a common arena for consciousness, the suggestions which we now have of a neuro-pilem covering the entire cortex and containing the finely divided fibers of centripetal and centrifugal nerves, which are merely closely associated without anastomosis, might seem to afford it without the necessity of setting aside the results of localization already given. If, on the other hand, we are justified in accepting the assurances of Wundt and Lotze that the concept of extension is out of place as applied to the soul, we still require from the standpoint of physiology some common ground for interaction such as would be furnished by the nerve-felt of the cortex.

Although Clarence Herrick twice visited Germany, he apparently never met Edinger personally, but an active correspondence passed between Edinger and the two Herricks. In a letter of 1908, Dr. Edinger wrote (to C. Judson Herrick, 1955): "Have you noticed that despite a strong aversion to giving personal names to fiber tracts, I cannot refrain from perpetuating your brother's name in some way? I have sponsored a Herrick's commissure in the fishes."

In 1915 the twenty-fifth volume of the *Journal of Comparative Neurology* was dedicated to Edinger "in commemoration of his fundamental researches in comparative neurology." In a letter of thanks to C. Judson Herrick (1954), Edinger replied:

I have just received the number of your Journal in which you have commemorated my sixtieth birthday in so distinctive a manner. Believe me, the sight of this number has moved me deeply. . . . I thought first of all of the time when your brother's influence was in action—one of the few men before whose eyes the ends for which we were working were clearly defined. . . . In those days we were but three or four workers. Only we—your brother and I—saw the farther goal, which was to learn to understand the genesis of the brain. When, from the start, others gradually came to our assistance, able men were added to our number, among whom you were in the first rank. The memory of your brother, however, must not perish, and it is, of course, natural, that I think of him today.

It is particularly appropriate to call attention to the Herricks' work at this Centennial Celebration, for the University of Chicago has been distinguished both by C. Judson Herrick's long and fruitful association as Professor of Neurology and, additionally, by the earlier and much briefer appointment of his brother, Clarence. It is not widely known that Clarence Herrick held a Professorship at Chicago for six months in 1891. He had joined the organizing University with the expectation of developing a venturesome "program of integrated teaching and research that would cross the usual departmental borders and involve co-operative relations with other associates in zoology, anatomy, physiology, genetics, animal behavior, psychology and philosophy in a search for scientifically acceptable principles of psychobiology." C. Judson Herrick very correctly remarks (1955): "If my brother had been permitted to develop his project, even in a limited way in the beginning, the innovation might have set a pattern that would have significantly changed the history of science in North America during the subsequent half-century and accelerated the movement toward the same objectives that is now in full flood."

DARWIN AND SPENCER

It seems clear that, to their contemporaries and early successors, the ideas of Spencer were fully as significant as those of Darwin in the development of concepts of evolution of the brain and behavior. While Darwin and his "prophet," Haeckel, were primary in influencing Edinger and the rise of comparative neuroanatomy, Spencer can be rated equivalently with Darwin in impressions made upon Pisarev and Sechenov and, through these latter, upon Pavlov and the growth of neurophysiology in Russia and the U.S.S.R. There can be no question, however, of the predominant influence of Spencer upon Hughlings Jackson and, through him, upon the formation of evolutionary views in Western neurological thinking.

Spencer's influence can be accounted for in part by its priority, for he applied evolutionary principles to an understanding of the brain earlier than Darwin and, indeed, before the latter's ideas were published at all. Additionally, the broad sweep of Spencer's interests was doubtless attractive, for he attempted to account for the whole range of neural function, from instincts to the most complex features of the mind, in keeping with his Comte-like propensity for global synthesis.

By contrast, Darwin's orientation was primarily toward those basic biological activities observable in animal behavior. Darwin recognized that each surviving organism adapted to its environment by staking out a territory, in which it gained food and so preserved its individual life, and sought a mate and reared its young and so preserved its race. The intrusion of predators or enemies into this territory was met by aggression or defense. So general were these activities that Darwin clearly saw the need for brain mechanisms concerned with the management of feeding, fighting, fleeing, and undertaking reproductive activity. So widely could these fundamental performances be identified in the animal kingdom that their neural representation must be sought in older, primitive, deep-lying, subcortical portions of the brain. Even in his *Descent of Man* (1871), Darwin's interest was obviously more preoccupied with principles of sexual selection than by considering the development of neocortical capacities for association, abstractions, and communication, which are easily the most distinctive features of human evolution.

Other contrasting features of their personalities and outlooks appeared to have led Darwin and Spencer to develop reservations about one another. An early phrenological characterization of Spencer (1904) concluded: "Such a head as this ought to be in the Church. The self-esteem is very large." While Darwin's own phrenology indicated that he "had the hump of reverence developed enough for ten priests," his tendency to self-deprecation seemed, on the other hand, to have amounted to a real sense of inferiority when comparing himself with Spencer. Each seemed also to have cultivated possibly willful difficulties in understanding the other's views. In a letter to Hooker in 1868, Darwin wrote: "I feel *Pangenesis* is stillborn. H. Spencer says the view is quite different from his (and this is a great relief to me, as I feared to be accused of plagiarism but utterly failed to be sure what he meant, so thought it safest to give my view as almost the same as his, and he says he is not sure he understands it" (Quoted in F. Darwin, *Life and Letters of Charles Darwin,* 1925).

In his letters, Darwin characteristically acknowledged Spencer's brilliance but usually expressed some question of the soundness or reliability of his views. In a note thanking Spencer for a present of his *Essays* in 1858, Darwin wrote: "Your remarks on the general argument of the so-called development theory seem to me admirable. I am at present preparing an Abstract of a larger work on the changes of species; but I treat the subject simply as a naturalist, and not from a general point of view; otherwise, in my opinion, your argument could not have been improved on, and might have been quoted by me with great advantage."

In a letter to Hooker in 1866, Darwin wrote:

> I have now read the last No. of H. Spencer (*Principles of Biology*). It is wonderfully clever and I daresay mostly true. I feel rather mean when I read him: I could bear and rather enjoy feeling that he was twice as ingenious and clever as myself; but when I feel he is almost a dozen times my superior, even in the master art of wriggling, I feel aggrieved. If he had trained himself to observe more, even at the expense, by the law of balancement, of some loss of thinking power, he would have been a wonderful man.

In a final judgment, in his *Autobiography* (1958, ed. Nora Barlow), Darwin commented:

> Herbert Spencer's conversation seemed to me very interesting but I did not like him particularly and did not feel that I could easily become inti-

mate with him. I think he was extremely egotistical. After reading any of Spencer's books, I generally feel enthusiastic admiration for his transcendent talents and have often wondered whether in the distant future he would rank with such great men as Descartes, Leibnitz, etc., about whom, however, I know very little. Nevertheless, I am not conscious of having profited in my own work by Spencer's writings. His deductive manner of treating every subject is wholly opposed to my frame of mind. His conclusions never convince me; and over and over again I have said to myself after reading one of his discussions, "Here would be a fine subject for half a dozen years' work." His fundamental generalizations (which have been compared in importance by some persons with Newton's Laws!)—which I daresay may be very valuable under a philosophical point of view—are of such a nature that they do not seem to me to be of any strictly scientific use. They partake more of the nature of definitions than of laws of nature. They do not aid one in predicting what will happen in any particular case. Anyhow, they have not been of any use to me.

Reciprocal comments on Darwin and his work by Spencer were made primarily from the point of view of their relations to Spencer's own ideas and interests. With respect to publication of the *Origin of Species,* Spencer later wrote (1904):

That reading it gave me great satisfaction may be safely inferred. Whether there was any set-off to this, I cannot now say; for I have quite forgotten the ideas and feelings I had. Up to that time, I held that the sole cause of organic evolution is the inheritance of functionally-produced modifications. The *Origin of Species* made it clear to me that I was wrong; and that the larger part of the facts cannot be due to any such cause. Whether proof that what I had supposed to be the sole cause could be at best but a part cause gave me any annoyance, I cannot remember; nor can I remember whether I was vexed by the thought that, in 1852, I had failed to carry further the idea then expressed that, among human beings, the survival of those who are the select of their generation is a cause of development. But I doubt not that any such feelings, if they arose, were overwhelmed in the gratification I felt at seeing the theory of organic evolution justified.

To have the theory of organic evolution justified was, of course, to get further support for that theory of evolution at large, with which, as we have seen, all my conceptions were bound up. Believing as I did, too, that right guidance, individual and social, depends upon acceptance of evolutionary views of mind and of society, I was hopeful that its effects would presently be seen on educational methods, political opinions and men's ideas about human life. Obviously, these hopes that beneficial results would presently be wrought, were too sanguine. My confidence in the rationality of mankind was much greater than it is now.

In 1872, Spencer acknowledged a copy of Darwin's work on *The Expression of the Emotions* as follows:

Dear Darwin: I have delayed somewhat longer than I intended acknowledging the copy of your new volume which you have been kind enough to send me. I delayed partly in the hope of being able to read more of it before writing to you; but my reading powers are so small, and they are at present so much employed in getting up materials for work in hand, that I have been unable to get on far with it. I have, however, read quite enough to see what an immense mass of evidence you have brought to bear in proof of your propositions.

I will comment only on one point, on which I see you differ from me. . . .

There were, of course, more than *points* of difference between these two influential figures, for their antipathies seem expressions of a generally contrasting orientation toward the advancement of knowledge. Darwin's remark—"His deductive manner of treating every subject is wholly opposed to my frame of mind"—emphasized the extreme nature of Spencer's deductive approach. Developing an interest in evolution from fossil-collecting, reading Lyell's *Principles of Geology,* and running across von Baer's concept of embryogenesis in the preparation of a book review, Spencer formulated his "development hypothesis" and, flagging a dilatatory cerebral circulation with which he was hypochondriacally preoccupied, devoted the balance of his career to its application to all fields of knowledge in a multi-volumed *System of Synthetic Philosophy.* Descartes, the father of the deductive method, could not have recommended more.

Darwin was also influenced by Lyell's *Geology,* a copy of which he had on the *Beagle,* with advice "to read, but not believe it." In South America, Darwin's study of fossils, Andean geology, and Galapagos speciation began his long, methodical, almost ponderous, accumulation of data. Upon return to England, his zoological and botanical studies, extended by plant and pigeon-breeding, continued at Down. After reading Malthus, a synthesis of this material was reached, but study and deliberation continued until the pressure of Wallace's paper (similarly formulated after reading Malthus, but more rapidly) forced publication of the *Origin of Species* after a quarter of a century of preparation. Darwin's approach was thus inductive to an extreme that even Bacon, the father of the method, did not advocate.

The value of Darwin's efforts and the principles he induced from them have been repeatedly and widely recognized, and the present centenary of his work has brought a crescendo of acclaim. By contrast, mention is rarely made today of Spencer. Although Spencer himself made no scientific discoveries, his deductive syntheses stimulated the interest and activity of many productive workers. One must conclude that the efforts of each of these individuals contributed significantly to evolutionary concepts of the organization and function of the brain which are recognized today.

References

DARWIN, C. 1839. "Narrative of the Surveying Voyages of His Majesty's Ships, Adventure and Beagle, Between the Years 1826–1836. Describing Their Examination of the Southern Shores of South America and the Beagle's Circumnavigation of the Globe." Vol. III, *Journal and Remarks, 1832–1836*. London.

———. 1859. *On the Origin of Species by Means of Natural Selection, or the Preservation of Favored Races in the Struggle for Life*. London.

———. 1871. *The Descent of Man, and Selection in Relation to Sex*. London.

———. 1872. *The Expression of the Emotions in Man and Animals*. London.

———. 1958. *The Autobiography of Charles Darwin (1809–1882)*. Edited by NORA BARLOW. London: Collins.

———. 1925. *Life and Letters of Charles Darwin*. Edited by FRANCIS DARWIN. 2 vols. New York and London: Appleton.

EDINGER, L. 1885. *Zehn Vorlesungen über den Ban der Nervösen Zentrolorgare*. Leipzig: Vogel.

———. 1899. *The Anatomy of the Central Nervous System of Man and of Vertebrates in General*. Translated by W. S. HALL. Chicago: F. A. Davis.

FREUD, S. 1933. *New Introductory Lectures on Psycho-Analysis*. Translated by W. J. H. SPROTT. New York: Norton.

———. 1950. *The Ego and the Id*. Translated by J. RIVIERE. London: Hogarth Press.

———. 1953. *On Aphasia*. New York: International Universities Press.

———. 1954. *The Origins of Psycho-Analysis*. New York: Basic Books.

HERRICK, C. J. 1954. "One Hundred Volumes of the *Journal of Comparative Neurology*," *J. Comp. Neurol.*, C, 717–56.

———. 1955. "Clarence Luther Herrick, Pioneer Naturalist, Teacher and Psychobiologist," *Trans. Amer. Philosoph. Soc.*, XLV, 1–85.

HERRICK, C. L. 1891*a*. "The Problems of Comparative Neurology," *J. Comp. Neurol.*, I, 93–105.

———. 1891*b*. "Neurology and Psychology," *J. Comp. Neurol.*, I, 183–200.

JACKSON, J. H. 1958. *Selected Writings of John Hughlings Jackson*. 2 vols. New York: Basic Books.

JONES, E. 1953–57. *The Life and Work of Sigmund Freud*. 3 vols. New York: Basic Books.

KRÜCKE, W. and SPATZ, H. 1959. Ans den Erinnerungen von Ludwig Edinger. Schriften der Wissenschaftlichen Gesellschaft. An der Johann

Wolfgang Goethe *Universität Frankfurt au Main Naturwissenschafliche Reihe*. No. 1, pp. 1–25.

LAYCOCK, T. 1869. *Mind and Brain.* 2 vols. New York: Appleton. 1869.

PAVLOV, I. P. 1928. *Lectures on Conditioned Reflexes,* Vol. 1. Translated and edited by W. H. GANTT. New York: International Publishers.

———. 1955. *Selected Works,* Edited by K. S. KOSHTOYANTS. Moscow: Foreign Languages Publishing House.

PISAREV, D. 1958. *Selected Philosophical, Social and Political Essays.* Translated by R. DIXON. Moscow: Foreign Languages Publishing House.

PLATONOV, G. 1955. *Kliment Arkadyevich Timiryazev.* Moscow: Foreign Languages Publishing House.

SECHENOV, I. 1935. *Selected Works.* Edited by A. A. SUBKOV. Moscow-Leningrad.

SPENCER, H. 1899. *The Principles of Psychology.* 2 vols.; 4th ed. London.

———. 1904. *An Autobiography.* 2 vols. London.

ALEXANDER VON MURALT

A DECISIVE STEP IN EVOLUTION: SALTATORY CONDUCTION

The conduction velocity of the nervous impulse is the determining factor for the behaviour of every animal. Locomotion, reaction, all the protective functions, the appraisal of a danger, and the defensive steps taken by the individual in a hostile world in order to maintain its own life and that of its offspring depend in the last instance on conduction velocity, reaction time, and instinctive protective behaviour. As long as life occurred only in the sea, hiding, absence of motion, and mimicry were very efficient defense measures because of the small visibility and the obscurity in sea water. Leaving this protective matrix meant a considerable danger which could be counterbalanced only by an equal increase in the velocity of locomotion and reaction.

The conduction velocity of a nerve was measured for the first time by Helmholtz in 1850. He found, for myelinated frog nerves, 25 m/sec at 20° C., and at lower temperatures a considerable decrease in the velocity. This was a first quantitative measurement, which discarded the vague notions of some physiologists of the eighteenth and nineteenth centuries that nervous conduction occurs "flashlike."

Comparison of conduction velocities of nerves in the animal kingdom reveals the fact that the speed increases, in general, with higher organisation, the lowest values being found in primitive forms, the highest (about 100 m/sec) in the mammalian body.

A very curious fact comes to light if a comparison of nerves with the *same* conduction velocity is made at various levels of evolution. In general, the morphological improvement of the body has been the main object of considerations in evolution, and the data have not been sufficient to extend them also to functional, i.e., physiological, aspects. This is a new line of approach to which I would like to con-

ALEXANDER VON MURALT is Professor of Physiology at the University of Berne. Educated at Zurich, Munich, Heidelberg, and Harvard, Dr. von Muralt's special field is neurophysiology, to which he has devoted several publications. He is president of the Swiss Academy of Sciences, of the Foundation for Fellowships in Medicine and Biology, and of the Swiss National Science Foundation.

tribute one very interesting example which has been studied by various investigators so carefully that one may safely derive conclusions.

The conduction velocity of a nerve fibre from the saphenous nerve of a cat (taken as an example of a mammal) with an outside diameter of 3.5 μ at 37° C. is 25 m/sec. All myelinated nerves show an increase in conduction velocity proportional to the outside diameter of the fibre,

$$v = k \cdot d \text{ (37° C.)},$$

where v is the conduction velocity in meters per second, d is the diameter of the fibre in microns, and k is a constant, which, for the saphenous nerve of the cat, is 7.4 (Gasser and Grundfest, 1939).

In the frog, a nerve fibre with the same conduction velocity of 25 m/sec has a diameter of 10 μ (20° C.), which means that the diameter has increased 3.5 fold, the biological material needed for the same performance 12 times, and $k = 2.5$.

If we now compare these data with the earthworm, in which a nerve fibre with 25 m/sec conduction velocity attains a diameter of 100 μ (giant fibre), it becomes apparent that the increase is 30 fold for the diameter and 900 fold for the material needed to build up such a nerve. And for *Loligo* (cephalopod) the giant fibre with a conduction velocity of 25 m/sec has a diameter of 650 μ, which corresponds to an increase of 185 fold in the diameter and 34,000 fold with respect to biological material.

In evolution the sequence was the opposite. Low forms of life had to develop giant nerve fibres in order to attain high conduction velocities, and, as their organisation developed, the diameters were reduced, material was economised, and the conduction velocity increased even to 100 m/sec in comparatively small nerve fibres. If mammalian nerves were built on the same principle as those of *Loligo,* our brachial nerves would form a bundle larger in diameter than our arm and our peripheral nervous system alone would have the volume of our whole body.

What is the reason for this improvement of construction economy and performance in the evolution of the peripheral nervous system? Conduction velocity depends in vivo on four factors: (*a*) mechanism of conduction, (*b*) diameter of the fibre, (*c*) myelinisation of the fibre, and (*d*) temperature.

Giant nerves without or with very little myelin surrounding the axone conduct the nervous impulse with a continuous velocity. From the active region local currents are flowing to the adjacent inactive regions, producing a stimulation by their depolarising action. Excitation is created in this region, and the nervous impulse perpetuates itself in this way continuously over the whole length of the fibre. In small fibres without myelin the velocity is low; the larger the diameter of these fibres, the higher is the velocity. If a small sheath of myelin—a myelotropic sheath—is present, the velocity increases, and the diameter can be smaller for the same performance. (This is the case in the giant fibre of the earthworm.) Fibres of shrimps and prawns are even heavily coated with myelin.

In vertebrates with the appearance of heavily myelinated nerves, a new mechanism of conduction appears, which was called "saltatory." The history of the discovery of this mechanism, which may be considered as being of the greatest importance for evolution, is rather fascinating.

In 1925, Lillie studied models of nervous conduction by immersing steel wires in nitric acid. He observed that an "impulse" traveled with a continuous velocity from one end to the other but that the velocity could be increased considerably by slipping pieces of glass capillaries (as a model of the myelin sheath) on the wire, leaving small open spaces between them (as a model of the nodes in myelinated nerves). He stated that rapidly conducting myelinated nerve fibres showed surprising similarities with his model, "the conducting element (axone) being enclosed by a tubular sheath of apparently high electrical resistance, the medullary sheath, which is constricted or interrupted at regular intervals." He concluded from comparison of non-medullated with medullated nerve fibres that the latter were transmitting impulses at about ten times higher velocities because, possibly, "a distance action effect, acting from internode to internode" could be a factor in this high-speed transmission.

> The electrical resistance between the surface of the axone and the surrounding medium may be assumed to be relatively low at the constrictions; diffusing substances (dyes) enter most rapidly at those regions, and the same is presumably true of ions. At least such a possibility must be considered; for, in the physical sense—if transmission is an effect of secondary stimulation by the currents of local circuits—the conditions in the model and in the nerve are of the same general kind [Stämpfli, 1954].

Erlanger (working with Blair), ten years later, concluded from his experiments on living nerves: "Such relevant evidence as we have been able to acquire all favors the view that progression is saltatory" (Erlanger and Gasser, 1937). These ideas, however, were not recognised by physiologists and were considered doubtful. New evidence appeared, when Tasaki (1939), with the technique of single isolated nerve fibres, a technique developed in Japan by Kato, showed that the myelin sheath of the internode acts as an insulator and that excitation can be produced only at the nodes. He concluded that it seemed highly probable that the excitation jumped from one node to the next, and he published a series of papers on this subject during the war (cf. Tasaki and Takeuchi, 1941). Tasaki studied the safety factor for this new type of conduction and found values of 5–7, which means that the action current of an excited node is sufficiently great to insure the stimulation of an adjacent node through the cable-like structure of the internode.

This was the situation in 1945. Erlanger had discussed the possibility of a stimulation from node to node by currents eddying around the segments (today we call them the "internodes") or excitation traveling from segment to segment across the node. Tasaki has shown that the nodes were the only sites in a myelinated nerve where stimulation from the outside is possible, owing to the insulation of the myelin sheath in the internode, the Pfaffmann (1940) had proved that the "action current" registered by electrodes lying in the region of the internode was due to spread from the neighbouring nodes.

After the war the work on this problem was intensified. In order to explain the results, something must be said about the structure of the myelinated nerve. The nerve cell is the origin of every axone, which extends as a protoplasmic thread in some long nerves over more than 1 meter. This axone is built by long threads of protein of a diameter of 100–200 A (Schmitt, 1950; Schmitt and Geren, 1950; Fernández-Morán, 1950) and shows birefrigence. At the nodes (Ranvier), which are regularly spaced at a distance L, the diameter of the axone is reduced to ½ and the myelin sheath is interrupted. For nerves from a bullfrog the internodal distance is

$$L = 146d$$

if d is the fibre diameter in microns. For a fibre of 10 μ, the distance L between two nodes is 1,460 μ (cf. Tasaki, 1953). But we have seen that there is also a correlation with the conduction velocity, v,

$$v = k \cdot d \ (k \text{ for frog} = 2.5).$$

We therefore obtain

$$\frac{L.}{v} = \frac{146}{2.5} = 59 \text{ sec} \cdot 10^{-6} \text{ or } 59 \ \mu\text{sec}.$$

This is the average conduction time for one internodal length and is independent of the size of the fibre diameter. (The internodal length L increases proportional to d, and so does v.)

The axone is surrounded by a well-insulating myelin sheath, which is built of spiral concentric layers of about 100–200 turns (Schmitt and Bear, 1937), in which protein and lipid layers alternate. Robertson (1955) has obtained electron micrographs of rare beauty showing these layers and their spiral structure. Betty Geren (1954) proved that the myelin sheath originates from a successive rolling of these layers by the Schwann cell. The nucleus of the cell remains on the outside. One Schwann cell always belongs to one internode, and at the node the Schwann cells of the adjacent internodes touch each other, without forming a syncytium but with an intimate interlocking. The axone has a specific resistance of 100Ωcm., whereas the myelin sheath, with 800 MegΩcm., is a very good insulator. The myelinated nerve can be compared with a cable and, from the electrical point of view, has many properties of such a model.

Is it possible to measure the spread of the nervous impulse on a myelinated nerve fibre with such accuracy as to insure whether it occurs in a continuous way or by a saltatory mechanism? This was done by Huxley and Stämpfli (1949*a*, and *b*) with a very ingenious technique, which cannot be described here (cf Von Muralt, 1958). They found that the radial membrane current in every part of the internode is always flowing from the axone through the myelin sheath to the outside, which corresponds to eddy currents flowing through a shunt in the cable insulation from the excited node through the axone, the sheath, and back to the node. Measurements at the neighbouring node revealed, however, that the current flow is to the outside for only a short time and that, with the "newborn" excitation, the membrane current reverses its sign and flows to the inside. This can be due only to an active process at the site of the nodal membrane due to excitation. In isolated myelinated nerves, conduction of the impulse is saltatory.

The question arose whether this might be due to damage to the fibre, acquired during preparation. Isolated fibres in vitro might show saltatory conduction, whereas living fibres in vivo might conduct in a continuous way. Stämpfli and Zotterman (1951) showed conclusively that, even in an uninjured nerve trunk, single fibres which respond to the stimulation of single-touch receptors show saltatory conduction; similar findings were reported by Frankenhäuser (1952).

Another problem was the question whether warm-blooded animals also showed saltatory conduction in their nerves and how the mechanism of conduction might be in the spinal cord and the brain. Gessler (1954) proved that the myelinated nerve trunk of the rat at 37° C. conducts with the saltatory mechanism and that there is no difference, except for the fact that these nerves conduct much faster at smaller diameters, as already discussed. The same statement, with a completely different technique, led Lussier and Rushton (1952) and Bishop and Levick (1956) to accept the view that the saltatory mechanism was a general phenomenon; Hodler, Stämpfli, and Tasaki (1952) showed that the longitudinal current flow spreads along the axone from the excited node to the neighbouring node in a manner similar to a signal traveling along a submarine cable at a finite rate. Although excitation "jumps" from node to node, a certain time lag occurs in the internode and contributes to the relatively low conduction velocity.

It is a mystery why in the last twenty years textbooks of the anatomy of the central nervous system have claimed that there are no structures similar to the nodes in the central pathways. It shows that not only were texts copied in the Middle Ages by uncritical copyists but that these are still alive. Cajal (1909) and many others have described nodes in the central pathways, and those who have looked for them were always able to demonstrate them. Hess and Young (1952) measured their internodal length in the rabbit and found a relation of

$$L = 120\text{–}\dot{x}\text{–}d$$

for the grown-up animal and

$$L = 80\text{–}\dot{x}\text{–}d$$

for young animals. Tasaki (1952) tried to stimulate fibres in the spinal cord of the frog by inserting microelectrodes and found responses only at certain sensitive spots, which were spaced at 0.2–0.4 mm. These indications are rather convincing that the pathways of the central nervous system of vertebrates also have saltatory conduction. Direct proof is missing, however.

And now let us return to evolution and let us try to evaluate what has been gained by the introduction of this new mechanism of nervous conduction. Saltatory conduction has made it possible to attain conduction velocities up to 100 m/sec in warm-blooded animals. A tremendous reduction of biological material was possible because of myelinisation and restriction of excitation to the nodal membranes. Sanders and Whitteridge (1946) found that the internodal length in fully regenerated mammalian nerves was only about half that in normal nerve, although the conduction velocity was the same. This finding was considered evidence against the saltatory theory. The relation between conduction velocity and node spacing shows a very flat maximum. Natural selection has apparently operated in such a way that the internodal length of myelinated nerves falls near this maximum, so that changes in node spacing, if they are not extreme, have no or very little effect on conduction velocity (Huxley and Stämpfli 1949*a*).

Myelinated nerves of the frog consume about 40 mm^3/gm per hour of oxygen at rest and produce about 240 mcal/gm per hour of heat.

Non-myelinated giant nerves of *Loligo* have an oxygen consumption of 100–150 mm^3/gm per hour and produce about a corresponding amount of heat. With activity, the maximum rate at which a frog nerve consumes oxygen or liberates heat is only about twice that at rest. In *Loligo* the increase is only 10 per cent. From this the following conclusions may be drawn. Per gram of nerve, the myelinated nerve is three to four times more efficient with regard to oxygen consumption. But this is not all. If one considers the relation between the biological material needed to convey a message in *Loligo* and in frog, it is apparent that a fibre of 10 μ is just as efficient as the giant 650 μ fibre of *Loligo,* and the relation of material needed for the same performance is 1 : 4,225. This figure shows better than many words the tremendous economy achieved in evolution through the development of the myelin sheath and the introduction of the saltatory mechanism for nervous conduction.

References

Bishop, P. O., and Levick, W. R. 1956. *Jour. Cell. Comp. Physiol.,* XLVIII, 1–34.

Cajal, S. R. 1909. *Histologie du système nerveux,* I, 269–75. Paris.

Erlanger, J., and Gasser, H. S. 1937. *Electrical Signs of Nervous Activity.* Philadelphia.

Fernández-Morán, H. 1950. *Exper. Cell. Res.* I, 309–40.

Frankenhäuser, B. 1952. *Jour. Physiol.* (London), CXVIII, 107–12.

Gasser, H. S., and Grundfest, H. 1939. *Amer. Jour. Physiol.,* CXXVII, 393–414.

Geren, B. B. 1954. *Exper. Cell. Res.,* VII, 558–62.

Gessler, U. 1954. *Pflügers Arch. ges. Physiol.,* CCLIX, 165–68.

Hess, A., and Young, J. Z. 1952. *Proc. Roy. Soc. London, B,* CXL, 301–20.

Hodler, J., Stämpfli, R., and Tasaki, I. 1952. *Amer. Jour. Physiol.,* CLXX, 375–89.

Huxley, A. F., and Stämpfli, R. 1949*a. Jour. Physiol.* (London), CVIII, 315–39.

———. 1949*b. Arch. Sci. Physiol.,* III, 435–47.

Lillie, R. S. 1925. *Jour. Gen. Physiol.,* VII, 473–507.

Lussier, J. J., and Rushton, W. A. H. 1952. *Jour. Physiol.* (London), CXVII, 87–108.

Muralt, A. von 1958. *Neue Ergebnisse der Nervenphysiologie.* Berlin, Göttingen, and Heidelberg: Springer.

Pfaffmann, C. 1940. *Jour. Cell. Comp. Physiol.,* XVI, 407–10.

Robertson, J. D. 1955. *Jour. Biophys. Biochem. Cytol.,* I, 271–78.

Sanders, F. K., and Whitteridge, D. 1946. *Jour. Physiol.* (London), CV, 152–74.

SCHMITT, F. O. 1950. *Jour. exper. Zoöl.,* CXIII, 499–515.
SCHMITT, F. O., and BEAR, R. S. 1937. *Jour. Cell. Comp. Physiol.,* IX, 261–73.
SCHMITT, F. O., and GEREN, B. B. 1950. *Jour. exper. Med.,* XCI, 499–504.
STÄMPFLI, R. 1954. *Physiol. Rev.,* XXXIV, 101–12.
STÄMPFLI, R., and ZOTTERMAN, Y. 1951. *Helvet.* physiol. pharmacol. acta, IX, 208–13.
TASAKI, I. 1939*a*. *Amer. Jour. Physiol.,* CXXV, 367–79.
———. 1939*b*. *ibid.,* pp. 380–95.
———. 1939*c*. *ibid.,* CXXVII, 211–27.
———. 1952. *Jap. Jour. Physiol.,* III, 13–94.
———. 1953. *Nervous Transmission.* Springfield, Ill.: Charles C Thomas.
TASAKI, I., and TAKEUCHI, T. 1941. *Pflügers Arch. ges. Physiol.,* CCXLIV, 696–711.

W. HORSLEY GANTT

PAVLOV AND DARWIN

The lives of Pavlov and Darwin overlapped. When Darwin was producing the great work which we now celebrate, Pavlov was a stripling lad of ten years, romping and scuffling with the urchins on the streets of Ryazan in central Russia. They both lived in the great age of the adolescence of science, in the century when science, like a rambunctious youth, felt the cocksureness of the teen-ager.

Darwin's theory of evolution liberated thinking among the masses. He gave to science a freedom from authority; he justified its right to stand in a new field upon facts. Pavlov was perhaps a more militant and conscientious champion of science than Darwin. The liberalization of science for which Darwin was responsible arose more from the impact of the theory of evolution than from any missionary zeal on Darwin's part. But Pavlov had the ardor of the reformer. He felt very much the prevalence of subjective thinking, the vague, confused arguments that permeated the psychology of that period. And it was against this kind of reasoning and false explanations that Pavlov struggled rather than against the existence and the importance of our subjective living.

Although Darwin and Pavlov were unlike in personality and in methods of working, in one major way they resembled each other. Darwin's concepts were compounded from numerous detailed observations—observations made under many conditions, in many organisms, in many environments. Pavlov, too, was a keen observer. Over the portals of the new building for research in Koltushi, erected for him a few years before his death in 1936, he had inscribed the words "Observation and Observation." Not only did Pavlov rely on this attribute in his own experiments, but in epitomizing the characteristics of the scientist, he emphasized the strict collection of facts, the method which

W. HORSLEY GANTT has been Director of the Pavlovian Laboratory of the Johns Hopkins University School of Medicine since 1929. He is also principal research scientist at the Veterans Administration Hospital in Perry Point, Maryland. After earning his M.D. degree at the University of Virginia in 1920, he spent a decade studying abroad, mostly with Pavlov at the Institute of Experimental Medicine in Petrograd. Dr. Gantt's writings and research frequently involve an extension of Pavlovian concepts to abnormal behavior and to physiology, such as the work described here on the cardiovascular conditional reflex.

Darwin used as the basis for his theory of evolution (Pavlov, 1941, p. 189).

Darwin's theory of evolution, although perhaps less firmly established as a law than are some of the Pavlovian principles, has nevertheless had a more profound effect upon popular thinking than have the discoveries of Pavlov. This is because Darwin's facts seemingly conflicted with the teachings of "orthodox" religion. Darwin has had the effect not so much of confuting the basis for religious beliefs as of defining the proper domains of religion and science.

Pavlov accomplished what was of equal importance for our mental life. He no more settled the age-old riddles of the fundamental nature of mind and of our spiritual life than Darwin solved the fundamentals of religion. But he did show that mental phenomena—the elements of our psychical life—have a physiological component and that in many of their aspects they should be studied by strictly scientific methods, the same methods that had been successfully used in the study of the digestive juices—for which Pavlov received the Nobel Prize in 1903.

This discovery was new in its field, but it was not new in principle for the function of the living organism. The study of the laws of nervous activity by physiological methods involved no principles that had not been employed, e.g., for the study of muscular activity: when we move our arm it can be explained as a voluntary action or as a system of levers and fulcrums according to Archimedes or as a utilization of chemical energy. For the thinking of the general public, this was nothing revolutionary.

But whenever science deals with the most complex functions of the nervous system and its contact with the subjective life and mental life, whether this be in psychology, psychophysiology, or psychiatry, Pavlov's discoveries have a profound influence, an influence which at present is barely realized. The "battle for the mind" [1] initiated by Pavlov is only in its infancy; it is still enshrouded in a good deal of obscurity or even confusion.

Despite the general photographic resemblances between Darwin and Pavlov, their marked differences in personality led to the equally marked differences in their methods. Though both were accurate observers, Darwin's observations were as a naturalist, Pavlov's as an experimenter. Darwin laboriously accumulated facts over a long period and methodically assembled them until they convincingly supported the theory of evolution, which had been known beforehand. Pavlov's

[1] The phrase "battle for the mind," though dramatizing the struggle for a scientific and objective point of view, perhaps gives the wrong slant. It is more properly a battle for the right of scientific investigation of what is available to "scientific methods," even if these are phenomena usually dealt with exclusively by subjective and vague formulations.

observations of laboratory facts were collected with equal care, but he then elaborated them into novel and original theories of brain action, often hypothetical. Also, in theorizing, Pavlov occasionally allowed himself great latitude of generalization, as when he spoke of the "reflex of freedom" and the "reflex of purpose."

Pavlov on Inheritance of Acquired Characteristics

Both Darwin and Pavlov considered the question of the inheritance of acquired characteristics. Before Darwin, Lamarck had given a dogmatic formulation for the mechanism, and Darwin felt that new traits were passed on to the progeny through the mixing of blood of the parents. Darwin's view, of course, was later superseded by the work of Mendel and subsequently by the principles of the gene theory (Weissmann, Morgan, *et al.*).

This belief of Darwin's is perhaps one reason why he is so popular currently in Russia. As his views on inheritance are generally known, I shall quote him only briefly:

> It seems probable that some actions, which were at first performed consciously, have become through habit and association converted into reflex actions, and are now so firmly fixed and inherited, that they are performed, even when not of the least use (Darwin, 1872, p. 39).

Chauncy Leake quotes a Russian physiologist as follows:

> Darwin means to us, perhaps, something a little different from what he means to Western Europeans or Americans. You think mostly of Darwin, I believe, or Darwinism as the ruthless struggle of nature, in which the strong conquer and eliminate the weak; in which nature rends, tooth and claw, and so on. We do not think that is Darwinism. That is Nietzsche and Huxley. But, to us, the important aspects of Darwinism is his principle of the survival of living things on the basis of their adaptations to a changing environment (Leake, 1959, p. 155).

The position of Pavlov in supporting Lamarckianism is surrounded by confusion, chiefly from two sources. First, although Pavlov often asserted in the 1920's that some of the educational efforts of the Bolsheviki were misdirected—because they considered as conditional reflexes what were really unconditional reflexes and therefore unmodifiable through education—he has been used as a champion of environment versus heredity by the dominating Lysenko school in the U.S.S.R. in its emphasis on modification by environment and in its struggle against the geneticists.

Second, this seeming contradiction in Pavlov's beliefs stems from a paper he read in 1923 concluding that conditional reflexes established experimentally in mice could be inherited—a position that Kleitman and Razran emphasize he did not refute (Razran, 1958). In the speech, delivered to the Eleventh International Physiological Congress meeting in Edinburgh, Pavlov said, apropos of inheritance:

> The latest experiments (which are not yet finished) show that the conditioned reflexes, i.e., the highest nervous activity, are inherited. At present some experiments on white mice have been completed. Conditioned reflexes to electric bells are formed, so that the animals are trained to run to their feeding place on the ringing of the bell. The following results have been obtained:
>
> The first generation of white mice required 300 lessons. Three hundred times was it necessary to combine the feeding of the mice with the ringing of the bell in order to accustom them to run to the feeding place on hearing the bell ring. The second generation required, for the same result, only 100 lessons. The third generation learned after 30 lessons, the fourth generation required only 10. The last generation which I saw before leaving Petrograd learned after 5 repetitions. The sixth generation will be tested after my return. I think it very probable that after some time a new generation of mice will run to the feeding place on hearing the bell with no previous training.
>
> It is well known that a chicken when it just comes from the egg immediately begins to pick up any black spot on the floor trying to find some grain, thus showing that it was an inborn reflex from the eye to the food. Why should we not build up the same reaction, not from the eye but from the ear as indicated in the case of the white mice? (quoted in Razran, 1958).

Razran points out in his paper, "Pavlov and Lamarck," that the theme of this address was not primarily inheritance of acquired characteristics, that this was "really only a small, but striking, aside." Previously, as Razran mentions, Pavlov had inclined to the view that it was possible for some habits to become fixed by heredity, viz., the unconditioned reflexes. For example, Pavlov said in 1913: "One may suppose that some of the conditional temporary connections may be later transformed into unconditional reflexes by heredity" (1928, p. 236). Again, in 1914, he said, "It is highly probable (and there are to this effect some factual indications) that, when the same conditions of life are maintained in series of successive generations, newly formed conditional reflexes uninterruptedly become constant unconditioned reflexes" (1928, p. 242).

Razran says, "No mention whatsoever was made of the problem in any of Pavlov's subsequent writings before 1923."

Much attention has been devoted to the postwar conflict in the U.S.S.R. between Lysenko and Michurin, on the one hand—the champions of environmental influence—and the geneticists represented by Vavilov and I believe earlier by the views of Kozlov. There has been so much written in this country concerning Lysenko and his claims that it is unnecessary to give details. Although the question of environment and heredity still has its different advocates, the opinions of competent authorities such as H. J. Muller, who spent eighteen months in Russia and knew both sides of the Vavilov-Lysenko controversy, indicate that Lysenko's claims are political rather than scientific (Zirkle, 1950).

The question of Lysenko's false claims is not so much of interest to us as is his effect in Russia on the views attributed to Pavlov there. Pavlov has been set up twenty years after his death as in violent opposition to Morgan and the Western geneticists, whereas Morgan's statement in regard to the 1923 address is by no means the virulent, anti-Pavlovian attack which the Russians have attributed to him. Morgan said, "There was some consternation in 1923 when the great Russian physiologist, Pawlow, reported the results of experiments that go far beyond what most Lamarckians have dared hope. Pawlow's conclusions—and as yet we have only his conclusions—are very surprising" (1925, p. 157).

I was in Pavlov's laboratory at the time that the experiments in question were being conducted by Studentsov; the purpose was to determine whether successive generations of mice would form conditional reflexes more quickly then their forebears. There was an ingenious apparatus, designed by Professor Hanike, by means of which laboratory mice were given all their feedings preceded by a conditional stimulus. At this signal the mice went from one cage into another, where they received food, and the number of times before they learned to go when the bell rang was recorded. Studentsov reported the facts of the lesser number of trials required for successive generations of mice to learn, as mentioned in Pavlov's Edinburgh address.

Studentsov died before the end of the experiments, I believe after they had been carried through 11 generations. When, with Pavlov's emphasis on thorough control of experiments and his habit of giving the same theme to at least two, sometimes three, collaborators working in different institutions, this problem was assigned to another investigator, the results did not show that successive generations formed the conditional reflex of going into the food cage at the signal any more quickly than did their forebears. The repetition of these experiments was done, as I recall, about 1924–26.

Since the later experiments turned out negatively, i.e., did not support inheritance of conditional reflexes, and since Pavlov had never read a paper the main theme of which was the inheritance of acquired characteristics, no significance is to be attached to the fact that he did not write a special article retracting these views. In fact, in the bibliography of his *Lectures on Conditioned Reflexes,* there is no record of any article by Studentsov or on the subject of inheritance.

I have described Pavlov's revised experiments as follows:

> The apparatus for forming conditioned reflexes in mice is very ingenious; it was perfected by Professor Hanike in 1925. When a given bell sounds, the mice run to a certain place to get food, and in going there they have to cross a platform attached to springs. When they step on this platform the act is registered on a revolving drum. A revision of the former work on inheritance of conditioned reflexes is being carried out with an entirely new apparatus. The mice now receive all their food (twenty times during each night) preceded by the bell. It is all done mechanically, so that the presence of the operator is not required. A clock arrangement makes twenty electrical contacts during the night (the natural time for the feeding of mice), rings the bell, and a few moments later opens a valve which allows grain to drop into a certain compartment of each cage. When the bell rings there is a general migration of the mice into the "dining room." The mice never get food without the bell. Males and females are kept in separate cages, so that the number of oncoming generations can be carefully regulated. When an experiment is made the mice are removed to a special cage where the results can be registered automatically (Gantt, 1928).

In a discussion which I had in 1926 with Pavlov about this work, he told me that his conclusions about the inheritance of acquired characteristics was one of the biggest errors of his scientific career. He attributed the mistake to the fact that he accepted—contrary to his usual custom—the results of Studentsov without personally supervising the experiments. He told me then that he had given the problem to another collaborator, whose results did not confirm those of Studentsov.

Neither in this case nor in any other is there evidence that Pavlov tended to cling to his theories or his concepts when further facts did not support them. Like other scientists, he made errors. One of these was his statement that the regulation of the pancreatic secretion was effected solely through the nerves. The discovery by Bayliss and Starling of the hormone secretin overthrew Pavlov's theory of nerve control as the *sole* regulator. To give up this conviction was difficult, but Pavlov stated to his collaborators after he had confirmed the experiments of Bayliss and Starling: "Of course they are right. We cannot claim to a monopoly on all scientific truth."

Pavlov and Types

Darwin's teachings emphasize the almost imperceptible gradations occurring in nature between different organisms in the process of evolution. Ernst Mayr (1959) emphasizes how the Darwinian cannot be a typologist:

Darwin introduced into the scientific literature a new way of thinking, "population thinking." What is this population thinking and how does it differ from typological thinking, the then prevailing mode of thinking? Typological thinking no doubt had its roots in the earliest efforts of primitive man to classify the bewildering diversity of nature into categories. The eidos of Plato is the formal philosophical codification of this form of thinking. According to it there are a limited number of fixed, unchangeable "ideas" underlying the observed variability, with the eidos (idea) being the only thing that is fixed and real while the observed variability has no more reality than the shadows of an object on a cave wall, as it is stated in Plato's allegory. The discontinuities between these natural "ideas" (types), it was believed, account for the frequency of gaps in nature. Most of the great philosophers of the 17th, 18th and 19th centuries were influenced by the idealistic philosophy of Plato, and the thinking of this school dominated the thinking of the period. Since there is no gradation between types, gradual evolution is basically a logical impossibility for the typologist.

The ultimate conclusions of the population thinker and of the typologist are precisely the opposite. For the typologist, the type (eidos) is real and the variation an illusion, while for the populationist the type (average) is an abstraction and only the variation is real. No two ways of looking at nature could be more different.

Pavlov said that the great difference between the human and the subhuman animals was in the language function, which he called the "second signaling system." He recognized that language represents a new function, present only in the human, upon which is based the quality responsible for the superior advances of *Homo sapiens,* viz., the capacity for symbolization and abstraction. Pavlov considered words not only as secondary signals—the signals of signals—but as a distinct and new function of the brain.

Here is introduced a new principle of higher activity (abstraction—and at the same time the generalisation of the multitude of signals of the former system, in its turn again with the analysis and synthesis of these new generalised signals), the principle of the conditioning limitless orientation in the surrounding world and of creating the highest adaptation of the human—science both in the form of a humanitarian empiricism as well as in its specialised form (1941, p. 114).

Pavlov divided the human being (perhaps without sufficient justification) into two main types according to the extent of the development of the second signaling system: In the first group were those who used mainly the first signaling system (e.g., artists), and in the second group were those with a more highly developed second signaling system—the scientist, mathematicians, etc., who depend chiefly upon abstractions. That Pavlov considered this function a definite human one is attested to by his statement (1941, p. 162):

> Until the time when *Homo sapiens* appeared animals were connected with environment so that the direct impressions fell upon the different receptors and were conducted to the corresponding cells of the central nervous system. These impressions were the several signals of the external object. However there arises in the developing human an extraordinary perfection, the signals of the second order, the signals of the primary signals in the form of words—the spoken, the heard, the seen word. Finally it came about that through these new signals everything was designated that the human being perceived both from the environment and from his inner world, and these signals commenced to serve him not only in communicating with other men, but also when he was alone.

The theory of natural selection rests upon the breaking-down of types. Pavlov, on the other hand, found definite variations among his dogs which led him to put them into four categories based upon the Hippocratic division into four temperaments—the extremes of choleric and melancholic, and the middle types of sanguine and phlegmatic (1928, p. 370).

Although the division into four rigid types on the basis of the predominance of excitatory and inhibitory conditional reflexes, as well as on the general behavior, may not be entirely satisfactory, there does seem to be some division possible, derived from the circumscribed conditional reflex studies in the dogs. In my own studies, although I have not been able to substantiate the strict conformity to the four Hippocratic temperaments, I do find it profitable to divide the dogs into groups according to their susceptibility to stress (Gantt, 1943).

Recent work from Russia on the typology of dogs, as well as of the human, is even more promising. Thus the work of Krasusky on the reaction of dogs to certain drugs (e.g., caffeine) is claimed to give a more rational basis for a classification of types. Krasusky makes the important revision in describing the behavior that external behavior is not to be taken as an adequate criterion, since this depends to a great extent upon what the individual is accustomed to, such as the presence of the human being, its early training, and environment. This work depends upon concepts that seem to be opposed to those of Lysenko. Though heredity is emphasized in Krasusky's research, there is no definite mention made of its relation to Lysenkoism.

A biochemical basis for a study of types in the human being has been laid by the Protopopov School of Psychiatry in Kiev. Their results support the view that manic psychotic attacks can be predicted by biochemical and metabolic studies some months before they occur and can be prevented by appropriate therapy.

Validity of Darwin's Observations In the Light of Modern Research

In his book *The Expression of the Emotions in Man and Animals* (1872) Darwin made profound deductions from careful observations, but his formulations have been neglected in the century since publication. I must admit that I was also unaware of Darwin's concepts on these subjects until this year. However, his formulations, made without benefit of laboratory or equipment, are so close in many cases to what I have found over the past thirty years by the conditional reflex methods that I must insert a digest of his observations, gathered from various statements throughout his book and placed here under the headings by which we now express the topics.

SCHIZOKINESIS

As I have defined this principle, it involves (1) specifically, a lack of parallel between the general autonomic (respiratory, cardiac) patterns and the more voluntary motor ones and (2) the extension of this principle to a persistance of useless reactions to the environment, representing maladaptations. There are at least twelve separate references in Darwin's book to differences between cardiac and muscular responses, similar to those I have seen in the laboratory and which I call "schizokinesis." Thus on page 28 Darwin says, "Some actions ordinarily associated through habit with certain states of the mind may be partially repressed through the will, and in such cases the muscles which are least under the separate control of the will, [the cardio-respiratory] are the most liable to act, causing movements which we recognize as expressive."

He goes further and states: "In all cases there seems to exist a profound antagonism between the same movements, as directed by the will and by a reflex stimulant, in the force with which they are performed and in the facility with which they are excited." Darwin also invokes the support of Claude Bernard: "L'influence du cerveau tend donc à entraver les movements réflexes, a limiter leur force et leur étendue."

In regard to the principle of schizokinesis (maladaptation) in disease Darwin quotes Maudsley as saying that "reflex movements which commonly effect a useful end may, under the changed circumstances of disease, do great mischief, becoming even the occasion of violent suffering and of a most painful death." According to Darwin, many of these responses that he states become purposeless or even harmful do so through a relationship between another emotional state and the original one, so that the secondary emotional state may produce an entirely inappropriate response. Many examples are given, such as cats shaking their feet at the mere sound of flowing water, as if they were actually standing in it, and the pounding of kittens with their extended claws against numerous objects, a movement appropriate only for the mother's mammae.

In regard to the self-destructiveness of the cardiac reaction, he says, "The heart beats quickly, wildly, and violently; but whether it pumps the blood more efficiently through the body may be doubted, for the surface seems bloodless and the strength of the muscles soon fails." Darwin repeatedly emphasizes the maladaptability of certain habits (conditional reflexes): "My object is to show that certain movements were originally performed for a definite end, and that, under nearly the same circumstances, they are still pertinaciously performed through habit when *not of the least use*" (italics mine).

CARDIAC CONDITIONAL REFLEX

Although the existence of a cardiac component of the conditional reflex has not yet been recognized in physiology in this centenary year, Darwin repeatedly cites well-known instances of what could be called a *cardiac component of the conditional reflex:* "When a man or horse starts, his heart beats wildly against his ribs, and here it may be truly said we have an organ which has never been under the control of the will, partaking in the general reflex movements of the body." The relationship is clearly expressed by him on page 73: "When the heart is affected it reacts on the brain; and the state of the brain again reacts through the pneumo-gastric nerve on the heart; so that under any excitement there will be much mutual action and reaction between these, the two most important organs of the body. . . . We must not overlook the indirect effects of habit on the heart."

Again he points out that this is so even though the heart is not "under the control of the will." A clear statement of cardiac conditioning follows indicating that muscular exertion is not the cause of the increase in cardiac activity: "On the principle of association, of which so many instances have been given, we may feel nearly sure that any sensation or emotion, as great as pain or rage, which has habitually led to much muscular action, will immediately influence the flow of nerve-force to the heart, although there may not be at the time any muscular exertion."

He states further that when the cerebrospinal system is highly excited "violent movements follow," and he points out that "voluntary muscular exertion relieves pain. . . . The anticipation of a pleasure leads to purposeless and extravagant movement. . . . Persons suffering from grief seek relief from violent frantic movements" (p. 176). In infants, too, he observes that "screaming brings relief."

All these observations by Darwin are strikingly parallel to what we have observed in the laboratory—that a quiescent state follows intense activity (Cruet and Gantt, *The Bulletin of the Johns Hopkins Hospital,* December, 1959).

EFFECT OF "PERSON"

We have devoted considerable attention at our Laboratory to the marked influence on dogs of the presence of other dogs and of human beings. We have seen an especially strong effect during tactile stimulation, such as rubbing behind the ears; in some animals the heart rate is reduced from 160 to 40 or less.

Darwin notes the wide prevalence of the effect of tactile stimulation in nature, plus a "strong desire to touch the beloved person. . . . Dogs and cats manifestly take pleasure in rubbing against their master and in being rubbed. . . . Monkeys delight in fondling and in being fondled" (1872, p. 233). He mentions kissing, rubbing noses, patting of the arms, etc. as human expressions, regarding kissing as innate "insofar as it depends on the pleasure derived from contact with the beloved person" (p. 352). He also describes a patient with heart disease and an extremely irregular pulse which "invariably became regular as soon as my Father entered the room" (p. 339).

While not agreed as to whether this desire for social contact is innate and constitutes a homeostatic drive, psychologists are well aware of its importance to mental health; much clinical data is accumulating on the effects of personal isolation and the implications of family environment for subsequent intellectual and personality growth.

EARLY LEARNING

In recent years the sterile environment-heredity issue has been succeeded by studies of a number of more specific factors that determine learning. There is an awareness of "critical periods," at which the individual reaches a state of maturation that facilitates learning of a particular skill or where he "suddenly" begins to respond to a certain set of environmental stimuli.

The phenomenon of "imprinting," introduced by Heinroth in 1911 and since studied in many species using various behavioral items, occurs at an exceptionally early age and involves sudden learning of a very tenacious sort. Perhaps the best-known example is that of Konrad Lorenz' greylag geese who were hatched in isolation by him and thereafter, as their behavior clearly indicated, regarded him as their mother.

If the young individual at a readiness point is not given opportunity in the form of the necessary equipment or a reasonable substitute or if he is denied the freedom to act, the normal behavior involved may fail to appear. Young mammals at Liddell's "Animal Behavior Farm" at Cornell, separated from their mothers soon after birth, failed to develop skill in sucking. Such was probably the case leading to observations by Hippocrates and by Harvey and later quoted by Darwin in *The Expressions of the Emotions,* namely, that "a young animal [removed from its mother] forgets in the course of a few days the art of sucking, and cannot without some difficulty again acquire it." An interesting corollary is the oft-observed refusal by the mother to accept the return of the offspring which has been snatched away.

There is certainly an element of timing here, too. Soon after birth, or later at "weaning-time," the young animal readily adapts to substitutes for mammae. However, being separated at a point in-between from that to which he has become accustomed, may result in abnormal and even fatal behavior.

Another learned item, eating habits, play a prominent and often pernicious role. Darwin (1872) refers to various caterpillars that, having regularly fed on the leaves of a particular tree, refused to eat the leaves of another kind of tree, although it could have supplied all their nutritional needs. We have observed similar food habits in dogs, and Stefansson has told me that arctic dogs refuse meat, which is not customarily part of their diet.

Post-Pavlovian Research

The main concepts of Pavlov have stood firm against assaults, as have the principles he derived. Many of his statements which he regarded as working theories are still in the realm of theory: difficult differentiation as the source of experimental neuroses; the nature, perhaps even the existence, of the processes of induction; concentration and irradiation, as he conjectured them; the explanation of sleep as the spreading of internal inhibition. On the other hand, many of his theories have later been supported by facts; thus, long before the discovery of chemical transmitters such as acetylcholine and sympathin, Pavlov postulated that excitation and inhibition in the brain depend upon definite chemical substances.

EVOLUTION OF CONDITIONAL REFLEX FUNCTION

Using two criteria for the extension of the conditional reflex, (1) formation of the positive reflex and (2) its differentiation, Pickenhain

(1959) concludes that in Hydra, Infusoria, and other organisms up to the echinoderms, there is a summation of excitation rather than new formation of the conditional reflex.

A central nervous system is necessary for the conditional reflex. Boycott and Young formed food and defense, excitatory and inhibitory, conditional reflexes in goldfish. Through extirpation of the phylogenetically most recent formations, viz., the cerebral ganglia, the conditional reflex formation was impaired or destroyed. Voronin formed conditional reflexes to light and to defense in crabs; these could be extinguished but were spontaneously restored the next day. In many insects conditional reflexes have been elaborated. Nikitina produced secretory fibers based on temperature in silk worms after 10–15 reinforcements, using change in light as stimulus. In bees Von Frisch has formed many diverse conditional reflexes reinforced by sugar syrup. Voskresenskaya determined that the head ganglion in the bee was required for the conditional reflex. Fankhauser and Vernon showed in salamandors that the number of ganglion cells is related to the conditional reflex formation. Simple formation of conditional reflexes, as well as differentiation, occurs after virtually the same number of reinforcements throughout the animal kingdom. Thus Angyan of Budapest states that he was able to form conditional reflexes in worms after 7 reinforcements. But how the experiments are performed is also important, as shown by the work of Voronin; he obtained formation in crabs after 25–50 reinforcements; in fish, after 30–45; in birds, after 40–120; in rabbits, after 47–107; in dogs, after 3–36; in chimpanzees, after 4–6 reinforcements.

For optimal formation of the conditional reflex in the various genera, it is necessary to observe the natural living conditions and not to base conclusions too strictly on the artificial laboratory environment, e.g., fish are grossly disturbed by an environment in which birds, rabbits, and higher animals are undisturbed. Also fish can form conditional reflexes to such complex stimuli as light plus tone better than to these separately. Nor can fish form trace reflexes where the interval is longer than 5–10 seconds. Trace reflexes are difficult in fowls, rabbits, and dogs, and unstable, while in apes inhibition is elaborated after 3 reinforcements and can be retained, as shown by Voronin, for 8 years.

The development of a centralized nervous system is expressed (1) in the ganglionic chain of the insects and (2) in the central nervous system of the chordates, with its opportunity for plasticity. There is added another mechanism, that of language, Pavlov's second signaling system, in the human being. A primitive form of abstraction is present in subhuman species, some of which can differentiate 7 different kinds of optical stimuli, e.g., in some birds 2 and in others 7. These abstractions seem to be limited in various analyzers to 7 possibilities, and even in the human being with tachistoscopic stimuli it is said that there can be only 7 good differentiations (Pickenhain, 1959).

It is interesting to emphasize that the simple formation of the conditional reflex occurs with nearly the same speed in all animals, viz., after a few combinations of the environmental situation and the unconditional stimulus. But the conditional reflex, though formed, may not be overtly expressed, as I have demonstrated with the cardiac conditional reflex.

The cardiac conditional reflex may be formed in the dog after 1 reinforcement, while the motor conditional is not seen until after 30 or 40 reinforcements, a phenomenon which I call *schizokinesis.*

The difference among the species of animals is not in the speed of the formation of the simple conditional reflex, but in the complexity, elaborations, and extension of the symbolizations that are possible. In the area where there can be formation, however, there is comparable plasticity in adaptation, which means that the lower animals can adapt to those situations within their limitations with the same readiness as can those animals higher on the zoological ladder.

The comparable speed of formation of the conditional reflex throughout the animal scale may rest upon a basic property of nervous tissue. This finds a parallel in the statement of von Muralt that everywhere in nature the passage of the nerve impulse is mediated through acetylcholine or nor-adrenalin, and everywhere the conversion of energy into movement occurs by one mechanism, viz., the chemical change involving adenosine triphosphate (ATP).

EXTENSION OF PAVLOVIAN CONCEPTS

Many other responses and reactions in the organism have been brought within the CR methodology since Pavlov. Bykov (1957) in Russia has done more than anyone else in this field. He has shown that renal secretion of urine, metabolic exchanges, thermal regulation of the body, hormonal secretions, ovulation, electrophysical components, and many others can be readily elaborated as CR's. In the Pavlovian Laboratory of the Johns Hopkins Medical School, we have also extended this field to include vestibular reactions of equilibration (Gantt, Löwenbach, and Brown, 1953), respiratory and cardiovascular responses, including heart rate and blood pressure (Dykman and Gantt, 1958; Gantt and Dykman, 1957), as well as responses to stimuli placed within the central nervous system (interoceptors) (Bykov, 1957). In the human being we have added to these the psychogalvanic response (Reese, Doss, and Gantt, 1953).

The cardiovascular conditional reflex.—Our chief interest since

1939 has been a study of the cardiac responses. I began this with Dr. W. C. Hoffmann, who came from Norway as a Rockefeller Fellow to work in my laboratory in September, 1939. The cardiovascular responses have both advantages and disadvantages for this type of work. The disadvantage lies in the widespread connections of this system with nearly all events occurring within and outside the organism and the consequent difficulty of controls and isolation of the individual. The advantage of including the cardiovascular system is that, on the theoretical side, it reveals mechanisms, to be pointed out subsequently, which we could never discover from the conventional secretory or motor components. Moreover, the cardiac response, unlike the motor and secretory, can give a measure below zero, as it were, viz., by a decrease in heart rate (HR) or blood pressure below the control, whereas with movement or with secretion, we cannot get such a negative measure, since the absence of secretion or movement is as far as we can go at present without complicated procedures, e.g., muscle potentials.

The first question that concerned us was whether there really existed a cardiovascular conditional reflex—which, for the sake of brevity, I will call the "HR-CR." The majority of physiologists and cardiologists whose opinions we sought were inclined to the view that the cardiovascular system would not participate to a measurable extent in the CR. There is, of course, another aspect of the question, and that is whether there is a cardiac CR to an adequate stimulus to the heart, as well as whether the cardiac system participates as a component in responses that are not primarily specific to the heart, viz., food and slight faradic stimulation. This has not been investigated so thoroughly, although there is sufficient evidence of a specific cardiovascular response to adequate stimuli.

To epitomize two decades of work from my laboratory on the cardiac reactions: the cardiac component of the CR's are, in general, parallel to the secretory and motor; there is a quantitative relationship with the intensity of the excitatory CR, a marked difference between the cardiac component of the excitatory and inhibitory CR's, a precise cardiac time reflex, etc. The inhibitory CR is characterized by a slight rise in HR, with a marked subsequent decrease below normal. Here we have in the cardiac response a measure of inhibition which gives an explanation to the quiescent phase and sleep, which Pavlov found resulting from inhibitions.

More important, however, than the resemblances between the specific CR movements and secretions and the cardiac rate (HR) component of the CR are the differences. First, the cardiac response is often the more sensitive of the measures we have employed. Second, contrary to our expectation, the HR-CR appears more quickly than either the motor or the secretory component. Thus it is often necessary to give 50–100 reinforcements before we see the elaboration of the CR to the motor or to secretory stimuli. But with the cardiac measure, we often see that a CR is formed after one reinforcement with the unconditional stimulus (food or pain) (Pinto, Newton, and Gantt, 1957). This quick formation of the CR explains much that was not evident previously; thus from laboratory studies it appeared that many repetitions were necessary to produce the CR, while the experiences of life showed that a CR was frequently formed after one coincidence of a physiological stimulus and its symbol, e.g., a single coincidence of a strong emotional experience often resulted in, at a later time in life, the same feelings being reproduced, although the original and real physiological unconditional stimulus was subsequently, as at first, lacking (Gantt, 1957).

Schizokinesis.—Now another unexpected event appeared to us. If the heart was sensitive enough to respond so quickly, would these earliest-formed CR's also disappear more quickly? Ask yourself the same question, and see whether you do not predict, as we did, that the cardiac CR's would be unstable and disappear more quickly than other components. In many, though not all, individuals, however, the HR-CR's are extremely stable; once formed, they outlast the more specific components, such as secretion and movement. In many dogs, even after repeated attempts to extinguish, the cardiac CR persisted as strong as ever and for as long as one to four years without practice, while the secretory and motor components remained extinguished!

This marked difference—the early formation and the extreme durability, in contrast to the greater plasticity of the specific secretory or motor responses—is what we have called *schizokinesis* (Gantt, 1953). In this term I intend to include not only in a narrow sense a difference between the general emotional components of the acquired responses and the specific ones, but in a broader way the lack of perfect adaptation that exists in our biological systems. The heart is doing one thing, out of adaptation to present reality; superficially, the individual may be in repose and undisturbed, but beneath, in the autonomic components of the response, there may be violent turbulence. Here may lie the explanation for the persistence of psychogenic hypertension to past experiences long forgotten (Gantt, 1957).

The tremendous advantage of the conditional reflex is the adaptability of the individual dependent upon experience. This adaptability consists in a readiness to act to the signals of an event and a certain plasticity added to the more stereotyped inborn reactions inherent in structure. But the increasing complexity of the conditional reflex function, seen especially in the human being, carries with it a great liability. First, on account of the complexity, there is a greater possibility for malfunction, just as there is a greater probability of a complex machine going wrong. Second, as we amass conditional reflexes, in order to preserve adaptability, there are of necessity more and more inhibitions with their consequent stresses. Third, as pointed out in connection with schizokinesis, these inhibitions are frequently only partial and imperfect adaptations. Thus, through an accumulation of only partially adaptive conditional reflexes, the individual becomes increasingly a museum of antiquities.

Autokinesis.—In several decades of studying neurotic as well as normal dogs, I have been struck by the changes occurring over a long period, based on past experiences but developing in the absence of the repetition of the original experience. Thus in "Nick" appeared a whole train of neurotic symptoms related to the original stress, but developing and becoming worse during 3 years when the animal was removed from this environment. Even more severe symptoms developed in "V3." These are examples of negative *autokinesis,* but there is also evidence of a positive autokinesis, e.g., when a single therapeutic conference or some single experience in the life of an individual has a profound and lasting effect for good. In the normal animal, autokinesis can be seen in the elaboration of new relationships among the original excitatory foci, modifying or completely changing the relationship between the conditional reflexes.

This is a circumscribed view of autokinesis, but one may conclude that there is a normally occurring basic principle of inner development—that this is a basic physiological law. Besides the examples from my laboratory, which provide striking contrasts because we have quantitative measurements for comparison and precise stimuli, there are a host of other examples from the laboratories of other workers, as well as from ordinary life. Embryology itself is an example of development determined from within, changes depending upon the internal structure more than on the external environment.

Although the peripheral impulses go into the cortex in specific patterns according to the receptor with definite mosaic arrangements, as shown by Vernon Mountcastle (personal communication), beyond that point is the possibility for integration, combination, and change. The neurological basis for such changes is described by Eccles (1958) in regard to synapses: "The initial activation of the synapses brings about a lasting improvement in the efficacy of these functions. . . . One explanation is that the synaptic knobs grow in size, another that the synaptic transmitter substance is increased." When he says that "usage enhances synaptic efficacy for days or months," here is evidence of central, internal change. Pavlov's action of induction over periods far beyond the action of the stimulus, his interaction of cortex and subcortex, are reinforced by the more recent "reverberating circuits" of the electrophysiologists. The effects of cerebral trauma on memory of events within 20 minutes of the trauma indicate that demonstrable changes are proceeding in the brain for at least this length of time after the stimulation has ceased.

These facts constitute the sketchy outlines of a shifting inner structure, with remodeling and rebuilding going on within. This function and capacity to change from within is what I mean by *autokinesis*.

This is something new for the organism. It apparently involves a changed relationship between centers in the central nervous system, and it occurs as the result primarily of *new relationships within* the organism. Since the organism is a little universe of its own, why should there not be possible changes occurring between the units of this organism on the basis of internal stimulation? Besides these facts, is there anything more impossible for reciprocal relations to be changed internally, among the inner centers of excitation, than for the organism to be capable of changing in respect to the external environment?

There is too much evidence to refute this idea. Its recognition will open up an entirely new field for exploration, a field which requires special emphasis for its future scientific development. This is an endeavor in which psychiatrists as well as physiologists should join.

The principle of homeostasis, so well developed by the genius of Claude Bernard and amplified by Cannon, now needs to be joined by the principle of schizokinesis and autokinesis, equally cogent for the understanding of normal behavior and psychopathology.

Conclusions

The conditional reflex function—at the apex of the evolutionary process—is itself not only an asset but a liability. Even more so is the supreme development of this function as represented by the second signaling system of the human. Not only do our great successes in science, mathematics, and literature rest on this quality but also our tragic failures—prejudices—and often our cataclysms of destruction—individual, national, international. Not only can the individual be categorized, branded, and persecuted through the function of such words as "capitalist," "Nazi," "Communist," and racial designations, but wars, with their annihilation of millions of lives, can be waged on the same basis.

Darwin's observations reveal the plasticity inherent in living organisms, the ability to cope with the environment, and the marvelous function for adaptations throughout the ages, while Pavlov's work on the conditional reflexes emphasizes the ability of the individual to adapt during its life.

To criticize Darwin and Pavlov because they did not discover what could not be discovered until their original contributions had been made and understood is not justifiable. Such criticism cannot be made by those who understand the history of science. Because Darwin did not enunciate the principles of Mendelism and of the later geneticists and because Pavlov did not demonstrate the role of acetylcholine or reveal the exact mechanism of inhibition or arrive at a final classification of types—these do not constitute scientific errors. Science has to take one definite step before it can attempt the next, and the next step will often give a point of view which will reveal the previous one in a perspective which was not possible before. Voltaire, in discussing Descartes, pointed out that he should no more be expected to make all the necessary elaborations of his theories which were made later than could Columbus be expected to describe in detail all the mountains and rivers of America.

Though Darwin and Pavlov differed in their methods, temperaments, and personalities, they were alike in their qualities of thoroughness, consistency, and patience in collecting facts on which to base their principles. Their ability to listen to the voice of nature, whether expressed in life or through the planned laboratory experiment, their capacity for laborious accumulation of data, their scientific zeal, their insight, their establishment of scientific laws by the facts—these are reasons why they both are foremost pioneers in the long procession of scientific explorers throughout the ages.

REFERENCES

BYKOV, K. M. 1957. *The Cerebral Cortex and the Internal Organs.* Edited and translated by W. H. GANTT. New York: The Chemical Publishing Co.

DARWIN, CHARLES. 1872. *The Expression of the Emotions in Man and Animals.* London. (Reprinted New York: Philosophical Library, 1955).

DYKMAN, R. A., and W. H. GANTT. 1958. "Cardiovascular Conditioning in Dogs and in Humans," in *Physiological Bases of Psychiatry,* ed. W. H. GANTT. Springfield, Ill.: Charles C Thomas.

ECCLES, J. C. 1958. "Physiology of Imagination," *Scientific American* (September, 1958) pp. 135–46. (Also data from lecture at the Johns Hopkins University, October, 1958.)

GANTT, W. H. 1928. *Medical Review of Soviet Russia.* London: British Medical Association. 112 pp.

———. 1943. "Measures of Susceptibility in Nervous Breakdown," *Amer. Jour. Psychiat.,* XCIX, No. 5.

GANTT, W. H. 1952. "Pavlovian Methods in Psychiatry," in *Progress in Psychiatry*. Edited by J. L. MORENO and J. MASSERMAN. New York: Grune & Stratton.

———. 1953. "Principles of Nervous Breakdown—Schizokinesis and Autokinesis," *Ann. New York Acad. Sci.*, LVI, 143–63.

———. 1956. "What the Laboratory Can Teach Us about Nervous Breakdown," in *Medicine in a Changing Society*. Edited by I. GALDSTON. New York: International Universities Press.

GANTT, W. H., and R. A. DYKHAM. 1957. "Experimental Psychogenic Tachycardia," pp. 12–19 in *Experimental Psychopathology*. Edited by P. H. HOCH and J. ZUBIN. New York: Grune & Stratton.

GANTT, W. H., H. LOWENBACH, and C. N. BROWN. 1953. "Acquired Vestibular Balancing Responses," *Trans. Amer. Neurol. Assoc.*, pp. 212–15.

LEAKE, CHAUNCEY. 1959. *Central Nervous System and Behavior*. New York: Macy Foundation.

MAYR, ERNST. 1959. In *Evolution and Anthropology: A Centennial Appraisal,* p. 2. Washington, D.C.: Anthropological Society of Washington.

MORGAN, T. H. 1925. *Evolution and Genetics*. Princeton, N.J.: Princeton University Press.

PAVLOV, IVAN. 1928. *Lectures on Conditioned Reflexes, Vol. 1*. Edited and translated by W. H. GANTT. New York: International Publishers.

———. 1941. *Lectures on Conditioned Reflexes, Vol. 2*. Edited and translated by W. H. GANTT. New York: International Publishers.

PICKENHAIN, LOTHAR. 1959. *Grundries der Physiologie der hoheren Nerventatigkeit,* pp. 109–16. Berlin.

PINTO, TERESA, J. W. NEWTON, and W. H. GANTT. 1957. "Comparative Speed Formation: Cardiovascular and Motor Conditioning," *Fed. Amer. Soc. Exper. Biol.* (April, 1957), pp. 15–19.

RAZRAN, GREGORY. 1958. "Pavlov and Lamarck," *Science,* CXXVIII, 758–60.

REESE, W. G., RICHARD DOSS, and W. H. GANTT. 1953. "Autonomic Responses in Differential Diagnosis of Organic and Psychogenic Psychoses," *Arch. Neurol. and Psychiat.*, LXX, 778–93.

ZIRKLE, CONWAY. 1950. *Death of a Science in Russia*. Philadelphia: University of Pennsylvania Press.

LESLIE A. WHITE

FOUR STAGES IN THE EVOLUTION OF MINDING

By "minding" we mean the reaction of a living organism to some thing or event in the external world.[1] It is therefore a process of interaction between an organism and a thing or event lying outside it. Minding is a function of the thing or event to which the organism reacts as well as of the organism itself: $O \times E = M$, in which O is the organism, E is the thing or event to which the organism reacts, and M is minding. Minding varies as either O or E varies.

Interaction between two bodies means a relationship between them. Minding may therefore be understood in terms of relationships between organisms, on the one hand, and events in the external world, on the other. We may deal with these relationships in terms of the *meanings* that things and events in the external world have for organisms: an organism approaches, withdraws from, or remains neutral toward some object in its vicinity, depending upon its meaning or significance to the organism. If it is beneficial (food), it may approach; if it is injurious, it may withdraw; and if it is neutral, the organism will remain indifferent. The concepts with which minding may be analyzed and interpreted are, therefore, *action, reaction, interaction, relationship,* and *meaning*. How are these relationships established? How are meanings determined?

Let us begin with inanimate bodies. According to the theory of gravitation, every particle of matter in the universe attracts every other particle, i.e., a relationship obtains between them; each has meaning for all the others. These meanings are determined by their respective masses and by the distances which intervene between them: "directly

LESLIE A. WHITE is Professor of Anthropology at the University of Michigan. After receiving his doctorate at the University of Chicago in 1927, he became Curator of Anthropology at the Buffalo Museum of Sciences; at the same time he began a series of field trips to study the Pueblo Indians, a study that extended over a quarter-century. His lengthy list of publications includes the widely reprinted *Science of Culture* (1949) and, most recently, *The Evolution of Culture* (1959).

[1] Cf. Leslie A. White, "Mind Is *Minding,*" *Scient. Monthly,* XLVIII (1939), 169–71; reprinted in *ETC., a Review of General Semantics,* I, No. 2 (1943–44), 86–91, and in White, *The Science of Culture* (1949; paperback, 1958), pp. 49–54.

as the mass, inversely as the square of the distance." Material particles attract or repel each other in such phenomena as capillary attraction, surface tension, and electromagnetic events. But all relationships among inanimate bodies can probably be reduced to three kinds: attraction, repulsion, and indifference. And in all instances, no doubt, these relationships are determined by the inherent properties of the bodies concerned, their topological relations, and their settings (presence or absence of catalysts).

When we cross the line that divides inanimate and animate bodies and come to living organisms, we find that the simplest reactions are precisely like those of inanimate bodies. The organism's reaction is positive (+), negative (—), or neutral (0).[2] That is, it approaches, withdraws, or does nothing, depending upon the meaning that the object has for it.

In this simplest type of reaction, which we may for convenience call Type I (Fig. 1), the meaning that the thing or event has to the or-

FIG. 1.—The simple reflex: Type I behavior

ganism is determined by the intrinsic properties of both organism and thing or event. Or, to put the matter otherwise, the relationship between organism and thing-or-event is determined by their respective intrinsic properties. (For the sake of completeness and precision, we ought, perhaps, to say: the relationship is determined by the intrinsic properties of organism and thing-or-event as conditioned by their setting and the factors—positive, or negative, or neutral catalysts—which it contains.) The organism approaches if the stimulus is positive (e.g., food), withdraws if it is injurious, and remains indifferent if it is neutral. But, whether a thing is *food* or not depends upon the intrinsic properties of the organism as well as of the thing; edibility is a function of the eater as well as of the thing eaten; what is food to one organism may be not-food to another. And so it is with injurious things or things neutral. In every instance in this simplest type of interaction the relationship between organism and thing-or-event is determined by the intrinsic properties of both.

The next stage in the evolution of minding is characterized by the conditioned reflex, and we may use the classic experiment of Pavlov with the dog and the electric bell to illustrate it. A hungry dog salivated

[2] If rest is a form of motion in which the velocity is zero, as Alfred North Whitehead says (*Introduction to Mathematics* [1st American ed.; New York, 1948], p. 29), then we may say that indifference or neutrality of an organism toward a body is a response in which the action is zero.

when he smelled food; he was indifferent to the sound of an electric bell. But when stimulated by odor and bell simultaneously for a number of times, the sound of the bell alone was sufficient to excite his salivary glands and evoke the response. We may call this kind of behavior Type II, and represent it diagrammatically in Figure 2.

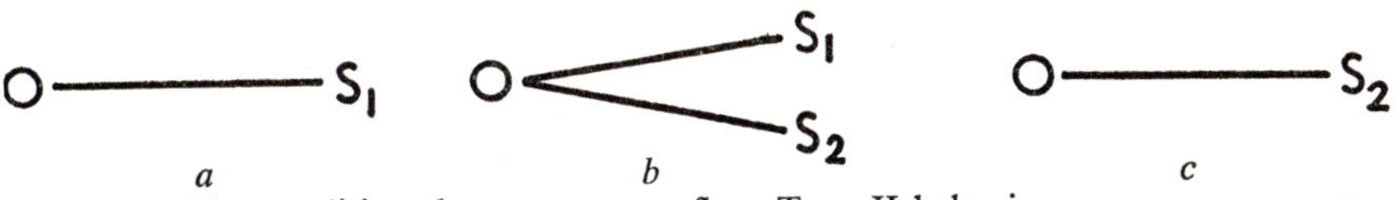

FIG. 2.—The conditioned response or reflex: Type II behavior

The process of condition takes place in three stages. In the first stage, *a,* in Figure 2, we have the same kind of situation that we have in Type I: the organism, *O,* and a significant stimulus, S_1, the odor of food, with a simple relationship between them in terms of their respective intrinsic properties. In stage *b,* we introduce S_2, the sound of the bell, which becomes related to S_1, the odor of food, on the one hand, and to *O,* the dog, on the other. Initially, S_2 is related to S_1 in time and in space, and, as a consequence of association, S_2 and S_1 become related to each other through the neurosensory-glandular system of the dog. A relationship between S_2 and the dog (*O*) is established at the same time and in the same way. When the relationship between S_2 and the dog has been established, S_1 may drop out. In stage *c* we again have a simple, direct relationship between the organism and a single stimulus.

Type II resembles Type I and grows out of it. It begins with a simple Type I reaction, and it ends with the *form* of Type I reaction. But Type II differs from Type I in a fundamental respect: Type II is characterized by a relationship between organism and stimulus which is *not* dependent upon their intrinsic properties. To be sure, the substitution of one stimulus for another could not have been effected, had not the dog been an organism capable of this kind of behavior. But the salivary-gland-meaning of the electric bell is in no sense intrinsic in the sound waves that it emits. Type II behavior ends with the *form* of Type I: the reaction takes place *as if* the relationship between dog and bell were intrinsic in them. But the response is fundamentally different in kind.

The next stage in the evolution of minding, Type III, may be illustrated by the example of a chimpanzee using a stick to knock down a banana which is suspended from the roof of his cage beyond the reach of his hand. We illustrate it in Figure 3, in which $O =$ chimpanzee, *E,* is the banana, and E_2 is the stick.

Type III minding is like Type II in one respect: the organism is

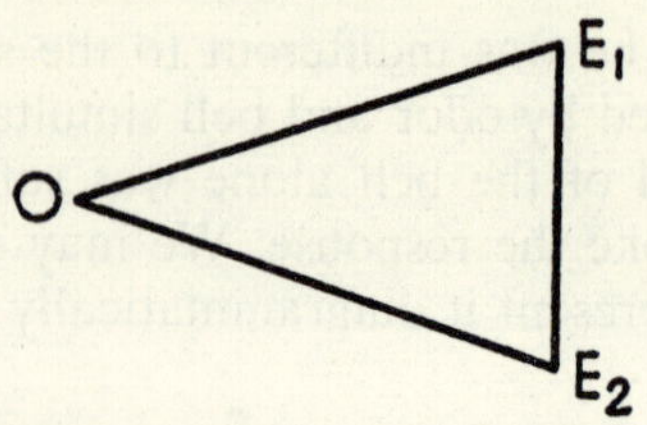

FIG. 3.—Type III behavior

related simultaneously to two things-or-events in the external world (as in Fig. 2, *b*). But Type III differs from Type II in a number of important respects. In the first place, the two things, E_1 and E_2, are significant from start to finish in Type III, whereas they are significant in only one of the three stages of the process in Type II (Fig. 2, *b*). Second, the relationships established in Type III are dependent in their entirety upon their respective intrinsic properties: the chimpanzee is a banana-eating, stick-wielding animal; the banana is a knock-downable-with-a-stick thing and eatable by a chimpanzee; the stick is wieldable by a chimpanzee and can be used to knock down a banana. Third, the relationship between E_1 and E_2 in Type III is established directly and extraorganismically, whereas they are related indirectly and intra-organismically (within the neurosensory-glandular system of the dog) in Type II. And, finally, the relationships established in Type III are determined intra-organismically, i.e., by the chimpanzee himself, "of his own free will and choice," so to speak, whereas the relationships established in Type II are not determined by the organism but by its relationships to other factors—the experimenter, or circumstances, such as chance association.

We may distinguish two kinds of roles of organisms in the process of minding. Either the organism determines the configuration of behavior which it executes, or it does not; it plays either a dominant or a subordinate role. Thus, in Type I, it is not the organism alone that determines its behavior. It behaves as it does because (1) of its own intrinsic properties and (2) because of the intrinsic properties of the stimulus, *E*. It has neither alternatives nor choice. The flower turns its face to the sun because it must; it can do nothing else. It is something that it undergoes as well as something that it does. Its behavior is subordinate to the intrinsic properties of itself and its stimulus.

The organism plays a subordinate role in Type II, also. The dog "has nothing to say" about how he shall respond to the sound of the bell; this is determined by the experimenter (chance associations may be the determining factor in other processes of conditioning). Here, also, the organism has neither initiative nor choice.

It is different with Type III minding. Here the organism plays a dominant role. It is the chimpanzee who decides what to do and how to do it. He has initiative, alternatives, and choice. He may use the stick to reach and knock down the food, or he may, as they sometimes do, use it to pole-vault ceilingward and snatch the food when it comes within his reach. Or he may decide to build a tower of boxes from whose summit he can reach his prize. This is what we mean when we say that the pattern of action, the configuration of behavior, is determined intra-organismically: the chimpanzee solves his problem by insight and understanding, formulates a plan, then puts it into execution. He is a sublingual architect and builder.

In Type III, then, we are again dealing with relationships determined by the intrinsic properties of organisms and things, but here the organism plays a dominant, instead of a subordinate, role in the formulation and execution of patterns of behavior.

Type IV minding is well illustrated by articulate speech and may be diagrammed as in Figure 4. *O* is again the organism, this time a human

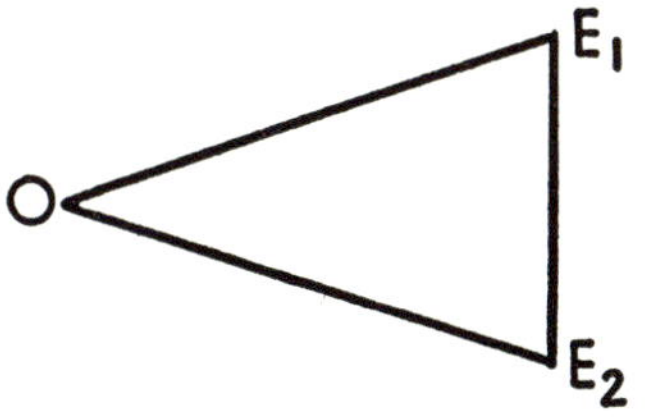

FIG. 4.—Type IV: symboling

being; E_1 is a hat; and E_2 is the word "hat." Again we have a triangular configuration as in Type III: there is a mutual and simultaneous relationship between the organism and two things-or-events in the external world. And, as in Type III, the configuration is determined intra-organismically, by the organism itself, of its own will and choice; Type IV, like Type III, is characterized by alternatives and choice. But Type IV differs from Type III in that in the former the relationships are not dependent upon the intrinsic properties of the elements involved, as they are in Type III. That is to say, there is no necessary or inherent relationship between the object hat and the combination of sounds *hat*. In this respect, Type IV resembles Type II: both are independent of the intrinsic properties of the factors involved. But Type IV differs from Type II in a fundamental way: the organism plays a dominant role in Type IV, a subordinate role in Type II.

Looking back over our four types of minding, we notice similarities and differences among them. Types I and III are dependent upon the intrinsic properties of the elements involved in the configurations of behavior; Types II and IV are not so dependent. In Types I and II, the organism plays a subordinate role in the formulation and execution of patterns of behavior; in Types III and IV, it plays a dominant role. We may summarize these facts diagrammatically as follows:

	Organism plays a subordinate role	Organism plays a dominant role
Dependent upon intrinsic properties	Type I	Type III
Independent of intrinsic properties	Type II	Type IV

FIG. 5.—Comparison of four stages of minding

Another feature of our series of stages is that our types are *kinds,* not *degrees,* of minding. An organism is either capable of conditioned reflex behavior, Type II, or it is not; there are no gradations between Types I and II. Similarly, an organism is capable of Type III or Type IV behavior, or it is not. We are confronted by a series of leaps, not by an ascending continuum.

The question might be raised at this point, Does our series of stages constitute a biological evolutionary sequence or merely a logical one? It has been derived deductively, in a sense, from a consideration of a basic concept: relationship. It has been postulated that the relationship established in the process of minding is either dependent upon the intrinsic properties of organism and things-and-events in the external world, or it is not. Second, the relationship or, more specifically, the pattern of behavior in which the relationship is expressed, is determined by the organism or it is not. This gives us four categories of minding, and it might appear at first glance that our series of stages is more artificial than real.

But this is not the whole story. We did not begin with factors selected at random. Our premises and postulates were in fact derived from a careful scrutiny and analysis of the behavior of very real organisms. The fact that the series proceeds from the simple to the complex would suggest that it constitutes an evolutionary, as well as a logical, sequence. But there are other facts that make it quite clear that we do indeed have here an evolutionary sequence in a biological sense.

Let us begin, first of all, by classifying all living species with reference to our four types of minding as far as our information will permit (Fig. 6). All organisms are capable of Type I: simple reflexes or tropisms. We know from observation that some species are capable of Type I only. Type II has grown out of Type I, as we have seen; therefore, it may be assumed that organisms capable of Type II are capable of Type I also; this assumption is validated by observation.

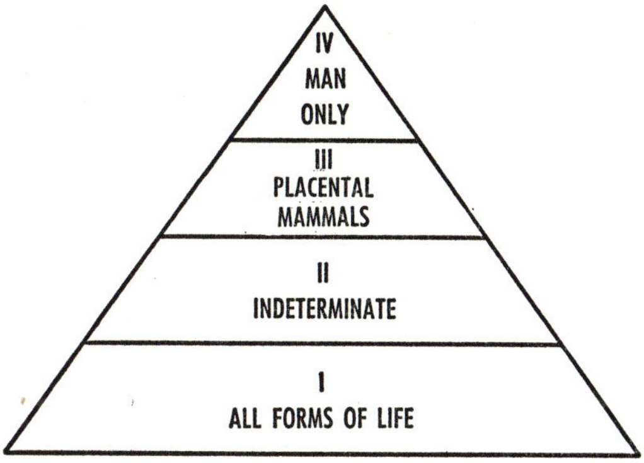

FIG. 6.—Classification of organisms with respect to types of minding

But, obviously, there are fewer species capable of Type II than are capable of Type I. We know, also from observation, that there are organisms capable of Type II that are incapable of Type III, but all organisms capable of Type III are capable of Types II and I also. Only one species, man, is capable of Type IV, and this species is capable of Types III, II, and I, also. Hence the following generalizations: (1) our series of stages is incremental and cumulative, a new stage being added to the one, or ones, that preceded it; (2) the number of species capable of a given type of behavior diminishes as we proceed from Type I to Type IV; and (3) organisms classified according to our series of types of minding are thereby arranged in a biological evolutionary series, the lowest and simplest organisms being at the bottom, the higher and more complex at the top.

Specifically, we cannot assign each and every species to one or another stage in our series of types of minding for the simple reason that we do not have the requisite information. But we have every reason to believe that we could do this if we possessed full information as to their behavior and that no species would fail to be accommodated by our series.

We do not know at what point in the evolutionary scale organisms become capable of Type II behavior. Snails, it appears, are capable of conditioned reflex behavior, which puts this ability fairly low in the scale of biological evolution, but what other species belong here is a question that we cannot answer.

We know that apes are capable of Type III minding, and there appears to be much evidence that dogs and elephants also possess this ability. It seems reasonable to suppose that all placental mammals fall within this class, but whether marsupials and monotremes belong here or not is a question to which comparative psychology provides us with no answers.

Incidentally, although we have illustrated Type III minding with an instance of tool-using (Fig. 3), it must be made clear that the ability to wield tools is not essential to this kind of behavior. The characteristic of Type III minding is the ability to formulate intra-organismically—i.e., by means of insight and comprehension and in terms of alternatives and choice—a configuration of behavior in which the organism is related simultaneously to two or more things or events in the external world in such a way that all elements of the configuration are related to one another in terms of their intrinsic properties, as in the case of the chimpanzee, the stick, and the banana.

But an organism need not wield tools to exhibit this kind of behavior; it may move itself, rather than a thing (tool), with reference to two (or more) things, all of which are related to one another in a configuration of behavior in terms of their intrinsic properties. Let us illustrate with an example.

The dog (*D*) in Figure 7 has formulated a plan of action, intra-organismically, in which he has related himself, the food, and the open gate to one another in terms of their respective intrinsic properties, and he then executes this plan in overt behavior. He has done, in effect, what the chimpanzee did in reaching the banana with the stick, except that the dog moved his body with reference to the two other factors instead of moving one of those external factors. Thus tool-wielding may be the most characteristic form of expression of Type III minding, but it is not essential to it. In this example (Fig. 7) we are reporting the results of some experiments in which it was demonstrated that a dog, but not a chicken, was capable of solving this problem by insight and comprehension. The chicken, placed in this situation, simply ran back and forth aimlessly along the fence at point *P*.

Our series of types of minding constitute a progressive series of advantages for the life-process, for living organisms: they emancipate organisms from limitations imposed upon them by their environments, on the one hand, and confer positive control over the environment, on the other. This is another significant indication of the biological evolutionary character of our sequence of stages.

Every living organism exists in a setting which imposes limitations upon its behavior in many ways: gravitation, temperature, atmospheric pressure, and humidity are universal factors that have to be reckoned with and which circumscribe the behavior of the organism. Food, enemies, and incidental obstacles further condition behavior. As we have seen, organisms capable of only Type I minding are to a great extent subordinate in their behavior to factors of their environment: the petals of a flower close at night to reopen after sunrise; the

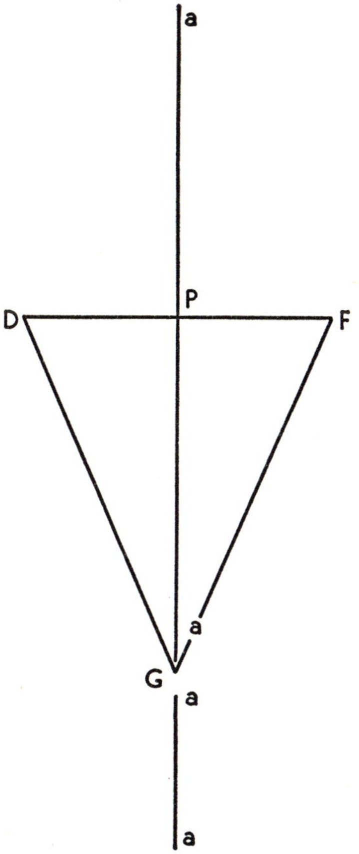

FIG. 7—*D* is a dog; *a, a* is a fence; *F* is dogfood; *G* is an open gate. The dog can see and smell the food. But he cannot approach and obtain it directly because of the intervening fence which he cannot jump over or crawl through or under at *P.* But he knows, either by previous experience or by observation, that there is that open gate, *G,* down the line. He therefore promptly runs to the gate, passes through it, and on to the food.

paramecium approaches or withdraws from a stimulus; and so on. The organism has neither alternatives nor choice; it must do what its own inherent properties and those of its environment dictate. All living organisms are dynamic (thermodynamic) systems, to be sure; but the behavior of organisms capable of only Type I is something that they *undergo* as well as something that they do.

The advent of the conditioned reflex brought about a revolution in minding. At a single stroke it emancipated organisms from many limitations imposed upon them by their environment. To be sure, the organism can live only with boundaries circumscribed by such factors as gravity and temperature, but it need no longer limit its behavior to responses to stimuli as determined by their intrinsic properties. A stimulus (a thing or event) can have only one meaning in Type I minding; it may mean any one of many things in Type II. Thus the sound of the electric bell has only auditory meaning in Type I; it may mean food, danger, sex, or something else in Type II. The conditioned reflex was, so to speak, an emancipation proclamation for evolving life-forms: it emancipated organisms from limitations imposed upon them by the natural properties of things in their environment; it multiplied the number of kinds of responses that could be made to a given thing as a stimulus.

But, under Type II, the organism was still subordinate to outside things and circumstances. With a conditioned reflex, the organism can acquire new meanings for things or events (Pavlov's dog acquired a new meaning for the sound of the bell), *but it cannot determine what this new meaning shall be;* this is done by an experimenter or by other circumstances. The organism still plays a passive role with reference to his environment. The advent of Type III gives him positive control over it. The extent of this control is not, of course, complete or absolute, *but it is control,* and it was this that was lacking in Types I and II. In Type III the organism rises above his environment. It now has alternatives, and it can make choices. It is the ape himself who decides how he is to reach the banana and with what means. In Type III the organism plays a dominant role in his interaction with his environment.

In Type IV we have the emancipation from the limitations of intrinsic properties of external things or events which was won in Type II, combined with the ascendance to control over environment that was achieved in Type III. In Type IV, it is the organism, man, who determines what meaning the sound of the electric bell shall have, and he can give it any meaning he chooses. Emancipation *and* control are thus united in Type IV.

The career of life is a struggle between thermodynamic processes: a building-up process coping with a running-down, breaking-down process. Life is simply the name that we give to a thermodynamic process that moves in a direction opposite to that specified by the Second Law of Thermodynamics for the universe as a whole. Life sustains itself by the capture and utilization of free energy, obtained in the form of food. Life is enabled to extend itself by the multiplication of numbers and to evolve new and higher forms because of its ability to capture and utilize free energy in increasing amounts.[3] To do this, it must overcome the obstacles of its natural environment. The eman-

[3] See L. A. White, *The Evolution of Culture* (1959), pp. 33–38, for fuller discussion of this subject.

cipation from limitations imposed by the intrinsic properties of things, on the one hand, and the assertion of positive control over environment, on the other, that are won by Types II and III, respectively, and which are combined to produce Type IV have been the ways and means by which biological evolution has been achieved.

Symboling

Type IV behavior, or minding, is characterized by freely and arbitrarily bestowing meaning upon a thing or an event and in grasping meanings thus bestowed. These meanings cannot be comprehended by the senses alone. Holy water is not the same kind of thing, from the standpoint of human experience, as mere H_2O; it has a distinctive quality, or attribute, in addition to hydrogen and oxygen in molecular organization. This new quality, or meaning, that holy water has, which distinguishes it from ordinary water, was bestowed upon it by human beings, and this meaning can be grasped and appreciated by other human beings. It is this ability to originate and bestow, on the one hand, and to grasp, on the other, meanings that cannot be comprehended with the senses that we have termed "symboling." [4] Symboling is a kind of behavior that is characterized by traffic in non-sensory meanings.

To be sure, a symbol must have a physical basis or form, otherwise it could not enter our experience. But the *meaning* of the symbol and its physical basis or form are two quite different things, and there is no inherently necessary relationship between the two. The combination of sounds *s-ee* may be the vehicle of an indefinite number of meanings—to use the eyes, seat of power of a bishop, yes, or anything else that we please. Articulate speech is the most important and characteristic form of symboling. A symbol may have any kind of physical basis: an object (keepsake or fetish), an act (tipping one's hat), a color (royal purple), or a sound (a spoken word).

Things that we call "symbols" have often been confused with things that we call "signs." Some psychologists who have worked with rats or apes have described their behavior in terms of symbols. The red tri-

[4] We have discussed this subject at some length in "The Symbol: The Origin and Basis of Human Behavior," *Philosophy of Science,* VII (1940), 451–63. This essay was reprinted, in slightly revised form, in Leslie A. White, *The Science of Culture* (New York: Farrar, Straus & Co., 1949; New York: Grove Press, 1958). It has also been reprinted in *ETC., a Review of General Semantics,* I (1954), 229–37; *Language, Meaning, and Maturity,* ed. S. I. Hayakawa (New York, 1954); *Readings in Anthropology,* ed. E. Adamson Hoebel *et al.* (New York, 1955); *Readings in Introductory Anthropology,* ed. Elman R. Service (Ann Arbor, Mich., 1956); *Sociological Theory,* ed. Lewis A. Coser and Bernard Rosenberg (New York, 1957); and *Readings in the Ways of Mankind,* ed. Walter Goldschmidt (1957).

angles that mean food, the green circles that mean an electric shock, the blue or yellow poker chips that are used in the "chimp-o-mat," etc., have been called symbols because their meaning or significance is not inherent in them; their meanings have been assigned to them. Therefore, they conclude, they are just like our symbols—holy water, words, crucifixes, fetishes, etc. This reasoning is unsound because it has allowed a similarity to obscure a fundamental difference between these two kinds of situations.

The red circles and the blue poker chips have indeed acquired a meaning which is not inherent in their physical structure and composition, just as holy water or the combination of sounds *see* has acquired a meaning. This is the similarity. But there is a fundamental difference also. It is not the rats and apes that have determined the meanings which the red circles or the blue poker chips have acquired, and, what is more, *they are incapable of doing this.* Only man is capable of freely originating such kinds of meanings and of bestowing them upon things or events. The apes and rats can *acquire* meanings for things, but they cannot originate or determine them. They are acquired by the mechanism of the conditioned reflex, on the level of Type II; the ability to originate and to bestow meanings is found only on the level of Type IV. The behavior of the rats responding to red circles or the poker-chip-using apes is not, therefore, symbol behavior but *sign* behavior.

A sign is a thing that indicates something else. Its meaning may be either inherent in it and its context (dark clouds a sign of rain) or extrinsic to it (yellow quarantine flag). In either case, the meaning of the sign has become identified with its physical form and context through experience and the mechanism of the conditioned reflex (or, in the case of man, by means of observation and reason, good or bad). The meaning of the sign, having become identified with its physical form, may be grasped and appreciated with the senses: we distinguish blue from red poker chips with our eyes.

But a thing or event which is significant in a context of symboling may be translated and become significant in a context of sign behavior. Thus we create a symbol: *boko*. We make it mean "hop on your left foot." The processes of originating and bestowing this meaning and the act of grasping it by non-sensory means constitute, as we have said before, an instance of Type IV behavior, of symboling, a kind of behavior peculiar to the human species. But, after we have used this word as a command several times, its meaning becomes identified with its physical (phonetic) form, becomes a sign, and we grasp and comprehend its meaning with our senses: we distinguish it from *loko,* "hop on your right foot," with our ears.

The Principle of Regression

The fact that a meaningful thing or event may have its origin on the level of Type IV but subsequently descends to the level of Type II brings us to an important principle in behavior, namely, the tendency for behavior to regress from higher to lower levels. We have just analyzed an instance of the translation of a thing from a symbol context (Type IV) to a sign context (Type II). But it may go even lower and assume the level of Type I. Let us illustrate with an example.

A word is a physical thing or event, a sound or a combination of sounds, or a visible mark on paper or some other substance. A word may be significant in a variety of contexts. It may be significant and have meaning on all levels of minding that we have distinguished. Words and their meanings originate only on the level of Type IV, where they function as symbols. But they may function in Type III contexts also. The chimpanzee reaches his banana with a stick, but a human being can use the word to obtain the same result: he need merely say "bananas" in the store, and the clerk brings him some. When the word descends from the symbol level to the sign level, it has descended to the level of Type II, as we have just seen. But it may go even lower and become significant as a thing or event in terms of its intrinsic properties. We use words because of their phonetic qualities in poetry: "charms" rhymes with "arms," not "legs." Some names are thought to be pretty (Sylvia); others ugly (Bridget). Primitive peoples use words in some situations as if their properties were intrinsic in their phonetic structure (spells, incantations). Members of a certain tribe sought prescriptions from a European physician, but, instead of taking the prescription to the pharmacist and exchanging it for medicine, they soaked it in water until the ink was dissolved and then drank the water; they availed themselves of the therapeutic values of the words by drinking them. And even in our culture we treat certain four-letter words of Anglo-Saxon derivation as if their meanings were inherent in their phonetic structure (their Latin cognates may, however, be used with propriety). Words are born as symbols, but they may descend to the level of things in themselves.

One reason for regression of behavior to lower forms is that an economy of effort or time is effected thereby. If one had to stop and think, when driving an automobile, whether the red light symbol means stop or go, he would probably violate a traffic rule or possibly kill someone. Traffic lights originated as symbols. Meanings were freely and arbitrarily bestowed upon colored lights (any colors would do). But these meanings quickly become identified with their physical structures (wave lengths), descended to the level of Type II, and became signs. But we react to the red light *as if* its meaning were inherent in it. Thus we end up with the *form* of Type I minding: a simple, straight-line reaction of stop or go (Fig. 1). This fact is recognized and expressed sometimes by saying "I instinctively did so and so."

Each stage in the evolution of minding has given life a new dimension: Type II, the conditioned reflex, emancipated living organisms from limitations imposed upon them by the intrinsic properties of the environment; Type III gave organisms positive control over things and events in the external world. But Type IV, symboling, has been especially significant, at least as far as human beings are concerned; indeed, it was symboling that made human beings of certain primates.

Both human behavior and culture are expressions and products of symboling. Human behavior consists of acts and things, dependent upon symboling, considered in terms of their relationship to the human organism. Tipping one's hat is an example of human behavior. And the hat, too, may be considered as a form, or product, of human behavior; it is human behavior locked up in a form and a fabric. But all these things and events that are dependent upon symboling may be considered in another context, also: an extra-somatic context. That is, instead of regarding them in relationship to human organisms, we can consider them in their relationships to one another and without reference to the human organism. Thus tipping one's hat may be thought of as a ritual and in terms of its relationship to other rituals, to customs of kinship, to social or class structure, and so on. In short, things and events dependent upon symboling may be thought of as constituting a continuum, a flow of culture traits from one individual and generation to another. We may think of words as items of human behavior, as acts of the human organism, as things and events in terms of their relationship to the human organism. But we may think of words as constituting a class of things and events *sui generis,* and we may study them without reference to the human organism. This is the science of linguistics and is concerned with such things as phonemics, grammar, syntax, word order, etc., whereas the scientific study of words in their relationship to the human organism is the science of the psychology of speech and is concerned with such things as habit formation, imagination, conception, attitude, and so on.

Now, just as we may think of words in a somatic context (i.e., in terms of their relationship to the human organism) or as a self-inclosed continuum, as a process *sui generis,* so we may consider all things and events dependent upon symboling in terms of their relationship to the human organism, in which case they are human behavior; or we may consider them as a self-inclosed flow, a process *sui generis,* in which case they are *culture.* "Culture" is the name of a flow of things and events dependent upon symboling considered in an extra-somatic context.[5]

Symboling has brought a certain kind of things and events into existence. They constitute a continuum, a flow of tools, customs, and beliefs, down through the ages. Into this flow, this extra-somatic continuum called "culture," every human individual and group is born. And the behavior of these human beings is a function of this extra-somatic continuum: an organism born into Tibetan culture behaves in one way (as a Tibetan); an individual born into Scandinavian culture behaves in another way. Thus the determinants of human behavior, insofar as the individuals may be considered as typical or average, are no longer the properties of the biological organism; the determinants are to be found in the extra-somatic tradition (culture). It is not the nature of the lips, palate, teeth, tongue, etc., that determine whether the human organism will speak Tibetan or Swedish; it is the linguistic tradition that determines this. Therefore, in contrast with all other kinds of living organisms, if we wish to learn why a typical individual—a typical Crow Indian or a typical Englishman—behaves as he does, we must concern ourselves not with their bodies, their neuromuscular-sensory-glandular systems, but with the cultures into which they have been born and to which they respond. Similarly, if we wish to learn why the Japanese behave differently in 1960 than in 1860, we must concern ourselves with the changes that have taken place in their culture.

In the scientific study of behavior, or minding, therefore, we must concern ourselves with organisms when we are dealing with Types I, II, and III. But when we come to Type IV and man, a new world has been created, an extra-somatic, cultural environment, and it is this which determines the behavior of peoples [6] living within it and not their bodily structures. This means that we must have a new science: a science of culture rather than a science of psychology if we are to understand the determinants of *human* behavior. But this is not the place to go into that, and, besides, we have touched upon it elsewhere.[7]

[5] See Leslie A. White, "The Concept of Culture," *Amer. Anthropologist,* LXI (1959), 227–51, for a fuller discussion of this point.

[6] The behavior of *individuals* is, of course, determined, or conditioned, by their biological make-up as well as by cultural factors. But we know of no biological differences among *peoples*—tribes, races, etc.—that would produce corresponding differences in human cultural behavior. *Peoples,* therefore, may be considered as a constant biologically, as an independent variable; it is the exosomatic cultural factor that varies.

[7] "The Science of Culture" and other essays in *The Science of Culture* (New York, 1949).

R. W. GERARD

BECOMING: THE RESIDUE OF CHANGE

FIXING EXPERIENCE

The enormous impact of Darwin's work on the thought of man derived from, but far transcended, its biological focus. The *Origin of Species* kindled emotions by its challenge of dogma concerning the genesis of man and animal; but the light generated in time illumined a vast territory of human interest. Darwin, in effect, crystallized the problem of cumulative change. A century later we celebrate the ramifications of his contribution in all sectors of knowledge, from astronomy and chemistry to linguistics and the study of culture. For each system—a molecule or cell or individual, a galaxy or society or species—presents aspects of the same basic problem of "becoming."

At any moment, or cross-section of time, an org (material system) possesses some structure or inhomogeneity. Certain aspects of this recur essentially unchanged in different time-moments and constitute the enduring architecture or "being" of the entity. Upon this "being" impinges the environment and its turbulences, the stimuli and loads ("stresses") to which the org responds. The responses to any particular environment input are determined by the architecture at that particular time. Mostly, responses are adaptive (homeostatic) and transient (reversible) and constitute the "behaving" of the system. But reversibility is perhaps never complete; and, under appropriate conditions of intensity or repetition or "meaningfulness," the response to experience becomes mainly irreversible. This secular or cumulative residue of change in the longitudinal time section constitutes the development or history or evolution of the system—its "becoming." Darwin channeled thought on the problem of becoming, a far wider one than the evolution of taxa.

The problem of fixing experience is universal; the mechanisms involved are highly particular. The mitotic dealing of genotypes and the environmental sieving of the resulting phenotypes, which helps pro-

RALPH W. GERARD is Professor of Neurophysiology at the Mental Health Research Institute, University of Michigan, specializing in neural metabolism, experimental psychiatry, and interrelated problems of human structure, function, and behavior. He has written half-a-dozen textbooks and over 300 research papers.

duce species of well-adapted organisms, is patently different from the residues left by activity in neurons and synapses, which underlie learning and well-adapted behaviors. The formation of antibodies or adaptive enzymes at the molecular level is different from the differentiation of muscle or bone at the cellular level, the callousing or wrinkling of skin (or the laying-down of engrams) at the organ level, the acquiring of percepts and skills and memories at the individual level, or the establishment of customs and edifices and languages at the social level. Yet even here some commonalties exist (Gerard, 1954).

If Darwin's natural selection is more than the tautology—"What survives, survives"—it is an emphasis on environment as the major variable; organisms are passively chosen or rejected by it—like souls on a Calvinistic judgment day. Lamarck emphasized the organism as the major variable; organisms actively achieve their adaptation—as salvation is won by good deeds. Both factors, of course, operate; and several essayists in the present volume have emphasized the continuum between extreme positions. For each kind of system, it deserves inquiry as to the relative importance of extrinsic and intrinsic factors and as to how these change from level to level and, especially, from time to time or stage to stage in its "becoming."

Evolutionists have explored with great success the operation of gene mutation and distribution in presenting blindly the arrays of organisms from which isolation, drift, and selection forge new species. Yet even here, as Waddington (1957) and Dobzhansky (1959) especially have insisted, there is more interaction than has been generally recognized between the production of novelty by the organism and its fixing by the environment; the process has its eyes at least partly open. The environment can mold only the malleable; and selection operates to enhance not only particular adaptive mutations but also mutability in general. More variable organisms, those that more easily acquire new characteristics, ones with mutable genes or mutation-inducing genes—these are the prime stuff for selection and have been favored by it.

The same is true more generally. Adaptive enzymes appear when the challenging substrate is present, but only when there is also the genetic capacity for such mutation. Muscle cells hypertrophy with use, but the capacity for so responding is inherent. Neurones and junctions are altered by the flow of nerve impulses, but this learning from experience is a contribution of the learning ability of the organism's components. The salivation of puppies at the smell of meat is conditioned, but this reflex develops especially easily. One society adapts its culture rapidly to altered environments, another remains rigidly unresponsive.

The problem can now be reformulated in terms of change and

changeability or of learning and learning to learn. Most generally, (1) as a system advances through time, how is its (relatively) permanent architecture—its internal heterogeneities and its boundary with the environment—altered in a cumulative and, therefore, irreversible manner? This is the fixation of experience, evolution, or learning, and implies the appearance of a different response to an identical stimulus. And (2) does the system, over time, fix experience more nimbly, and, if so, how does this come about? This is increasing malleability, autocatalysis, evolving the ability to evolve, learning to learn.

Whatever the mechanisms—a series of analytic subquestions that will receive attention—the answer to the main descriptive question is that living systems do exhibit, at all levels and in especially high degree, the ability to develop secular change and to do so at an accelerating rate. This is, in fact, the great invention of life-stuff, the epigenetic mode ! As I wrote elsewhere (1960*b*):

> The epigenetic mode allows organisms to respond ever more rapidly and adaptively to the environmental challenge; it similarly enables mindful organisms to meet their environmental problems with greater skill and speeds; and it has brought about that accelerating cultural change in civilizations which seems almost to have reached an explosive point. Epigenesis was enhanced by increased gene mutability, by the development of chromosomes and sex assortment, by adaptive changes in individual characteristics (in themselves or as a richer array for the action of simple natural selection), by the invention of a nervous system and of highly differentiated or coded responses.
>
> Environment operates upon a system at all levels, differentially selecting for survival particular genes or gene arrays, cells and cell aggregates, organs and organ systems, and individuals and groups of various sizes. The environment operates not so much on the finished product as on the formative process. It supplies the physico-chemical milieu determining molecular changes, the electro-chemico-mechanical field guiding cellular changes, the neuro-chemico-mechanical influences modulating organ development, the material and biotic stimuli that guide the maturation of the individual, and the coded and meaning-laden signs and symbols which are added to these in the course of enculturating an individual into his group. The stresses applied by the environment determine the direction of development of the individual and the selection of the individual in the group. It may determine the adaptations of the body, the behavior of the individual, the norms of the culture, and, in general, the goals or values that guide the course of future change of a system.

And, from another angle:

> At each stage and at each level, the system or sub-system presents to the environment a structure which has at least some aspects of a template, and so can lead to the production of more of itself; and at least some aspects

of a program or set of operation rules, so that the kinds of responses it will make to certain situations are roughly indicated. The outcomes are never identical and never foreseeably deterministic, because the fine details of the particular template and program, even in identical twins, are not absolutely identical and, even more, because the environmental conditions to which these are exposed are never even roughly identical. Despite relative constancy in "beings," therefore, outcomes are always more variable, the exact one in each case depending on the particular, often chance, details of the individual-environment interaction. Clearly, the line between heredity and individual experience becomes vague indeed. A gene array is a template and a program; so is an engram [Gerard, 1960*b*].

Problem-Solving

All organisms "track" through life, attempting to bring their actual state into congruence with a "desired" state. The heedless ones correct for the present error, the mindful ones correct for the projected error. Heliotropic caterpillars eat upward on a shoot, whether long or short; cats pounce where the dashing rat is expected to be, not at where it is. The nervous system is the prime device for learning, and for learning to learn; it links organism to environment in the most sensitive and specific manner, via information flow; it is highly valued by organisms that possess it and most highly by those that possess it in the most developed state; [1] it is the organ, and supports the function, which most clearly, in evolutionary efflorescence, marks the directional change—I would still call it "progress"—from primitive and simple to advanced and complex in form and behavior. The evolution of neural structures and functions plays a central role in animal evolution; surely, behavior (in securing food, escaping predators, finding mates, calling for as-

[1] "Man's value judgment in ranking animals on a behavior scale is supported by a seeming value judgment of the body in cherishing its nervous system. This organ is wrapped in multiple membranes, floated in liquid (and the composition of CSF suggests a purely supportive function), and encased in bony armor—all giving maximal protection against mechanical damage. Carotid receptors help insure a constant supply of blood of proper composition at the portal of the brain, and the state of the bathing fluid is further under precise regulation from central receptors for osmotic pressure, temperature, carbon dioxide, etc., which supplement the peripheral regulators. Local vascular adjustments protect against oxygen excess or deficit. In addition, a special permeability barrier exists; so that the intercellular milieu of the neurones is doubly protected from outside perturbations. Only in the vicinity of the special chemoreceptor areas is the blood-brain barrier breached. Moreover, glia cells closely invest neurone elements and must contribute to the fine control of their environment; indeed, there may be essentially no true extracellular space. Finally, a fifth or more (a half in a five-year-old) of the resting metabolism of the body is allotted to a nervous system constituting, even in man, only a fiftieth of his weight; and this expensive organ is maintained relatively well through a starvation period, all other organs except the heart being used as fuel. The nervous system is indeed a well-buffered black box, protected from all inputs save those external and internal ones for which it is specifically coded" (Gerard, 1960a. Citations in the original are here omitted).

sistance) is more important to survival than is structure alone (thick fur, heavy armor, horns) or even physiology (high temperature, quick clotting, strong contraction); and the laying-down of engrams is central to the development of individual behavior. In both aspects of "becoming," the nervous system must star in the epic of life and change.

The "tracking" by organisms mentioned above is equally true for individuals and for species. For the former, the "error" is mostly the result of a transient environmental load and the attendant displacement (or stress) of some subsystem from its equilibrium ("desired") state; and the "correction" is a compensatory behavior; also mostly transient, but with learning, and involving homeostatic mechanisms, but with environmental adjustments. For the species the "error" is the result of the lag between secular change in the environment and adaptive ("desired") change in successive generations of descendants; the "solution" is evolutionary adjustment or innovation. In both cases, but involving vastly different times and mechanisms, a problem is posed by the environment and solved by the org, mostly by reversible adjustments of behaving but most significantly by the irreversible ones of becoming. Successful irreversible solutions involve enduring structural and functional alterations of individuals, singly or in populations.

Some of the key problems met and solved by the stream of the living seem clear. Almost before life could be called such, self-reproducing molecules had to exist. Autocatalysis and heterocatalysis are basic to the specific synthesis of growth and reproduction and to the dynamic equilibrium of metabolism and the turnover of substance and energy. Perhaps the first spectacular success of living things was the invention of chlorophyll, which tapped the largest omnipresent energy source. The equally important, but greatly variegated, oxidizing systems arose either earlier or later, depending on the actual succession of reducing and oxidizing atmospheres on earth and the time of biogenesis. Another molecular triumph was in gene mutability, which added to the stability of self-reproduction a needed plasticity. This variability was fully exploited by the spectacular device of mitosis-meiosis and of sex, the major accelerator of evolution that enabled taxa to "learn to learn."

Other innovations were associated with the invasion and exploitation of new habitats, especially the great emergence from ocean. Multicellularity, with the division of labor between organ systems, is involved here, and the goodly array of specializations appears. Integument to control desiccation and organs of intake and output to regulate the constant internal chemical environment; rigid material for support, and muscles to move the skeletal structure so available; systems for defense (mechanical and chemical), for interval co-ordination (circulatory and endocrine and neural), and for reproduction; and the information-handling neural receptor instruments all flourished.

For ever faster performance came warm-bloodedness, and the high metabolic rate which is both its antecedent and its consequent, made possible by the further molecular invention of oxygen-carrying hemoglobin. And for ever better-aimed performance came action plans, associated with a brain and variable behavior, and then imagery and symbolism and full-fledged foresight and purpose, associated with the cortex and its wonderful capacities. Finally, with growing multi-individuality, came epiorganisms (societies, ecosystems), with further division of labor and co-operative reintegration made possible by a new interval environment—culture—and with new integrating mechanisms, involving language (see Emerson, 1959; Gerard, 1956). In the evolution of the higher animal phyla, the nervous system and its functions are thus of dominant importance. This topic is my main one; but first some attention to the mechanisms for fixing experience.

Learning and Communication

Experience—racial evolution, social change, individual learning—is fixed in material traces at all levels, from molecules upward. Repetitive or cyclic production of the same entities is, of course, necessary to biological existence, but it is not sufficient for biological change. Gene duplication without gene mutants (which then continue the duplication mode) would allow no evolution at the molecular level and only limited reassortment above this. Cell division without (irreversible) differentiation would largely blot out ontogeny and, by obstructing phenotypic expression, also sharply constrict evolution. Stereotyped behavior and fixed views do not favor cultural development; the new idea, a sort of social mutant (Gerard, Kluckhohn, and Rapoport, 1956), is needed.

Fixing experience in the nervous system also involves many levels (for fuller details and references see Gerard, 1960*a, b*). Neurones reproduce their protoplasm rapidly, as fast as three times a day, and so manufacture nucleotides, proteins, and other components. Activity increases the turnover and can leave enduring changes in amount and, at present only by inference, kind of these molecules. The polynucleotides, especially, may well assemble their mononucleotide units in altered sequences in response to different patterns of message input from neurone to neurone or period to period—by transducer linkages that are only dimly even guessed at. Such a neural engram, a nucleotide change induced by factors external to the molecule and then held in further molecular duplications, would be highly congruent with the gene mutation, a sort of racial engram.

Cellular changes, from new molecules, are also involved in neural memories; as they are, via mitosis, in species change. A well-used reflex arc transmits more rapidly and easily across its synaptic hurdles than does a little-used one. This also is a memory residue and involves such quantitative shifts as the magnitude of the membrane potential (or the somatic potential), the size and number of presynaptic endings, perhaps the rate of formation and release of certain threshold-altering chemicals, and so on. Still further, there is growing evidence that engrams may result from the concurrent activity of groups of neurones, which thereafter are more likely to work together than before their chance linkage. Individual neurones may participate in many separate groups, thus being involved in many discrete memories at the organ level. And, of course, changes in receptors (thickened or inflamed skin and cutaneous sensations, drugs, etc.) and even more in effectors (muscle hypertrophy with use), as well as in the nervous system, alter the total individual's perception of and behavior in the environment, just as somatic as well as genetic changes alter the total individual's selection by the environment.

The residues of change at the level of taxa or ecosystem involve individual differentiation and interaction and, a special aspect of these, population genetics. Individuals may have become morphologically specialized, as in colonial insects or echinoderms, or functionally specialized for performing given roles, as in flock, pack, or other group leaders. Structural change commonly follows functional commitment—the cowboy's legs bow, the dancer's calves swell, the farmer's face weathers, and the office worker develops a slouch and pot belly. Indeed, tool use is now suggested as the trigger, rather than the result, of brain, hand, and other structural developments in hominid evolution (Washburn, 1959). There is mounting evidence for a similar sequence of evanescent memory and dynamic neurone activity, perhaps reverberating neurone circuits, during minutes or hours, before an irreversible material engram is formed. Here, again, the same tools are at the service of evolution, giving racial change, and of experience, altering the individual.

Interactions and the machinery of communication between individuals have also evolved, necessarily, with the evolution of epiorganisms; and this is explicitly dependent on advanced nervous systems. An animal emits a signal only to elicit some response to it. Bats use echoes of their own high-frequency sounds, as electric fish use distortions in their own high-voltage fields, for information about their material environment. But the vast majority of signals are intended to convey meaningful messages to other organisms as so to alter their behaviors. Not only has the larynx evolved to produce varied and loud sounds; no less has the involuntary grunt or cry of pain evolved to exploit this resource. The whole complex of sound-emitting and -detecting organs (and of those of smell; perhaps even of light, particularly in the ocean deeps) is mainly concerned with the biotic environment, not with the material surround. Some information is concerned with food and is not by intended signal, as in smelling or seeing some dead or fixed morsel; in hearing, smelling, or seeing some prey and attacking it; or in similarly detecting some predator and fleeing or freezing. But, beyond this and increasingly so with the evolving social groups, is deliberate communication for mutual benefit.

At least four areas are involved: the warning of danger, as the starling call; the guidance to food or other good, as the bee's pollen dance or the deer's tail flick to the faun; the call for help, as the cry of pain or anguish of more advanced vertebrates or as the kitten's meow; and the invitation to mate, the sounds and smells and gyrations widely emitted by animals. The studies of ethologists are exhibiting a rich array of detailed cues and responses that guide animal behaviors, most of them involving specific, rather than generalized, communication and the transfer of coded information. Whether inborn or imprinted or imparted, such coded transmission involves particular neural patterns in both sender and receiver—either inherited structural engrams or those laid down by experience. And the greater the degree of individual rather than of racial learning in the code, the more complex and malleable is the required nervous system.

Imprinting and true social teaching and learning, which demand some symbolism, are now established for a number of birds and, especially taught behavior, of mammals; so the subhuman rudiments of language exist. But it remains true that the full, rich, coded transfer of percepts and concepts, the evocation of feelings and thoughts, by symbols—the phonemes and morphemes and the ideograms or alphabets of language—is the unique privilege of man. For this capacity of behavior, there are required extra refinements of the nervous system. What these may be, and the order of their appearance, now deserves attention (see Gerard, 1959, for fuller discussion).

Neural Evolution

All organisms interact with their environment, if only at the level of chemical exchange, as they also all maintain internal co-ordination, exhibit metabolic flow and equilibrium, and perform growth and reproduction. But, whereas the chemical traffic and controls of metabolism and the machinery of reproduction are closely similar from microbe to man, the capacities and mechanisms of experience and behavior have gained enormously. Sensory thresholds have fallen, by a million, million fold, and sensory range and discrimination have risen; and responses have similarly vastly improved in speed and power and discreteness. These improvements were achieved primarily at the cellular level, by better receptor and effector elements, and were largely or entirely completed with the appearance of arthropods and vertebrates. Even the basic conducting nerve fiber, transmitting synapse, and integrating neurone had reached their asymptote in the lowest vertebrate, with fast all-or-none conduction, irreciprocal transmission, and reflex facilitation and inhibition. Further gains, mainly in the patterning and generalization of experience and in the variability and modifiability of behavior, came from increased numbers of elements—the neurones of the central nervous system.

As given transistors, capacitances, and resistors can be wired with low or high sophistication to form a radio with inferior or superior performance, so can neurones be interconnected for better or worse. Many factors are already recognized (see Gerard, 1960*a*) and include the following: the presence of parallel paths, giving redundancy and reducing error; the associated convergence, or funneling, from larger input channels to smaller output ones, favoring interaction and selection; the superposition of higher-level loops on local ones, with greater centralization and integration; the appropriate interplay of excitatory and inhibitory components to modulate output and insure that incompatible responses do not blend but rather succeed one another as functional wholes; the closed-loop neurone chains that permit activity to reverberate and so gain temporal dissociation of output from input; the feedback loops that serve as volume controls throughout the nervous system—on input, output, and thruput or inner activity; the synchronization of neurone rhythms, involving threshold level and discharge rate, brought about more by field effects, such as steady potentials, than by nerve impulses; the rhythms themselves and other spontaneous, irregular variations in the physiological parameters of neurones; the growing ease of fixing changes induced by activity and the consequent learning; the organization of neurone sheets that favors regulated traveling waves that interact to give engrams and conditioning and recall; a temporary information-storage device that allows grouping and sorting of different kinds of bits before feeding into attention and fixing. These and like inventions and improvements in circuitry have enhanced the performance of advanced nervous systems. But this is still insufficient.

The comparative anatomy and physiology of such neural properties are insufficiently known to permit solid generalizations; but it seems highly probable that these advances were essentially complete with the arrival of modern mammals. The further gains in capacity, seen most strikingly in the primate line and culminating in man, are due to simple increase in numbers rather than to improvement in units or patterns. That increasing brain size parallels richer performance, even for particular regions and functions (e.g., tongue motor area and speech), is a commonplace (see Spuhler, 1959); how this operates is less clear. Sheer increase in number, without secondary specification (which does also occur), might seem unable to generate new capacities but only to intensify old ones; but this is not the case. Examples have been given elsewhere (Gerard, 1959*b*, 1960*a*) of qualitative novelty with quantitative gain: at the genetic level, chromosome reassortment becomes possible only when two or more pairs are present. In the brain, an increase in the anatomical neurone population raises the limit on the physiological neurone reserve and so allows greater variety of selection and greater richness of analyses and combination, expressed in modifiable and insightful behavior. The relation of decreased physiological reserve to the behavioral changes of age, anxiety, and schizophrenia, as well as its normal role in thought and information handling, has been discussed in other articles (see Gerard, 1960*a*, and references).

Epigenesis

Learning to learn, accelerating evolution, the rise of modifiability, the epigenetic building of the new upon the old—this attribute has characterized living organisms from their inception. If genes and mutation were the great inventions for speeding change at the molecular level, and gamete formation and fusion at the cellular level, so were the nervous system and the engram at the individual level, and language and other information flow—and the cerebral mechanisms that mediate this—at the social level. Learning by insight or by instruction increases vastly in the neurally better-endowed animals, as does speed of acquisition and fulness and duration of retention, in general. (In relatively simple situations, a monkey may do less well than a rat—or an expert professional less well than a naïve subject—perhaps from worrying about too many trees!) The mechanisms for fixation of experience are presumably more powerful: synapse change, altered dendritic potentials (the entire non-specific afferent system—associated with conscious level, attention, affect, and the like—may affect learning and retention via this mechanism), modified neurone assemblies, and other factors in engram formation have appeared or improved during neural evolution.

Excessive ease and firmness of fixation might leave behavior too rote and too entirely at the caprice of the current environment, as in imprinting. Besides increase in passive receptivity, there is an increase in innovative and selective capacity, probably related to the rhythmic and random threshold changes of single neurones and to the increased numbers and combinations of all neural elements. Neurone groups or assemblies can be formed for each component of a perception or an act, and these can be combined in multiple ways to give unlimited patterns of wider integration—the ideas and plans of high-level functioning. So few kinds of letters, with a great quantity of each kind, form many words and sentences, each in lesser quantity, and these form longer passages, often unique but of limitless variety. How imagination is fed by recombinations of the known and how action choices become modified as experience changes the probabilities of various outcomes are but dimly understood as mechanisms, though well established as phenomena.

Such capacities improve individual functioning, including adjustment to and increasing control of the environment. No less do they open wide the way to richer and more effective collective functioning. With language, and the rich conceptual clarification and communication it permits and generates, the minds of men interact in great numbers—as do the neurones of the brain. Differentiation of roles and enhancement of skills follow; and co-operating experts can then create new techniques and develop new knowledge. Collectively generated instruments have extended man's receptors (scopes of all sorts) and effectors (building machinery wave emittors) and are rapidly supplementing his mental processes (computers), even at the level of learning to learn. The pursuit of science has increased exponentially from its start three centuries ago (Price, 1956). With such a premium, even by the minimal criterion of survival, on differentiation and reintegration, it is not surprising that co-operation also appeared early in evolution and accelerated in impact. From interactions of molecules and cells to co-operation of individuals and altruism in family and social groups, working together paid off and was selected as fitting.

The resulting societies and related multi-individual epiorganisms, evolving from multicellular organisms, have deployed widely and rapidly. The more developed ones have increased their territory and influence, and many species have disappeared along the way. The survivors are great and complex, with adjoining territories and different goals. It may be too much to expect that biologically identical or similar groups will remain indefinitely in ecological balance; if not, a world state may encompass all men. Then, as present life precludes the origin of new life from the non-living, new species of society may arise only as do new species of organisms. The factors making for change—mutation or innovation, drift, migration or diffusion, and selection—will then operate sluggishly and "becoming" will slow, if not halt.

We honor the centenary of Darwin's masterpiece. This is proper tribute to the greatest pinnacle of biology. But other men have risen toward his height, and it is the entire range of such peaks of individual achievement that has given us the elevated, yet substantial, view available in the mid-twentieth century. It is appropriate that Darwin's unifying theme, of cumulative and enduring change dependent on system-environment interaction, has integrated knowledge from such disparate fields as molecular biology, cytology, breeding, systematics, and sociology and has fed insights back to these and others. Here is a prime example of collective functioning at the epiorganismic level and of the social learning to learn.

References

DOBZHANSKY, T. 1959. "Variation and Evolution," *Proc. Amer. Phil. Soc.,* CIII, 252–63.

EMERSON, A. E. 1959. "Dynamic Homeostosis: A Unifying Principle in Organic, Social, and Ethical Evolution," *Scient. Monthly,* LXXVIII, 67–84.

GERARD, R. W. 1954. "Experiments in Microevolution," *Science,* CXX, 727–32.

———. 1956. "A Biologist's View of Society." In *General Systems Yearbook,* edited by L. VON BERTELANFY and A. RAPOPORT, I, 155–62.

———. 1959*a*. "Brains and Behavior." In *The Evolution of Man's Capacity for Culture,* edited by J. N. SPUHLER, pp. 14–20. Detroit: Wayne State University Press.

———. 1959*b*. "Neurophysiology: Brain and Behavior." In *American Handbook of Psychiatry,* edited by S. ARIETI, II, 1620–38. New York: Basic Books, Inc.

———. 1960*a*. "Neurophysiology: An Integration." In *Handbook of Physiology,* Section 1: *Neurophysiology,* Vol I (in press).

———. 1960*b*. "The Fixing of Experience." In *Brain Mechanisms and Learning,* edited by DELAFRESNEYE (in press).

GERARD, R. W. KLUCKHOHN, C., and RAPOPORT, A. 1956. "Biological and Cultural Evolution," *Behavioral Sci.,* I, 6–34.

PRICE, D. J. 1956. "The Exponential Curve of Science," *Discovery,* XVII, 240–43.

SPUHLER, J. N. 1959. "Somatic Paths to Culture." In *The Evolution of*

Man's Capacity for Culture, edited by J. N. SPUHLER, pp. 1–14. Detroit: Wayne State University Press.

WADDINGTON, C. H. 1957. *The Strategy of the Genes.* New York: Macmillan Co.

WASHBURN, S. T. 1959. "Speculations on the Interrelations of the History of Tools and of Biological Evolution." In *The Evolution of Man's Capacity for Culture,* edited by J. N. SPUHLER, pp. 20–31. Detroit: Wayne State University Press.

ERNEST R. HILGARD

PSYCHOLOGY AFTER DARWIN

Men who are judged truly great for their contribution to the history of ideas seldom belong to one discipline alone; particularly as time goes on they become recognized for their contributions to several fields outside their own. While Darwin would not have thought of himself as a psychologist, much that he wrote would today be accepted as psychology—partly because he changed the direction in which psychology was moving.

The psychological climate for an evolutionary doctrine was prepared by Herbert Spencer, whose *Principles of Psychology* appeared in 1855, four years before the *Origin of Species.* Spencer had a broad scheme of evolution which helped keep psychology in a biological setting and emphasized the adaptation of the organism to its special environment. While the flavor was Lamarckian, Spencer's views were quoted by Darwin in support of his own opinions respecting the inheritance of acquired modifications. Darwin also quoted Spencer favorably on the notion of an equilibrium between the forces of action and reaction in nature. Darwin stood right at the threshold of psychology's efforts to become empirical, and his own empiricism provided a marked impetus for these efforts.

In estimating Darwin's influence upon psychology, and in attempting to look ahead as to where psychology is going, I wish first of all to recognize Darwin's influence upon empirical developments within specified topics of psychological science, and then to consider how he has influenced the current trends of psychology.

Darwin's Influence on Modern Psychology

Three topics, or "chapters," of modern psychology stand out as tributes to the direct influence of Darwin: comparative psychology,

ERNEST R. HILGARD is Professor of Psychology and Education at Stanford University and Associate Director of the Laboratory of Human Development, where he is currently directing a research program concerned with human motivation, psychodynamics, and hypnosis. He is a past president of the American Psychological Association (1948–49) and author of two widely used texts: *Conditioning and Learning,* with D. G. Marquis (New York: Appleton-Century, 1940); and *Theories of Learning* (2d ed.; New York: Appleton-Century-Crofts, 1956).

the study of emotional expression, and the study of individual differences among men.

COMPARATIVE PSYCHOLOGY

Darwin's book *The Descent of Man* (1871) is essentially a comparative psychology. It attempts to show, among other things, that there are but very slight *qualitative* differences between the mental activities of man and the lower animals. Two chapters are devoted specifically to a comparison of the mental powers of man and the lower animals, including such topics as curiosity, attention, memory, imagination, reason, moral sense, sociability.

The evolutionary point of view that there is a continuity between lower animals and man led to divergent tendencies in interpretation. One interpretation was that we must gain increased respect for the intelligence of animals. (We know how clever man is; because there is this close relationship to man, animals are probably more clever than we have heretofore thought them to be.) This tendency to raise the level of animal intelligence was found particularly in Romanes (1882) who collected many anecdotes about rather remarkable animal performances.

The opposite point of view also gained some currency. (Let us not be anthropomorphic; we are little better than our near neighbors; perhaps much that we have thought of as rational conduct in man is merely the operation of conditioned reflexes similar to those in lower animals.) Thus we can close the gap between man and animals in one of two ways: raise up the animal or lower man. Lloyd Morgan, in his reaction to the excesses of romanticizing the animal mind, coined his equivalent of Occam's Razor, known to psychologists as "Morgan's Canon" (1894): "In no case may we interpret an action as the outcome of the exercise of a higher psychical faculty if it can be interpreted as the outcome of the exercise of one which stands lower in the psychological scale."

It was Lloyd Morgan who more than any other single person established among psychologists the evolutionary point of view in the empirical study of animal behavior. Morgan's books were widely read (e.g., *Animal Life and Intelligence,* 1891; *An Introduction to Comparative Psychology,* 1894), and in the 1890's he lectured at Clark University and at Harvard, encouraging studies in those centers. The early work of Thorndike (*Animal Intelligence,* 1898) was undertaken at Harvard, and the introduction of the rat-maze experiment was at Clark by Small in 1899. Harvard's laboratory came under the direction of Hugo Münsterberg in 1892. Münsterberg was much impressed by evolution. The story is told that he kept his fingernails filed square at their tips so that when he drummed on the desk the relationship between man's nails and a horse's hoof would not be overlooked. It was at Harvard that Robert Yerkes began his animal experiments that ranged widely over the animal kingdom, from earthworms through turtles, frogs, pigs, monkeys, and man. And, of course, it was out of comparative psychology that John B. Watson's *behaviorism* emerged—America's most distinctive contribution to psychological theory. Behaviorism went all the way in applying Morgan's Canon to man and made it fashionable for a time to bridge the gap between animals and man by refusing to use the concept of mind in dealing with man (Waters, 1939). The tendency still exists, among some of those whose experimentation has been confined to lower animals, to play down any uniqueness in man's capacities, for example, in the things that he can learn (Harlow, 1958).

The province of comparative studies.—I shall not attempt to outline comparative psychology as it has developed and flourished. There are two major interpretations of its province. The *first* of these is that the purpose of the study of animal behavior is to do for behavior what comparative morphology does for structure, that is, to examine behavior from an evolutionary point of view, to describe it in detail, and to point out similarities and differences that will permit a historical and functional understanding of heredity, development, variation, adaptation, and species differentiation. While there are many who would accept this statement as an ideal, comparative psychologists with few exceptions have not taken the evolutionary task very seriously, especially in its historical aspects, and have contented themselves with studying a few species in considerable detail. This is in part because of the *second* interpretation of the province of comparative psychology: using animal subjects to study topics of psychological interest primarily because animals often make convenient subjects. This interpretation is similar to that which leads to the use of animal subjects in nutritional investigations, or in the study of disease. Just as the guinea pig is a small animal with some of the characteristics of a cow (for nutritional purposes) it is also an animal useful to psychologists in the study of audition because of the convenient anatomy of its ear. The ear is enough like a human ear to make electrophysiological investigation useful in determining how the human ear works, and the evolutionary problem does not have to be faced any more concretely than this. It does not learn as quickly as a rat; hence, when the psychologist does a learning experiment he is more likely to use a rat, without necessarily asking about the relative position of rat and guinea pig on an evolutionary scale. I am inclined to believe that the term "comparative psychology" should be confined to studies carried out in evolutionary spirit, but contemporary practice does not distinguish sharply between those who use animals in the one way or the other.

The concept of instinct.—Darwin used the word "instinct" freely. It was quickly picked up within psychology, but during this century has had a peculiar history of recurrent popularity and taboo. I cannot leave the topic of comparative psychology without saying a few words about it.

It is common to say that instinct is such a confusing and poorly defined concept that it is better to drop it (Beach, 1955). This began in the anti-instinct period of the early 1920's, in which American psychologists (e.g., Dunlap, 1919–1920; Kuo, 1921) and sociologists (e.g., Bernard, 1924) became antagonistic to McDougall's instinct psychology (McDougall, 1908, 1923). Bernard was able to find some 894 uses of the term "instinct" reducible at best to 325 uses. When an argument of this kind arises, semantic considerations often take the blame for many other aspects of the controversy. The objection to McDougall was as much to his vitalistic and hereditarian doctrines as to his definition of instinct; the battle against instincts came to be carried by the behaviorists and environmentalists of the time.

I wish to defy the common assertion that instinct is poorly defined by asserting that I believe it to be as well defined (when used carefully) as any of the terms that compete with it, such as tendency, disposition, drive, maturation, motivation, need. Of course these are ambiguities in definition, but I do not believe that these ambiguities are the root of the trouble. It is always possible to undercut any definition by pointing to marginal cases. For example, I have tried for some years to define "learning," but I have stopped trying. You and I know what I mean when I pick clear examples, such as learning to operate a typewriter, or memorizing a poem. But how about the blacksmith's apprentice whose biceps become stronger as he learns to wield his hammer? Where does learning stop and growth begin? How about the typist, tired at the end of the day? Are her errors a sign that she has forgotten to type? How do we distinguish between fatigue and forgetting? We really do not have any great trouble because we find it hard to distinguish by definition between what is learned and what shows up in the actual performances from which learning is inferred. The issue is a little more clouded when it comes to problem-solving, for some problem-solving does not involve any new learning at all, depending merely upon the utilization of learned habits, while other kinds of problem-solving require inventive learning.

I believe the issue to be similar with respect to instinct. The clear illustrations of instinct are easily pointed out. The behavior of a hive of bees that is characteristic of that species of bee, building the web characteristic of the spider, building the nest characteristic of the bird —that is what we mean by instinctive behavior. It is characterized by behavior specific to the species, a behavior that appears in its essentials with a minimum of learning. That it employs past learning or can be improved by learning, or that there will be some improvising if ordinary materials are lacking (e.g. string substituted in nest-building by chipping sparrows that formerly used horsehair), is really immaterial to the definition. It is not essential that instinct be 100 per cent innate any more than that learning be 100 per cent acquired.

There are difficulties with the concept of instinct, but I am tired of hearing that the difficulty is one of definition. The difficulty rests primarily upon the use of the term as an explanation rather than as a mere label for a class of behavior. We do not explain any behavior by calling it instinctive, but by so naming it we do call attention to some of the problems involved. There is nothing circular in saying that we know a given species of pigeon has a homing instinct because its members come home, for not all species show this behavior. Of course, we could talk about homing without talking about instinct; we could also talk about words typed per minute without talking about learning, if we had an aversion to the use of the word learning for something so complex (and mysterious) as the smooth operation of a typewriter.

Following the anti-instinct period of the 1920's we tried to get along by dividing the topic of instinct into two new topics, which psychologists called *motivation* and *maturation*. The new terms were helpful in that they called attention to sets of experimental problems that did not have in them all the controversies that had grown up around instincts. Motivation research became primarily the study of *drives* in animals: the effects of depriving an animal of food or water or sex, or, in the case of females, of their young. When a study is made of the amount of electric shock an animal will suffer in order to reach a sex object, or in order to retrieve young, we are studying the strength of what was formerly included in "instinct" without raising the question of whether or not the behavior is "instinctive." *Maturation* became the study of sequences of behavior in the development of the unborn or immature organism and the effects of environmental changes on such development. We stopped asking the futile question: Is walking instinctive? Instead the question became: At what age does the infant shift from crawling to walking, and what are the in-between stages?

There was a certain vigor to this development, and many new observations were made because of the new framework used in the studies. But there was some loss, too, especially because the emphasis on quantitative laboratory study led away from naturalistic observation. The *new naturalism,* represented in this symposium by Dr. Tinbergen, brought instinct back into the foreground, but attacks on the word are again mounting, and the ethologists are now using the word somewhat less (e.g. Hinde and Tinbergen, 1958). From my point of view it is of little moment whether or not they use the word; their correctives to American comparative psychology have been very welcome.

So much for comparative psychology.

THE STUDY OF EMOTIONAL EXPRESSION

Emotional expression is well within the province of psychology, so that Darwin's book *Expression of the Emotions in Man and Animals* (1872) is clearly a contribution to the literature of psychology. It has been accepted as such, and his three principles of interpretation are commonly cited in our contemporary textbooks. These three principles bear repetition. They are (1) the principle of serviceable associated movements, (2) the principle of antithesis, and (3) the principle of direct action of the nervous system.

Classifying expressions.—The first of these accepts emotional expression as the residual of behavior once appropriate, now used by analogy, as when we bare our teeth in anger though we have no intention to bite, or when we purse our lips in disgust, although we do not spit anything from our mouths. This principle is more widely accepted than the second one, which sees opposite emotions expressed by opposite actions. Thus a gesture of distrust is that of clenched fist with arms held close to the body (as in defense against attack), while the gesture of welcome will be the defenseless opposite—outstretched arms. The corners of the mouth are turned up in happiness, down in sadness, etc. It is now considered somewhat far-fetched to look for all these opposites; too often opposite reactions may mean the same thing, as in either blanching or getting red-faced in anger.

The third point—action depending on the way that the nervous system is made—shows Darwin's modesty and honesty. His followers try to see adaptive significance in all behavior: he is willing to accept what he finds. If behavior sometimes seems just disorganized in emotion, then he accepts that as one form of emotional expression.

For psychologists to quote Darwin is one thing; it is another to follow his leads and to make some progress in the description of emotional expression. The field has, in fact, been a bit dreary. Some success has been achieved in the classification of human facial expression in emotion, yielding a surface on which expressions can be plotted in the two dimensions of pleasantness-unpleasantness and, at right angles, attention-rejection. Thus the diagonals include contempt as relatively pleasant but rejecting, and surprise as pleasant and attentive, while opposite to them lie fear as unpleasant and attentive, and disgust as unpleasant and rejecting. A third dimension is provided by the intensity of feeling (Schlosberg, 1954).

Theories of emotion.—The physiology of emotions has coursed from the famous James-Lange theory, announced in 1884 and 1885, through the criticisms and emendations of this theory by Walter B. Cannon, Bard, and others (e.g., Cannon, 1927). The issue of the James-Lange theory—what it is that constitutes the subjective "feel" of the emotion, and how this is timed in relation to the physiological state—is a relatively sterile one, and it is little surprising that the issue never gets settled.

In two little-known papers, John Dewey set out to harmonize Darwin's views with the James-Lange theory. Dewey was much impressed by Darwin, perhaps by way of G. Stanley Hall, under whom he had studied, and William James, whom he admired. The two papers, parts I and II of a total called "The Theory of Emotion," appeared in the *Psychological Review* in 1894 and in 1895. They present a conflict theory of emotion. The bear (of whom James said we were afraid because we ran) is again brought into the picture. According to Dewey we are afraid, not because we run, but because we are in conflict as to what to do—whether to keep an eye on the bear to see whether or not he is coming after us, or to give full energy to running away. It is out of the uncertainty, the confusion, the paralysis, that the fear comes. This is, in fact, a very appealing theory, and features of it appear from time to time in the descriptions of anxiety by later writers. But for some reason or other the theory did not "catch on." There are few mentions of it in the later textbooks, although two psychologists chose to recount its virtues in serious articles in 1927 and 1928—over thirty years after it was written (Angier, 1927; Howard, 1928). I regret to say that little more has been heard in the additional thirty years since they attempted to awaken interest in it.

It is difficult to guess why some theories gain a great deal of attention and others do not, when in retrospect one seems as inviting as the other. Certainly Dewey's style has none of the felicity of James's, but this has not prevented attention to other Dewey writings. I would venture a guess that may be wide of its mark but is nevertheless consistent with other tendencies in assimilating what a writer stands for: Dewey was well known for a conflict theory of consciousness that was clearly consistent with an evolutionary interpretation of consciousness as adaptive. When we meet a situation that we cannot handle by smooth-running habit, he said, consciousness becomes vivid; we reconstitute the situation cognitively and then proceed to make an adaptive response. Hence his idea of conscious awareness fit neatly into a scheme of adaptive behavior. The emotion theory did not fit so neatly. Those who were fondest of quoting Dewey were fond of finding all behavior adaptive. When Dewey then presented a theory of emotion as arising out of conflict but not leading to anything but chaotic disorganization, they believed that they had not understood him. Surely the emotional shock must be good for something! It is these same readers who forget that Darwin's third principle of emotional expression was exactly of this kind: the direct discharge of the nervous system, unexplained by any special adaptive significance. Readers forget those parts of Darwin that are inconsistent with their understanding of what Darwin stood for: I am inclined to believe that they dismissed Dewey's theory of emotion because they could not fit it into the pattern of what they thought Dewey stood for.

The theory does, in fact, supplement Darwin and James in rather sensible ways. Darwin's theory of serviceable associated movements is reinterpreted to mean that the original movements, now mimicked in emotion, served the adaptive purposes of the organism, but not necessarily as earlier expressions of emotion. Thus the act of spitting a bitter substance out of the mouth was not necessarily accompanied originally by an emotion of disgust, as Darwin seemed to believe; it could be reflex action, pure and simple, according to Dewey. Only when there was some conflict would the emotion arise; one aspect of the conflict might very well be a minimal tendency to spit something out of the mouth; thus Darwin's origin of the expressive movement might be right, although the origin of the emotion as such might be different. The variation on James was that perception of the bear is not neutral until after the bodily discharge takes place, but rather that the constitution of the bear as a fearful object takes place at the same time as the bodily discharge.

The evolutionary emphasis in emotion theory came back more pronouncedly with Cannon's "emergency theory" of the action of the sympathetic system in its general antagonism to the parasympathetic system (Cannon, 1915). Some of the old problems of distinguishing between fear and anger have taken on new interest since the discovery that in addition to adrenalin there is noradrenalin, and some animals appear to use one of these more than the other. The timid ones that run when frightened, such as rabbits, show a predominance of adrenalin in their adrenal medullary secretions. Those that are more aggressive, such as lions, are said to show relatively high amounts of noradrenalin (Funkenstein, 1955). The temptation is strong to see some kind of selection taking place on the basis of principles of survival appropriate to the differentiated roles of the various species in their environments.

The advances in the anatomy and physiology of the limbic system and of the reticular formation yield a number of suggestions bearing on emotional activity, but this is not the place to review these.

I have given enough to show that Darwin has a definite place in the history of the study of emotion. To my knowledge, some of his work has not yet been followed up, and some of the questions he raised are still to be answered. For example, the mechanism of tear production in the young infant appears not yet to be satisfactorily explained, specifically the fact that the earliest crying is without tears—an observation that Darwin was careful to study and check.

INDIVIDUAL DIFFERENCES

Darwin was greatly concerned with the problems of heredity, as his theory demanded; his studies of hybridization as reported in *The Variation of Animals and Plants under Domestication* (1868) represent the closest he came to laboratory methods. This direction, as it applied to man, was picked up immediately by Francis Galton, who soon published his *Hereditary Genius* (1869), to be followed a few years later by his *Inquiries into Human Faculty* (1883). A number of references to Galton's *Hereditary Genius* appear in Darwin's *The Descent of Man* (1871).

The family history method as used by Galton is now considered a little unrefined in that it paid little attention to the environmental influences that would give sons of successful men access to opportunities lacking to equally able sons of the less successful. The fact that among the few presidents of the United States there have been two Adamses, two Harrisons, and two Roosevelts is hardly to be accounted for on the basis of genes alone.

The importance of Galton lies in his recognition and systematic study of the variations among men, so that, following him, individual differences could no longer be neglected. The subsequent development of intelligence tests and other measures of individual differences is well known. The continuity between Galton and modern psychology is recognized in the coefficient or correlation, which he invented and which was refined by his follower Karl Pearson, and in the methods of factor analysis, which we owe, in the first place, to Charles Spearman, who in turn owed much to Galton and Pearson.

The heredity-environment controversy.—One of the basic problems with which Galton began is still a source of study and of controversy within psychology. This is the issue over the relative contributions of heredity and of environment to intelligence, personality, and other differences. He gave these studies their name—studies of "nature versus nurture."

Modern genetics has taught us that identical twins are ideal subjects for the study of human inheritance, because with identical genetic constitutions all differences must be non-genetic in origin, either congenital or postnatal. Hence the study of twins reared in unlike environments furnishes a useful source of data. The best-known study is that of Newman, Freeman and Holzinger (1937). They found that identical twins reared in unlike environments tended to be somewhat less alike in intelligence than twins reared in the same homes, although identical twins reared apart resembled each other about as much as fraternal twins reared together. The number of subjects is too few and the measures of environment too crude for any very exact determination of the relative contribution of heredity and environment (Woodworth, 1941).

The fact that there is *some* contribution from environment and *some* contribution from heredity seems indisputable, but because there are remaining areas of quantitative disagreement, there is room for the expression of preference and prejudice.

Means and correlation coefficients.—How data can be selected to favor one's own position is well illustrated by those who are "environmentalists" on this issue and those who are "hereditarians" by preference. The environmentalists, in *reporting* the effect of home environment on the intelligence of foster children, stress differences in *mean* performance. They can then show that the mean intelligence level of foster children tends to be higher than would be predicted from the intellectual level of their true parents. This argues, of course, for the favorable results of being adopted into the environment of homes selected as good homes by placement agencies. The hereditarians, in analyzing the data from exactly the same studies, choose not means but *correlation coefficients*. If correlation coefficients alone are studied, it will be found that foster children show higher correlations with the true parents *with whom they have not lived* than with the foster parents who have provided their favorable environments. Thus the same data can give support to the environmentalists if *means* are studied and to hereditarians if *correlations* are studied! This interesting thing, from the point of view of the social history of science, is that the authors who stress the means often do not even bother to report the relevant correlation coefficients, and vice versa. The puzzling problem of finding some sort of model that will take into systematic account both kinds of changes at once has not been solved.

Lest what I have just said sound too paradoxical, consider only what has been happening to the height of American children relative to the heights of their parents. The *mean height* of children has been going up steadily, regularly surpassing that of their parents. This we tend to associate with diet, vitamins, and other improvements in environment. At the same time, the *correlation* between the heights of parents and children has not changed from Galton's day to ours: tall parents still have tall children; short parents still have short children (with the exceptions expected from low correlations). Hence the *hereditary* basis of height represented by correlations is not challenged by the *environmental* basis suggested for changes in the mean height of the next generation.

Stated in this way, there appears no mystery whatever in the inconsistencies in different results between means and correlation coefficients. But when one wishes to state the relative contributions of heredity and environment to height there is a serious problem created. It may be that the question of relative contribution is an improper one, especially if we try to answer it from the measurements of human beings growing up in a society as varied as ours. Consider what the problem would be if one produced a very heterogeneous population of rats by first breeding for some strains of different size on the same diet, and then confused the phenotypes by imposing another set of size differences through differential feeding in infancy. If mating were now according to some pattern, possibly selecting for phenotypical size, and the offspring were fed the diet that was used with their parents (or some compromise when the parents had differing diets), a very heterogeneous population would result. Out of these populations what would be the chance of unraveling the relative importance of diet and of heredity in the determination of size? One can conceive of setting up a model to work backwards from if one knew what had gone into this experiment, but in the human situation we are given data like this without very much information about what went in originally.

The method of identical twin studies has been used also in the assessment of the hereditary basis for mental disease, but in these studies there appears to be some sway of prejudice as well. While the impression is abroad that a number of twin pairs have been reared apart and subsequently became schizophrenic, only two pairs have actually been so studied, neither pair in contrasting environments (Jackson). In one case, for example, a girl was raised in the home of a bachelor brother of her schizophrenic mother, her twin raised in the bachelor home of another brother of this same schizophrenic mother. The possibility of some common environmental pressures upon these sensitive girls is surely not ruled out merely because they are separated identical twins.

I have paid as much attention as I have to unfinished business in this field because it is important to recognize that there is always a growing edge to science, and the relative quiescence of the nurture-nature problem in psychology today does not mean that the issues have been settled.

With this brief review of comparative psychology, the psychology of the emotions, and some persistent problems of accounting for individual differences, I have tried to show the contemporary influence of Darwin on some of the special topics within psychology.

Functional Psychology: A Psychology of Adaptive Response

In the early years of the then "new" psychology in America, two "schools" came to be contrasted: the *structuralist* school of Titchener, at Cornell, and the *functionalist* school of Angell at Chicago.[1] While functionalism as a school with a recognized leader and loyal adherents disappeared when behaviorism took over, in some ways the functional viewpoint has remained (with some changes) the most characteristic outlook of American psychologists. One need only mention such names as Dewey, Angell, Judd, Thorndike, Woodworth, Yerkes, Carr, Robinson, Thurstone: all influential writers who never embraced behaviorism even though they were active in behaviorism's heyday. Some men who went along and called themselves behaviorists, such as Lashley and Tolman, remained very close to functionalism. Related viewpoints have found expression in Europe through men such as Bartlett in England and Bühler and Brunswik in Vienna, the latter two later coming to America, although their positions were established abroad. Brunswik came to describe his position as a "probabilistic functionalism" (Brunswik, 1955).

That functionalism of sorts is the characteristic American psychology is acknowledged by both Woodworth (1948) and Boring (1950), senior psychologists whose careers have covered the period when controversies among the "schools" were most vehement.

Boring, whose own early experience was in Titchener's camp, has this to say in assessing American psychology: "It had inherited its physical body from German experimentalism, but it had got its mind from Darwin. American psychology was to deal with mind in use." (1950, p. 506).

Or again: "The apparatus was Wundt's, but the inspiration was Galton's." (1950, p. 507).

The history is clear enough. There was William James, whose two-volume *Principles of Psychology* of 1890 is a classic and still a joy to

[1] Note that "functional" in psychology contrasts with "structural." Among biologists the contrast is often between "functional" and "historical." Functionalism in psychology is *both* functional and historical in the biologist's sense.

read. James set the tone for a functional psychology. It is to be noted here that he was an enthusiastic Darwinian and took up the cudgel against the inheritance of acquired characters in favor of chance variation and selection. One of James' students was G. Stanley Hall, another enthusiastic evolutionist, who became the teacher of John Dewey at Hopkins. Dewey, much under James' influence, gets the credit for officially establishing functional psychology at the University of Chicago, where it was carried on under James Rowland Angell, and later under Harvey Carr. This is the historically recognizable functionalism, but there were many other relatives such as Ladd and Scripture at Yale, followed by Judd, both there and at Chicago; James Mark Baldwin, at Toronto, Princeton, and Hopkins; Cattell, Thorndike, and Woodworth at Columbia. The official interest in Chicago functionalism lessened when John B. Watson, a Chicago Ph.D. of 1903, announced behaviorism in 1913. But what I propose to talk about is the unofficial residue that is alive in contemporary psychology.

There are many meanings of the word functionalism in psychology. I do not intend to go through a dull catalog of all of them, but I prefer instead to delineate three meanings that have both historical and contemporary significance.

CONSCIOUSNESS

In the days when psychology was the science of consciousness, the functionalist tried to discover the survival value of consciousness. Hence the first meaning of functional psychology was *an interpretation of consciousness according to the function that consciousness serves* in the adaptive economy of the organism.

Here the evolutionary emphasis was clear: unless consciousness had some sort of survival value, there would be no consciousness. William James (1890) said: "Consciousness . . . has in all probability been evolved, like other functions, for a use—it is to the highest degree improbable *a priori* that it should have no use." When all runs smoothly, habit takes over, and consciousness lapses; only when there are obstacles or difficulties is consciousness intense. This view was elaborated by John Dewey, who put it that consciousness arises out of conflict, and it became a basic tenet of the early functional psychology associated with James Rowland Angell.

The problem of consciousness no longer has the central role in psychology that it once had, thanks to John Watson's vigorous insistence that you could write all of psychology without it. Whether or not you can write *all* of psychology without mentioning consciousness is less important than that you can write *much* of it without it;

as Boring has well pointed out, with modern operational logic one can handle problems of consciousness as problems of behavior, and not worry about the ontological status of awareness (Boring, 1950, p. 658). The *existence* of consciousness is a pseudoproblem, but the *efficacy* of conscious awareness, as reported, for example, by the problem-solver, can enter into laboratory data. Thus the essence of the functionalist's position that consciousness, or what is attributed to it, redirects perception and helps resolve conflict is still valid.

BEHAVIORAL FUNCTION; ADAPTATION

In the remaining definitions of functionalism we may drop the problem of consciousness and speak of behavioral functions, without any bias against consciousness. The second meaning of functionalism, one that also goes back to the early days, is expressed in the following familiar definition of psychology: *psychology is the study of the adaptation of the organism to its environment.* This makes psychology a branch of biology, perhaps especially of ecology, but it has a life of its own because of the problems that psychologists have chosen to work upon.

The adaptational point of view is another heritage of the evolutionary outlook, although the details may have little to do with developments in evolutionary theory. Thus studies of learning, habit-formation, memory, thinking, and problem-solving, are characteristic topics within a functional and adaptive psychology, although the references to survival value are minimal. Because many of these theories are also theories of need-satisfaction, attributing habit-strengthening to reward through drive-reduction that comes about basically through satisfying the needs that lead to survival, there is a roundabout connection with evolutionary thinking.

During the last decade or so this kind of thinking has increasingly pervaded the field of perception, through the interpretation that needs determine idiosyncratic interpretations of ambiguous stimuli. The extreme, is of course, the mirage in the desert in which the thirsty traveler sees an oasis. Similar results can, however, be found in the laboratory.

The distortion of perception through personal factors, and through habitual modes of adaptation, has been related to the functionalist tradition. One of the books about these developments has a foreword by John Dewey, in which he points out that the *transactional* viewpoint in perception is what he was talking about all his life. This, coming in 1947, was among the last of his writings, fifty years after his paper on the reflex arc, announcing this position (Kelley, 1947).

A side issue of the adaptational point of view—often cited in disparagement by the critics of functionalism—is that the study of adaptation leads to diverting much energy into applied fields. It is true that the functionalist tends to be broad-minded, and congenial to the practical applications of psychology. I am calling this a side issue, however, for in supporting applied research the functionalist is at the same time interested in pure research. It is my belief that a cult of pure science is as dangerous to progress in science as is an impatience with practical applications: what is needed is a proper balance, with division of labor and two-way communication.

QUANTIFYING BEHAVIOR

The third meaning of functionalism defines psychology as *the study of the functional relations between antecedent events and their consequents.* This is practically the mathematical meaning of function, in the simple form of $y = f(x)$. I may well be accused of a verbal dodge here, substituting the mathematical meaning of function for the adaptive meaning that it originally had. However, functionalism was always a "cause-and-effect" psychology; an antecedent-consequent relation is about all the scientist can mean by cause and effect.

If in the refinement of their work functionalists can be shown to have moved toward the study of mathematical functions, without any feeling that they have changed their orientation, I believe it is fair to attribute this choice to some natural affinity between their original functionalism and functionalism in its mathematical form.

I had already written along these lines several years ago, for the transition from one form of functionalism to another seemed to me very natural as I had witnessed it among functionalists of my acquaintance (Hilgard, 1956). I did not at that time offer any support, and I have been criticized for the sudden jump from functional adaptation to mathematical function. Therefore I am going to take this opportunity for a brief defense.

I had in mind the work of men like Carr, Robinson, McGeoch, and Melton, who were in the direct line of succession from Dewey and Angell at Chicago. Carr wrote against what he called "the quest for constants," urging the necessity for studying the detailed relationships between all sorts of antecedent events and the results to which they led (Carr, 1933). In a small book on association psychology, Robinson (1932) similarly urged the search for functional relationships that could be expressed in the form of continuous mathematical functions. His own attempt to do this was in connection with the role of similarity (Robinson, 1927). Later McGeoch (1936) espoused what he called "a dimensionality interpretation," in which all the dimensions of influence should be studied. Melton (1950) picked this up and made it the basis for his statement of the contemporary functional position with respect to learning. While none of these men was trained in a strongly mathematical tradition, the implications of their proposals are that the laws of psychology will be written as mathematical functions. It is to Thurstone (1930), another functionalist, to whom we owe one of the earliest mathematical formulations of a learning theory. Lest Thurstone may be thought of as a functionalist-after-the-fact, I should point out that his major work was in the field of individual differences—which, since Galton, has always been functional—and that in his book on intelligence (1924) he espoused a point of view toward stimulus and response that was virtually a restatement of Dewey's reflex-arc concept.

In a recent article on Lamarck and Darwin, the historian of science Gillispie (1958) points out that the real gain of Darwinism was that it opened up the mathematicization of biological science. His reasoning as applied to Darwin, much more eloquently than mine shows that when one begins with masses of concrete relations as Darwin does, and as contemporary functionalists in psychology do, the outcome is "a conception of biological order no different from the order assumed by contemporary atomic physics—an order of chance to be analyzed by the techniques appropriate to mathematical probability." While this is the course of events in biology proper, it is also the course of events in much of contemporary functional psychology, epitomized now in our stochastic learning models (Bush and Mosteller, 1955).

Thus, from the adaptive nature of consciousness, through the adaptiveness of behavior in general, functional psychology has moved to the more neutral ground of what Brunswik has called a "probabilistic functionalism." These trends in the evolution of scientific psychology appear entirely coherent with the trends in other branches of biological and social science—trends largely precipitated by the impetus of Darwin.

Final Remarks

In choosing to discuss certain lines of work within psychology that show a continuity with the ideas of Charles Darwin, I have omitted some large sectors of psychology, particularly those that are represented by modern social psychology and the psychology of personality. A complete discussion of the role of psychology vis-a-vis evolutionary theory would have to take up the problem of cultural factors in evolution, a problem upon which I have not touched. Two conferences participated in by psychologists and evolutionists, one in 1955 and one in 1956, gave serious attention to these problems, and the results of their deliberations are now available in book form (Roe and Simpson, 1958). There the optimistic outlook that man has been an unusually successful product of the evolutionary process is clouded by the fear that evolution may also have provided man with the tools for his own destruction. The authors who introduce that note do not care to predict that he will use those tools (Freedman and Roe, 1958). Let us hope that Darwin's optimism will prevail: that man has evolved enough in the way of superior intelligence and social sympathy to become self-regulating; and that man will survive to contribute to his own evolution.

References

ANGIER, R. P. 1927. "The Conflict Theory of Emotion," *Am. Jour. Psychol.*, XXXIX, 290–401.

BEACH, FRANK A. 1955. "The De-scent of Instinct," *Psychol. Rev.*, LXII, 402–410.

BERNARD, L. L. 1924. *Instinct, a Study in Social Psychology*. New York: Henry Holt & Co.

BORING, E. G. 1950. *A History of Experimental Psychology*. 2d ed. New York: Appleton-Century-Crofts.

BRUNSWIK, EGON. 1955. "Representative Design and Probabilistic Theory in a Functional Psychology," *Psychol. Rev.*, LXII, 193–217.

BUSH, R. R., and MOSTELLER, F. 1955. *Stochastic Models for Learning*. New York: John Wiley & Sons.

CANNON, W. B. 1915. *Bodily Changes in Pain, Hunger, Fear and Rage*. New York: Appleton.

———. 1927. "The James-Lange Theory of Emotion: a Critical Examination and an Alternative Theory," *Am. Jour. Psychol.*, XXXIX, 106–124.

CARR, HARVEY A. 1933. "The Quest for Constants," *Psychol. Rev.*, XL, 514–32.

DARWIN, CHARLES. 1859. *Origin of Species*. 6th ed., 1890. New York: Appleton.

———. 1868. *The Variation of Animals and Plants under Domestication*. New York: Judd.

———. 1871. *The Descent of Man*. New York: Appleton.

———. 1872. *The Expression of the Emotions in Man and Animals*. London: Murray.

DEWEY, JOHN. 1894. "The Theory of Emotion" (I), *Psychol. Rev.*, I, 553–69.

———. 1895. "The Theory of Emotion" (II), *Psychol. Rev.*, II, 13–32.

———. 1910. *The Influence of Darwin on Philosophy, and Other Essays in Contemporary Thought*. New York: Henry Holt & Co.

DUNLAP, KNIGHT. "Are There Any Instincts?" 1919–20. *Jour. Abn. and Soc. Psych.,* XIV, 35–50.

FREEDMAN, LAWRENCE Z., and ROE, ANNE. 1958. "Evolution and Human Behavior," in *Behavior and Evolution,* eds. ANNE ROE and G. G. SIMPSON. New Haven: Yale University Press.

FUNKENSTEIN, D. H. 1955. "The Physiology of Fear and Anger," *Sci. Amer.,* CXCII, 74–80.

GALTON, FRANCIS. 1869. *Hereditary Genius.* New York: Macmillan Co.

———. 1883. *Inquiries into Human Faculty.* London: Macmillan Co.

GILLISPIE, CHARLES C. 1958. "Lamarck and Darwin in the History of Science," *Amer. Scientist,* XLVI, 388–409.

HARLOW, HARRY F. 1958. "The Evolution of Learning," in *Behavior and Evolution,* eds. ANNE ROE and G. G. SIMPSON. New Haven: Yale University Press.

HILGARD, ERNEST R. 1956. *Theories of Learning.* 2d ed. New York: Appleton-Century-Crofts.

HINDE, R. A., and TINBERGEN, N. 1958. "The Comparative Study of Species-specific Behavior," in *Behavior and Evolution,* eds. ANNE ROE and G. G. SIMPSON. New Haven: Yale University Press.

HOWARD, D. T. 1928. "A Functional Theory of Emotions," in *Feelings and Emotions, the Wittenberg Symposium,* ed. M. L. REYMERT. Worcester: Clark University Press.

JAMES, WILLIAM. 1890. *Principles of Psychology.* New York: Henry Holt & Co.

KELLEY, E. C. 1947. *Education For What Is Real.* New York: Harper & Bros.

KUO, Z. Y. 1921. "Giving Up Instincts in Psychology," *J. Phil.,* XVIII, 645–64.

MCDOUGALL, WILLIAM. 1908. *An Introduction to Social Psychology.* London: Methuen. 16th ed., 1923. Boston: Luce.

———. 1923. *Outline of Psychology.* New York: Scribner.

MCGEOCH, JOHN A. 1936. "The Vertical Dimensions of the Mind," *Psychol. Rev.,* XLIII, 107–129.

MELTON, ARTHUR W. 1950. "Learning," in *Encyclopedia of Educational Research,* ed. W. S. MONROE. Rev. ed. New York: Macmillan Co.

MORGAN, C. LLOYD. 1891. *Animal Life and Intelligence.* London: Arnold.

———. 1894. *An Introduction to Comparative Psychology.* London: Scott.

NEWMAN, H. H., FREEMAN, F. N., and HOLZINGER, K. J. 1937. *Twins. A Study of Heredity and Environment.* Chicago: University of Chicago Press.

ROBINSON, EDWARD S. 1932. *Association Theory Today.* New York: Appleton-Century.

———. 1927. "The 'Similarity' Factor in Retroaction," *Am. Jour. Psychol.,* XXXIX, 297–312.

ROE, ANNE, and SIMPSON, GEORGE G. (eds.) 1958. *Behavior and Evolution.* New Haven: Yale University Press.

ROMANES, G. J. 1882. *Animal Intelligence.* New York: Appleton.

SCHLOSBERG, HAROLD. 1954. "Three Dimensions of Emotion," *Psychol. Rev.,* LXI, 81–88.

SMALL, W. S. 1899. "An Experimental Study of the Mental Processes of the Rat," *Am. Jour. Psychol.,* XI, 133–165.

SPENCER, HERBERT. 1855. *The Principles of Psychology.* 2d ed., 2 vols., 1871, 1873. New York: Appleton.

THORNDIKE, E. L. 1898. "Animal Intelligence, an Experimental Study of the Associative Processes in Animals," *Psychol. Rev. Monographs,* Suppl. ii (4).

THURSTONE, L. L. 1930. "The Learning Function," *J. gen. Psychol.,* III, 469–93.

———. 1924. *The Nature of Intelligence.* New York: Harcourt Brace.

WATERS, R. H. 1939. "Morgan's Canon and Anthropomorphism," *Psychol. Rev.,* XLVI, 534–40.

WOODWORTH, R. S. 1948. *Contemporary Schools of Psychology.* Rev. ed. New York: Ronald Press.

———. 1941. *Heredity and Environment: a Critical Survey of Recently Published Material on Twins and Foster Children.* (Soc. Sci. Res. Counc. Bull., No. 47.)

MACDONALD CRITCHLEY

THE EVOLUTION OF MAN'S CAPACITY FOR LANGUAGE

The spate of criticism which followed the publication a hundred years ago of the *Origin of Species* often included the protest that Darwin in his argument had ignored man's higher mental faculties. This was perhaps true in fact, though captious in spirit. A recent critic, Leslie Paul, has written: "There is therefore through the invention of speech the entry into and the exploration of a new dimension of human activity. I think it was rather provincial and dull-witted of Darwin not to have shown a glimmer of interest in all this." In any event the gap was filled four years later when his geological colleague, Lyell, devoted a chapter in his classical *Antiquity of Man* to a comparison between the origin and growth of languages and of species. Schleicher, a botanist as well as a professional philologist, had called attention, three years before ever reading Darwin's book, to the struggle for existence among words, the disappearance of primitive forms, and the immense expansion and differentiation which may be produced by ordinary causes in a single family of speech. He looked upon languages as natural organisms, which, according to definite physical influences and independently of human will, take origin and mature, grow old and die, and therefore manifest the series of phenomena to which are given the name of "life." In 1863, Schleicher issued his pamphlet *Die Darwinsche Theorie und die Sprachwissenschaft,* which was the expansion of a letter he had written to Professor Häckel acknowledging a copy of the *Origin of Species.* Herein he argued that the inception of species is notably paralleled in the genealogy of language, and particularly of the Aryan and Semitic tongues. Analogous with the struggle for life among the more or less favoured species in the animal and vegetable kingdoms, a struggle for survival occurs among individual languages.

This analogy between the evolution of species and of language was discussed in contemporary scientific literature. F. W. Farrar (1870) believed that comparative philology supported Darwin's hypotheses

MACDONALD CRITCHLEY is Senior Physician at the National Hospital, Queen Square, London, and formerly Senior Neurologist at King's College, London.

in two important respects, viz., the effect of infinitesimal modifications in gradually bringing about great changes; and the preservation of the best and strongest elements in the struggle for existence. Just as very many primordial cells, closely resembling each other, may have been the earliest rudiments of all living organisms, so in philology different linguistic families may have sprung from multitudes of "speech cells" or "sound cells," that is, the fundamental roots of language. Like an extinct species, a language—once extinct—can never reappear. Intermediate linguistic forms also die out. Thus external factors disturb the primitive relationship of languages, and consequently one may find radically different languages existing side by side. Farrar said, "All this, as every naturalist is well aware, represents a condition of things precisely similar to that which prevails in animated nature."

Darwin's contemporary, Max Müller, who occupied the Chair of Comparative Philology at Oxford, was also interested in this parallelism between the struggle for existence in the biological sense and in the case of languages. He laid stress on an important difference, however. It is not on account of inherent defects that languages gradually become extinct, but rather because of external causes; that is to say, the physical, moral, and political weaknesses of those who speak the languages concerned. Müller considered that a much more pertinent linguistic analogy with Darwinism lay in the struggle for survival among words and grammatical forms which is constantly going on in every language, whereby shorter and easier forms gain the upper hand.

Views of this kind were of topical interest a century ago, but since then philologists, with the possible exception of Jespersen, have been largely out of sympathy with the application of Darwinian ideas to their own subject.

In 1871 Darwin himself dealt in some detail with the human faculties which he had rightly omitted from his earlier monograph; in *Descent of Man* the problem of speech was specifically discussed. The faculty of articulate speech, he wrote, in itself offers no insuperable objection to the belief that man had evolved from some lower form. The mental powers in some early progenitor of man must have been more highly developed than in any existing ape, before even the most imperfect form of speech could have come about. The continued advancement of this power would have reacted on the mind itself, by enabling and encouraging it to pursue long trains of thought. Complex reflection can no more be carried on without the aid of words, whether spoken or silent, than can a long abstraction without the use of figures or algebra.

Darwinian theories and ideas pervaded every aspect of scientific and philosophic thought, and Max Müller was also caught up in the current excitement. Leaving aside his purely linguistic considerations, we may examine his views upon the evolution of the speech faculties in man. After detailing *seriatim* the characters of mind and body which are shared by man and animal, he went on to enquire in 1861:

> Where, then, is the difference between brute and man? What is it that man can do, and of which we find no signs, no rudiments, in the whole brute world? I answer without hesitation: the one great barrier between the brute and man is *language*. Man speaks, and no brute has ever uttered a word. Language is our Rubicon, and no brute will dare to cross it. This is our matter-of-fact answer to those who speak of development, who think they discover the rudiments at least of all human faculties in apes, and who would fain keep open the possibility that man is only a more favoured beast, the triumphant conqueror in the primeval struggle for life. Language is something more palpable than a fold of the brain or an angle of the skull. It admits of no cavilling, and no process of natural selection will ever distil significant words out of the notes of birds or the cries of beasts.

Professor Müller's views were, on the whole, opposed to those of Darwin, and their differences of opinion were studied and discussed at length by the German linguist Noiré. Strongly critical of "Darwinian foibles, incompleteness and one-sidedness," Noiré proclaimed in 1879 that Max Müller was the only equal, not to say superior, antagonist who had entered the arena against Darwin. "Here is reason, here language, here humanity. None shall pass here; none penetrate into the sanctuary who cannot tell me first how reason, how speech, was born. And the shouting bands of the assailants were struck dumb, for they could give no answer."

Although Professor Müller was, in the main, out of sympathy with Darwin, nevertheless he proclaimed in 1873: "In language, I was a Darwinian before Darwin." As early as 1861 he was trying to reconcile Darwin's doctrines with linguistic phenomena. He compared Darwin with Epicurus, and he spoke of the origins of language in terms of natural selection, or—as he preferred to call it—natural elimination.

Anatomical Basis for Language

Any discussion of the evolution of man's capacity for language must entail an enquiry, not only into the appropriate intellectual equipment, but also into the necessary and actual anatomical substratum. This latter is a twofold problem. In the first place, a physiological cerebral mechanism exists peculiar to man. In addition, the faculty of speech requires certain peripheral instrumentalities, which can fulfil a complex co-ordinated activity of the lips, palate, tongue, pharynx, larynx, and respiratory apparatus. Herein lies the structural basis for the achievement of an audible motor-skill of the utmost delicacy. In the course of both phylogeny and ontogeny it is often possible to observe that anatomical structures are present even before they are actually utilized. Structure, in other words, antedates function. Consequently, within the animal series it may be expected that both cerebral and peripheral mechanisms will stand ready for use, though not yet productive of mature speech.

Even in the anthropoid there is no valid morphological reason, at a peripheral level, why speech should not occur. At any rate, most authorities would agree with this opinion. The relative coarseness of the tissues would no doubt impart a certain unmusical quality to the articulation, but the phonemic range would probably be not inconsiderable. As Max Müller said, there is no letter of the alphabet which a parrot will not learn to pronounce, and the fact that the parrot is without a language of his own must be explained by a difference between the *mental,* not between the *physical,* faculties of animal and man.

It must be stressed that language is a function which can be looked upon as overlaid, or even parasitic. There are no specific cerebral structures which are peculiar to the faculty of speech. It would be difficult—if not, indeed, impossible—to decide merely from a study of the brain, however meticulous, whether the subject had been a polyglot, an orator, a writer, or even an illiterate or a deaf-mute. Simply by microscopical examination of the cerebral cortex it would not be easy to distinguish gorilla from man. In other words, no essentially human cerebral speech centre can yet be confidently identified as an anatomical entity. Speech likewise makes use of predetermined bucco-laryngeal structures which were primarily destined to serve for acts of feeding and respiration. Certain teleological advantages accrued when the function of communication took over structures which were also being utilized for other purposes. Man did not develop *de novo* some entirely novel means for subserving the novel faculty of language. Linguistic precursors, anatomical, physiological, psychological, and cultural, must obviously have existed in the subhuman animal series. In some creatures like bees, simple communicative acts operate by dint of global movements. In birds and primates, elaborate combinations of cries, intention movements, and pantomimic displays fulfil the role of primitive sign-making or communication. Some birds possess the faculty of mimicking human utterances in a plausible and even startling fashion, but it must be remembered that this is a learned, artificial performance and that their innate instinctive calls are crude, raucous, stereotyped—indeed, anything but human in quality.

We recall Buffon's speculation as to what would happen if the ape had been endowed with the voice of a parrot and its faculty of speech. The talking monkey would, he said, have struck dumb with astonishment the entire human race and would have so confounded the philosopher that he would have been hard put to prove—in the face of all these human attributes—that the monkey was still an animal. It is, therefore, just as well for our understanding that Nature has separated and relegated into two very different categories the mimicry of speech and the mimicry of our gestures.

Koehler has put the question why it is, if there are so many precursors of our own language in the animal kingdom, no known animal speaks like man. It is because no animal possesses all those *initia* of our language at one and the same time. They are distributed very diffusely, this species having one capacity, that species another. We alone possess all of them, and we are the only species using words.

Criteria of Language

Among the prerogatives of *Homo sapiens* the faculty of speech is the most obvious. Other members of the animal kingdom, not excluding the higher primates, are not so endowed, however vocal may be the individuals. By contrast, it can be asserted that no race of mankind is known, however lowly, which does not possess the power of speech. Nay more, the linguistic attainments may be subtle, complex, flexible, and eloquent—even though the cultural level be primitive in the extreme. It is indeed difficult to identify among the races of man anything which can be justly termed a "primitive" tongue.

At the very outset it is important to be clear in what way the cries, utterances, calls, and song of birds and subhuman mammals can so readily be deemed as lying outside the category of language attainment. On enquiry, it is found that no one touchstone of distinction is entailed, but rather a co-ordination of factors, some of which may be present in this or that animal, but which do not come together in integration until the stage of *Homo sapiens* is achieved.

Doubtless the most weighty single criterion of human speech is the use of symbols. Animals betray abrupt fluctuations in their emotional state by making sounds. To this extent they may be said to utilise signs. Whether the sign be perceived and identified as such by other members of the same species is arguable. An alarm-call may act as a signal of danger, and others within earshot may take flight. This effect may be an instance of direct signaling between one bird or mammal and another. Or, possibly, the frightened creature's cry may be interjectional rather than purposeful, and others within call may thereupon be made merely partners in alarm rather than the recipients of a directed message. Be that as it may—and the possibility that both types of concerted action occur in nature cannot be gainsaid—the animal's cry cannot strictly be looked upon either as language or as speech. At most, it is communication. The communicative act may be deliberate, willed, directed encoding, while the comprehending recipients who act upon the signal may be looked upon as decoders. Here, then, is communication in the accepted sense of the term. Or it may be that the communicative act is merely incidental, and no true encoding and decoding can be said to take place.

"Animal communication" is therefore the term which carries with it the fewest drawbacks. In essence it can be said to comprise a series of signs which refer to ideas or feelings within immediate awareness. They do not and cannot apply to circumstances within past or future time. Herein lies an all-important distinction. Man's utterances entail the use of symbols or signs of signs and consequently possess the superlative advantage of applying to events in time past, present, and future and to objects *in absentia*. This endowment has been called the "time-binding" property of human language. It also possesses the merit of beginning the process of storage of experience, a process which eventually reaches fruition with the subsequent introduction of writing.

The Evolution of Behaviour

Most of the early arguments concerning the problem of the origin of speech in man have been either theological or linguistic. In the former case the doctrine of a divine creation was accepted, but many controversies arose, including such questions as monogenesis versus polygenesis. The purely linguistic theories rejected altogether the idea of a special creation of a mature system of language, and while some process of transition between the communication of animals and the beginnings of speech in *Homo sapiens* was assumed, there was disagreement as to the *modus operandi*. The sources of argument included the relative importance of the role of imitation, of interjectional utterances, of associated motor-vocal phenomena, of gesture and still other factors, none of which nowadays excites serious comment or concern.

Attention became focussed more upon the mysterious evolutionary changes which are believed to have taken place between the behavioural systems of the highest primates and those of earliest man. The beginnings of speech, in the strict sense of the term, rank among these changes. However striking in character and fundamental in importance, speech certainly cannot be looked upon as man's sole perquisite, singling him out from the rest of the animal kingdom. Several other important developments took place at more or less the same period of evolution, any one or any combination of which may actually prove to be supremely significant in the genesis of speech.

The principal clash of opinion turns around the debate whether the difference between animal communication and the speech of early man entail factors which are qualitative or merely quantitative. Expressed somewhat differently, the question has been raised whether the distinction between man and animals is one of kind or merely of degree, as far as the communicative act is concerned.

Within that stage between the most complex of animal communication and the speech-efforts of earliest man lies the core of our problem. Obviously, this transition from animal cries to human articulation is but an item in a much bolder process of evolution. Instinctive responses no longer prove biologically adequate, and more and more complicated vocal reactions gradually emerge. Linguistics alone can never afford the whole solution, and other realms of thought and endeavour will need exploring.

SOCIAL FACTORS FAVOURING LANGUAGE

Communal living.—Attempts which have been made to identify these important steps in the evolution of human language fall roughly into two classes. Thus one can distinguish sociological from intellectual hypotheses, the former envisaging some modification in behaviour, the latter implying a change in the mode of thinking as between animals and man. These two attitudes are not mutually irreconcilable, and both types of change may well have operated together.

Many would agree that the ancestry of language lay within the prehominid stages, at the same time denying the existence of anything that can be strictly termed "animal language."

Révész spoke of "contact reactions" as being important in the genesis of speech in man. By this expression he understood the basic, innate tendency of social animals to approach one another, establish rapport, co-operate, and communicate. Contact reactions are a necessary precondition of linguistic communication. In a rather unconvincing fashion, Révész seems to have equated the essential differences between human speech and the cries and directed calls of the animal world with an elaboration of this "contact reaction" in the domain of articulate utterance.

Much earlier, Lord Monboddo realized the critical role of communal existence. To convert man into a speaking animal, the factor of society is essential. He posed the question: which is the more important—language for the institution of society, or society for the invention of language. In his view, society came first and had existed perhaps for ages before language developed, for man is by nature a political as well as a speaking animal.

Biologists realize that communal existence is an important factor in survival, which can be traced as a principle throughout the animal kingdom, even in the lowliest species (Espinas, Kropotkin). Indeed, the physiological value of coexistence can perhaps be better demonstrated in the invertebrate phyla. In the higher ranks of Mammalia, there is perhaps a greater co-ordination of group activity, whereby there is a limited degree of sharing of function and a deputing of special tasks increasingly becomes the rule. Allee has shrewdly asked at what stage can an animal group be said to have become truly social: is it at the point when animals behave differently in the presence of others than they would if alone? If this is the case, then we witness in an interesting fashion the first hint of ethical or moral factors in animal behaviour.

One of the principal functions of speech is to co-ordinate the behaviour of the individual members of a group. Grace de Laguna stressed the progressively elaborate communal life which synchronizes with the development of speech. Planned hunting forays, the need for securing safety by night, the indoctrination of the young—all these are among the activities of early *Homo sapiens,* and they must have been considerably assisted by the faculty of speech. The power of speech thus confers an important survival value upon its owner.

Tool-using: tool-making.—Allied to this notion is the role of an increasing utilization of tools, as an immediate precursor of speech. *Homo sapiens* has often been identified not only with *Homo loquens* but also with *Homo faber*. An animal achieves its purposes by modifying its own bodily structure, that is, by making a tool out of some part of itself. Man ventures further by making use of instruments outside his own body. As L. S. Amery said, man also began to employ a "sound-tool"; that is to say, he made use of differentiated sounds as an instrument of precision, in order to indicate not only emotions but also specific objects, qualities, actions, and judgments. Both language and tools are instruments which humans alone employ to achieve definite and concrete actions. "Language, like the tool, and unlike the limb, is something objective to, and independent of, the individual who uses it" (de Laguna).

We now approach a critical point in the argument. The term *Homo faber* is ambiguous, for it can be interpreted in two very different ways. It can be read as meaning either the "tool-maker" or the "tool-user." This distinction is important and is not to be glossed over. Mere tool-taking or tool-utilizing is quite consistent with anthropoid behaviour; tool-making is not. The higher apes are not infrequently to be seen making use of a convenient stick as an implement with which to draw a delicacy within reach. But deliberately to choose and to set it carefully aside, against the contingency of finding at some possible future date an edible morsel just inaccessible, is outside the capacity of the anthropoid. To select an instrument and keep it for future use can be reckoned as analogous to fashioning a tool out of sticks or stones to attain an immediate need or desire.

When the species can do these latter things, it steps over the frontier and qualifies as *Homo sapiens*. Similarly, in the most primitive communal groups of man's ancestors, a piece of sharp stone, a stick, a shell might have been picked up and used straightway as a weapon to fell an object of prey, as a weapon of self-defence, or as a tool for decorticating a tree-trunk or skinning a beast. This sort of activity is consistent with primate behaviour, and speech acquisition is unnecessary. But when the apelike creature breaks a stick in two or pulls it out of a bush or if he puts it aside for another occasion, it is beginning this apprenticeship for qualifying as *Homo sapiens,* and here the first beginnings of speech may be detected.

With the art of knapping of flint core-tools or flake-tools or by shaving down a stake, we have the unmistakable marks of attainment of man's stature, and speech can doubtless be assumed as a concomitant. For here we have the earliest mastery over purely perceptual thinking, the dawn of conceptual thought, and release from the shackles of time-present.

Delegation of labour.—Closely linked with an elaborate communal life and the construction of tools, delegation of labour can also be reckoned as a factor in the ancestry of speech. Greater efficiency in hunting and in the acquisition and preparation of food for the group follows upon the use of speech and leads to the beginnings of a simple form of specialization. This is an aspect of linguistics which has naturally appealed particularly to Soviet writers. Soviet philosophers of language believe that language began when man—a new species of animal—began to use tools and to co-operate with others in order to produce the means of subsistence. Human labour is a new form of social activity and gives rise to a new phenomenon, articulate speech, and to a new characteristic of the mind, the conscious reflection of objective reality. Stalin—himself a dabbler in linguistics—looked upon language not only as a tool of communication but also as a means of struggle and development of society. Language is connected with man's productive activity and also with every other human activity. Seppe, a Russian neurologist, went further. Work appeared to him as a main factor in the development of higher and abstract thinking of man. Speech functions are created from work. Furthermore, we find Stalin declaring: "In the history of mankind, a spoken language has been one of the forces which helped human beings to emerge from the animal world, unite into communities, develop their faculty of thinking, organize social production, wage a successful struggle against the forces of nature, and attain the stage of progress we have today" (*Pravda,* August 2, 1950).

Such were the progressive elaborations in animal behaviour which immediately antedated and perhaps accelerated the development of speech in man. The alternative group of hypotheses puts the emphasis more upon the elaboration of certain ways of thinking, as bridging the gap between animal communication and human speech.

INTELLECTUAL BEHAVIOUR

Abstract thinking.—Since Aristotle, many philosophers have stressed man's gift of conceptual thought, whereby he is enabled to deal with general ideas as well as the particular or the concrete. Man's unique power of coping with "abstractions," "universals," "generalisations," has been associated with his endowment of speech. Geiger, a contemporary of Darwin's, was one of the most eloquent advocates of conceptual thought as a human perquisite. In his *Ursprung der Sprache* (1869) he wrote:

> It is easy to see that blood is red and milk is white; but to abstract the redness of blood from the collective impression, to find the same notion again in a red berry and, in spite of its other differences, to include under the same head the red berry and the red blood—or the white milk and the white snow—this is something altogether different. No animal does this, for *this, and this only, is thinking.*

Noiré (1879) enquired how man's power of abstraction came about. He attributed it to man's manual dexterity coupled with his ingenuity. More than any other creature, man has the power of selecting objects from his environment and then modifying them to suit his own purposes. Thus he became master of his environment. He learned to create things, and these creations were for him the first "things." Such "things" became endowed with independent existence, and from this point to the endowment of names for the things was quite an easy step.

Terminology readily misleads, however. We now believe that the older, narrow views upon the essentially human nature of conceptual thinking are not warranted by the facts. As Darwin showed, it is not possible to deny that in some animals, as judged by their behaviour, indications of a kind of abstract thinking can at times be traced. Although perhaps an exceptional state of affairs, it occurs often enough to cast doubts upon any notion of a Rubicon separating the brute beasts from man.

Let us recall the very beginnings of philology as a science, which can be said to date from 1772, when J. G. Herder wrote his essay on the *Origin of Language*. Herder rejected the doctrine that language was a divine creation and also the idea that it might be a willed invention on the part of men. Nor did he believe that the difference between man and animals was one of degree. He considered that there had taken place in man as he emerged from the subhuman state a development of all of his powers, in a totally different direction. This abrupt exploration led to the appearance of speech. Language sprang, of necessity, out of man's innermost nature. Herder likened the birth of language to the irresistible strivings of the mature embryo within the egg. In particular, man possessed a keener faculty of "attention" than any other animal, and he was thereby enabled to seize hold of isolated impressions from out of a mass of detail surrounding him. In this way man became able to identify the most arresting feature within his environment. For example, the distinguishing property of a lamb would be its vocalization; that is, its bleating. Thereafter the lamb would be recollected, and referred to, as a "bleater." So, according to Herder, primitive nouns stemmed from verbs (as indeed we know to be the case in the sign-language of the deaf-and-dumb).

Perception and conception.—Herder's theory, couched in somewhat different terms by Noiré, reappeared a century later. And fifty-five years after that, contemporary animal psychologists state this theory anew. Professor E. S. Russell warned us not to assume that an animal's perceptual world would be like our own: on the contrary, judging from its behaviour, we must conclude that every animal has its own perceptual world, one which is very different from ours. Animals do not ordinarily perceive their environment in the same "articulated" fashion as we do. They perceive things only as ill-distinguished parts of a general complex. Isolated from its habitual context, an object may not be recognised for what it is. Animals respond only to perception-complexes and not to simple and solitary stimuli.

We have already referred to the philologist Max Müller as in many ways an antagonist of Darwin. This would be to do an injustice to both writers, for—as we have stated earlier—in 1861 we find Müller aligning Darwin with Epicurus. The latter believed that primitive man's uncouth instinctive ejaculations were fundamental in the origins of language. In addition, there must have been an important second stage, whereby agreement is made in associating certain words with certain conceptions. For the "agreement" of Epicurus, Müller would offer his doctrine of natural selection, or natural elimination. The phenomenon of the origin of language would then be visualized as follows: Sensuous impressions would produce a mental image or *perception;* a number of perceptions would bring about a general notion or *conception.* A number of sensuous impressions might also occasion a corresponding vocal expression—a cry, an interjection, or an imitation of the sound in question. A number of such vocal expressions might be merged with one general expression and leave behind the root as the sign belonging to a general notion. The gradual formation of roots and of natural cries or onomatopoeia is a product of rational control. Rational selection is natural selection, not only in nature but also in thought and language. "Not every random perception is raised to the dignity of a general notion, but only the constant recurring, the strongest, the most useful." Of the multitudinous general ideas, those and only those which are essential for carrying on the work of life survive and receive definite phonetic expression.

Symbolic Behaviour

Another way of looking upon the development of human speech out of animal vocalizations is to regard speech as the utilization of symbols. The sounds emitted by animals are in the nature of signs, while man's speech is made up of symbols. Signs *indicate* things, while symbols *represent* them. Signs are announcers of events; symbols are reminders. In other words, symbols are not restricted to the confines of immediate time and place. As "substitute" signs, symbols can refer to things out of sight and outside present experience. When an ape utters a cry of hunger, it can be looked upon as perhaps making a declaration, perhaps an imperative utterance, or even an exclamation of discomfort. No ape, however, has ever uttered the word "banana," for such a word is a concrete symbol, a tool of thought which only man can employ, and he can do so in a variety of ways, irrespective of the barriers of time and space. Man can refer to a banana in past or future tense, as well as the present. Man can talk about a banana *in absentia.* No animal can do these things, the task being far beyond its system of thought and therefore of expression. Likewise no monkey can emit a word meaning "hunger," for this term would constitute, or refer to, an abstract or universal idea.

In Pavlovian modes of thought the use of symbols is regarded as a hallmark of man's cerebral function—although other terminologies are used. Pavlov taught that when the developing animal world reached the human stage, an extremely important addition came about, namely, the functioning of speech. This signified a new principle in cerebral activity. Sensations and ideas from the outer world constitute the first system of signals (concrete signals; signals of reality). Speech, however, constitutes a second set of signals—or "signals of signals." These make possible the formation of generalisations, which, in turn, constitute the higher type of thinking, specific for man.

Man's capacity for dealing with symbols rather than signs or things has been visualized by Korzybski as a specific "time-binding" faculty, peculiar to mankind. Pumphrey has described as many as three considerations which are attached to the human employment of verbal symbols as opposed to the sign-making cries of animals. These properties are (1) *detachment,* whereby man is able to use language to describe events in a wholly dispassionate fashion, if he should so desire; (2) *extensibility,* whereby a proposition can be made and discussed in terms of past, present, or future time; and (3) *economy,* whereby symbols enable man to abbreviate what would otherwise be a long-winded description or declaration.

Many writers view the origins of speech as merely part of a large developing faculty, namely, the beginnings of symbolic behaviour. As Sapir put it, language is primarily a vocal actualisation of the tendency to see reality symbolically, a property which renders it a fitting medium for communication. The problem, therefore, really resolves itself into a search for the earliest indications of symbolic behaviour, as the immediate precursor of speech. S. Langer believes that these beginnings of symbolic thought can be detected when an animal—the highest of the primates, in fact—behaves as if significance were being attached to certain objects or sounds. This attitude may be seen in the anthropoid in captivity, in its attachment toward some inanimate and favoured plaything—a piece of wood, a toy, a rag, or a pebble. Here, then, we are attempting to discern the dawn of symbolic thought; and here, too, we may perhaps descry the remote ancestry of human speech. In other words, the chimpanzee, although devoid of speech, begins to show a rudimentary capacity for speech—an opinion which reminds us of Müller's uneasy feeling that the gorilla is "behind us, close on our heels."

Some of the nineteenth century philosophers who were critical of Darwin, compared the appearance of language in man with the beginnings of religious belief. Certainly, at a very early date in man's emergence we find that there are indications of primitive magical practices, with evidences of ritual or ceremonial. It can safely be concluded that, at such a cultural level, primitive man was surely endowed with the faculty of speech. Noiré believed that the two aptitudes grew up in concert, the rise of mythology being an important and necessary stage in the development of language. This can be looked upon as a period when objects began to mark themselves off from the indefiniteness of the total perceptual processes and to form themselves into independent existence.

Emergence of Speech

The intriguing question naturally arises At which point in the evolution of primitive man did speech, in the strictest sense of the word, first make its appearance? When did the ululations of the anthropoids give way to the use of verbal symbols, disciplined by phonetic and syntactical rules? Obviously, it is not possible—nor is it ever likely to be possible—to answer this question with confidence. The evidence, such as it is, is meagre, indirect, and oblique. But speculation on this interesting matter is quite permissible.

Anthropological data are of great importance here. They comprise arguments which are of a cultural order and which discriminate clearly between anthropoid and human communities. They also include the weighty evidence which lies within the domain of comparative anatomy. Here are to be found the impressive distinctions between ape and man in respect to the crania and the problems which arise from a study of the fossil skulls of man's immediate ancestors. Here, too, are marshaled the anatomical features in the crania which are to be regarded as specific for *Homo sapiens*. The size of the cranial cavity will naturally indicate brain-volume. This is an important point in that, ordinarily speaking, the human brain differs from anthropoid brain in its greater size, while the cranial capacity of prehistoric man occupies an intermediate position. But the rule is not invariable: one or two prehistoric specimens are characterized by megalencephaly. More valuable than sheer size is the question of the shape and proportions of the cranial cavity. In addition, the endocranial markings may be taken as a likely index of the convolutional pattern of the cerebral hemisphere. In assigning a fossil specimen to its evolutionary rank, the development of such specifically "human" areas of the brain as the frontal lobes and the parietal eminences are all-important. Obviously, clues such as these may be followed when discussing the problem of when in prehistory man developed speech.

L. S. Palmer, a dental surgeon as well as paleontologist, has approached the question of the development of speech in man from a somewhat unusual angle. His distinguishes between human speech and animal noises, the former being regarded as being effected by delicate and voluntary variation in the size and shape of the oral cavity. The power of articulation (as exemplified in human speech) depends upon a specific morphology of the jaws, and here man differs in an important manner from the ape. In man the two rami of the mandibles are splayed apart, whereas in apes they are parallel. There results a difference in the shape of the posterior ends of the mandibles, together with an increased width between the condyles at the upper ends of the rami. In man there is consequently ample space for the free movement of the tongue, in this way facilitating articulate speech.

Another difference between the jaws of apes and men consists in the presence of a bony ledge connecting the anterior ends of the mandibles in apes. This "simian shelf" serves as an attachment for the genioglossus muscles. The range of lingual movement is rather restricted. In man, however, the tongue muscles are attached to a series of small genial tubercles, which, taking up but little room, permit freer movement of the tongue within a broader intermandibular space.

The same author (Palmer) also associates himself with L. A. White (*The Science of Culture*) and believes that there was an important connection in prehistory between favourable climatic factors and cultural acceleration, which naturally includes the origin of the faculty of speech. Palmer set out the chain of causes as follows:

> A rigorous ameliorating climate → appropriate gene mutation → expansion of the skull → development of brain → increased mental ability → development of articulate speech and the introduction of written words.

Thus we can surmise with no little confidence that Cro-Magnon man must surely have been endowed with speech, even though no firm evidence exists that written language was ever in use at that period. The refinement of the skeletal structure and the large cranial capacity point to a quite highly evolved type of *Homo sapiens.* But the weightiest arguments are of a cultural order. The skilful cave-paintings of the Aurignacian, Solutrean, and Magdalenian periods obviously must have been the work of individuals endowed with symbolic and conceptual thinking. The frequency with which hand-prints occur on the cavern walls may also be taken as suggestive of individual personal awareness. Furthermore, the relative preponderance of left hands over right at El Castillo (4 to 1) must indicate that cerebral dominance obtained at that period. Perhaps, too, the appearance of obscure linear markings—red blobs and dots—adjacent to these hand-prints may be looked upon as the very first modest indications of written communication. The fashioning of elaborate tools, the use of fire and of clothing, and the evidence of ceremonial burials as well as religious or magical practices cannot be reconciled with a speechless state. Even the Negroid variant of Cro-Magnon civilization, known as the Grimaldi man, as well as the Eskimo-like Chancelate man, is no exception to these arguments. These fragmentary clues take the story of language back to the last Ice Age of late Paleolithic times, that is between 25,000 and 10,000 years B.C.

Can language be assumed in even earlier man? European *Homo neanderthalensis* (*or mousteriensis*) constitutes a less straightforward problem. Some anatomists are tempted to explain some of the contradictory characteristics of Neanderthal skulls by suggesting that they were out on a side line, away from the main stream of evolution. Thus the anthropoid characters of the supraciliary and occipital ridges, the massive jaw, and the wide orbits and nasal apertures contrast with the large cranial capacity, which actually exceeds that of average modern man. The African Neanderthaloids, including the *Homo rhodesiensis,* and the *Homo soloensis* of Java, present essentially the same problem.

A. Keith believed that the faculty of speech could be traced back as far as Neanderthal man, but no further. His evidence was wholly anatomical, and not very convincing. The left hemisphere was apparently more massive than the right, indicating cerebral dominance. Tilney also believed that Neanderthal man was "possessed of linguistic capabilities not far below the standard of *Homo sapiens.*" His conclusion was based upon the depth of the parietal fossae in the skulls, suggesting a well-developed "auditory area," i.e., the abutment of the parietal lobe upon the outer occipital and upper temporal lobes. This postero-inferior part of the parietal lobe is commonly regarded as a true and specific human perquisite. L. S. Palmer, however, is impressed by the poor temporal lobe development in the brain of Rhodesian man, and he doubts very much whether this specimen of Hominidae ever could speak.

Cultural evidence is more convincing than the morphological. The coexistence of eoliths in the way of sharpened flints and arrow-heads, and signs of the use of fire and the practice of cooking, all point to Neanderthal man's possessing a degree of conceptual and symbolic thinking consistent with the possession of language, just as in the case of Cro-Magnon man. If this argument is admitted, then the story of language can be taken back to about 50,000 years B.C., that is, to the post-Acheulian Paleolithic period, or the last glacial era.

Let us now turn to the early and middle Pleistocene periods. The hominid representatives of this time are exemplified by *Pithecanthropus pekinensis* (*erectus*) or *sinanthropus,* by *Homo heidel-*

bergensis, and perhaps also by *ternifine man.* Possibly, too, the Swanscombe and the Steinheim skulls belong here, though admittedly they may represent a transitional or intermediate type between *Pithecanthropus* and *Homo sapiens.* Later specimens within this same period include the skulls associated with Ehringsdorf, Fontéchevade, Florisbad, Krapina, and Mount Carmel.

According to Tilney, the left frontal area of the brain was larger than the right, a fact which he was tempted to associate with a state of right-handedness. The same author pointed out the development of the inferior frontal convolutions, a feature which suggested to him that *Pithecanthropus* could speak. He went on to assert: "Doubtless the linguistic attainments were extremely crude." It is difficult to comprehend exactly what Tilney implied by this statement, for present-day linguistics has no knowledge of a language system which can be designated as "extremely crude," even among the most primitive and uncultured communities.

Upon other grounds, too, there is evidence that speech was an endowment of *Pithecanthropus.* Implements of quartz have been found in the caves alongside the human remains, obviously fabricated with some skill. There is evidence also that *Pithecanthropus* knew how to produce fire and that at times he produced cooking. Again it is almost useless to conjecture what manner of speech was employed by *Pithecanthropus.* Oakley was merely guessing when he surmised that the earliest mode of expression of ideas was perhaps by gesticulation, mainly of mouth and hands, accompanied by cries and grunts to attract attention.

By such suggestive paleontological clues, we can refer the faculty of speech to a period at least as far back as 100,000 years B.C., that is, the middle Pleistocene period. On the evidence of the Javanese and Chinese skulls (*P. soloensis* and *pekinensis*) the date might even be relegated as far back as the early Pleistocene era, that is, perhaps 500,000 years B.C.

Few would venture to seek the pioneers of speech at any more remote period. There arises for serious discussion, however, an interesting series of fossil skulls found in South Africa, small in size and of an interesting morphology. These are associated with an extinct series of pygmy man-apes, originally called the fossil Taung's ape, but more often nowadays as *Australopithecus.* Where these specimens rightly belong is debatable; Le Gros Clark regards them as exceedingly primitive representatives of the family which includes modern and extinct types of man. Leakey called them "near-men." There is no sure evidence that such creatures fabricated tools; consequently, it is unlikely that the *Australopithecus* can be assigned to the genus

Homo faber vel sapiens and that it was capable of speech. The recent findings of a number of crude stone artifacts in proximity to the bones and the associated fractured skulls of the fossil bones makes it possible that *Australopithecus* utilized stones as weapons, even though it did not, strictly speaking, manufacture weapons.

L. S. Palmer believes, however, that *Australopithecus* was perhaps endowed with speech. He bases his opinion upon the anatomical characteristics of the mandible. The absence of a simian shelf and of diastema (or gap between the incisor and the canine) and the convergence angle of the teeth are all features which correspond with a hominoid morphological pattern. Whether this type of argument is sufficient to militate against such arguments as the small cranium and the lack of sure evidence of tool-making is very doubtful.

Recently it has been suggested that *Australopithecus* had the ability to make fire. If this is really the case, it should be taken as an additional piece of evidence to suggest that speech might have been within its capacity. Obviously, the answer to the question awaits the production of further findings.

The date of *Australopithecus* is remote indeed—probably beyond the earliest Pleistocene era and back into the end of the Pliocene. This means anything from one to fifteen million years B.C.

There is one aspect about the beginnings of speech in man which is only too often completely overlooked. Was speech a consistent endowment in the case of early man? When man first appeared on the earth, as in the middle and late Pleistocene periods, perhaps only some of the newly evolved *Homo sapiens* were endowed with speech. Speech in those remote times might have constituted an exceptional phenomenon or aptitude, one which was within the competency of only a few highly favoured individuals.

In the case of hand-skills, too, maybe only comparatively few members of *Pithecanthropus* were able to fashion arrow-heads or flints, this expertise being a rare and no doubt highly accorded accomplishment. Then again, skill in handicrafts might perhaps have correlated closely with the faculty of speech. Such especially gifted members of the community probably also had an expectation of life above the average, and therefore speech and the art of tool-making may well have had considerable survival value.

Evolution and the Origin of Language

On rereading these remarks, it appears that perhaps insufficient attention has been paid to the difficulties inherent in a purely Darwinian conception of the origin of language.

It was implicit in this particular hypothesis as to evolution that differences between human and animal structure and function are matters of degree. Were this principle to be firmly established, then it would be difficult to avoid the idea that animal communication leads by insensible gradations to the faculty of speech in man. There are numerous linguistic objections to this view, however. It is important to realise, too, that language does not stand alone in this matter and that there are other weighty considerations which lead to the well-nigh inescapable conclusion that some potent qualitative change occurs at a point somewhere between the anthropoid and *Homo sapiens.*

Animals, at best, may possess a limited store of vocal sounds. These are innate, instinctive, or "natural." Under appropriate circumstances, internal or external, they are emitted. They may happen to possess communicative action in respect to other animals mainly of the same species. It is doubtful, however, whether these cries are always communicative in intent. In ordinary circumstances the adolescent animal does not increase its vocabulary of sounds, except that in one or two strictly delimited circumstances the vocal repertoire may be extended. Thus some animals in states of domestication may amplify their stock of cries and calls. Other animals, particularly certain birds, may elaborate their performance by dint of imitation. In this way the innate and instinctive bird song or call is overlaid by dint of learning from other birds, not necessarily of the same species. Finally, certain animals, particularly a small group of birds and a few higher apes in captivity, may be taught to mimic human articulate speech, the specific cries of quadrupeds, or even inanimate noises.

The foregoing recounts the sum-total of achievement in the domain of animal sounds. Between these and human articulate speech lies a very considerable gulf. Even in the case of the most untutored, primitive, and savage human communities the language-system is so far removed in its complexity from the crude and simple utterances of the sagest of the primates as to be scarcely comparable. And nowhere and at no time has there been any hint of an approximation between these two extremes. No "missing link" between animal and human communication has yet been identified.

Can it be, therefore, that a veritable Rubicon does exist between animals and man after all, as Professor Müller insisted when discussing the origins of language? Has a new factor been abruptly introduced into the evolutionary stream at some point between the Hominoidea and the Hominidae, constituting a true "barrier"? Can it be that Darwin was in error when he regarded the differences between man and animals as differences merely in degree?

The lessons gained from comparative linguistics would certainly suggest that there are serious differences "in kind" which interpose themselves at a late stage in evolution. We have been told that the contrasting of differences in kind and in degree is in itself an outmoded attitude. However that may be—and such argument is not easy to follow—it is tempting to doubt whether anything like a smooth gradation has occurred. Outside the domain of language there are other human endowments which are not readily traced in the animal series. As such, they scarcely pertain to our present subject-matter, unless it can be shown that their very existence depends upon the presence of a language system. Here, for example, may be placed the advent in man of what we might loosely term the various "moral faculties." Darwin was not oblivious of this problem, and he believed that a moral sense had been evolved from prehuman ancestors. This aspect of evolution was not mediated by a process of natural selection, however, but it arose from man's newly acquired power of reasoning. So then it is a mechanism of evolution additional to the ordinary natural selection. When early man became endowed with reason and when to that mental accomplishment was added the power of speech, then the way lay open for the operation of conscious purpose (de Beer) or the psychosocial factor (J. Huxley). In this way there develops—again indirectly out of the beginnings of language—the beginnings of choice as to conduct. This also implies the power of doing harm as well as the power of doing good. So arise ethical and altruistic considerations. The earlier stages of these aspects of behaviour can be visualized in the animal kingdom in the instinct of maternal solicitude. This instinct is restricted—be it noted—in both time and place. With the achievement of adulthood, the young animal no longer receives maternal solicitude. The instinct, too, is limited to the immediate family group. Altruism extends from beyond the family circle to the clan only with the attainment of human status; and thence, with the growth of social conscience, it expands to embrace the tribe and eventually the nation. This act of stepping outside the strict family circle may doubtless be assisted—if not mediated—by the faculty of language.

A. IRVING HALLOWELL

SELF, SOCIETY, AND CULTURE IN PHYLOGENETIC PERSPECTIVE

Introduction

When man is considered in evolutionary perspective, primary emphasis is usually given to his biological attributes, the morphological features that can be dealt with comparatively in other members of the primate order, living or extinct, and in organisms more distantly related. In the intellectual climate of the post-Darwinian period, however, human evolution was conceptualized in much broader terms. The advent of Darwinism helped to define and shape the problems of modern psychology as it did those of anthropology. An evolution of "mind" within the natural world of living organisms was envisaged. Now a bridge could be built to span the deep and mysterious chasm that separated man from other animals and which, according to Descartian tradition, must forever remain unbridged. Darwin himself explicitly set processes of reasoning, long considered an exclusively human possession, in an evolutionary perspective; he also advanced an evolutionary interpretation of the facial and postural changes of man when expressing emotion.[1] He argued that mental differences in the animal series present gradations that are quantitative rather than qualitative in nature. A. R. Wallace, who had had more intimate contacts with primitive peoples than Darwin, was immensely impressed with their abilities. To him, the evolution of the brain represented a sharp mutational development in man that permitted adaptive capacities far beyond those he thought were necessary for survival on a primitive

[1] Darwin, *Descent* and *Expression of Emotions*. Cf. R. H. Waters in C. P. Stone (ed.), *Comparative Psychology*. In her Preface to a recently published edition of *The Expression of the Emotions,* Margaret Mead relates Darwin's work to the developing interest in "the non-verbal aspects of human communication—the new science of kinesics."

A. IRVING HALLOWELL is Professor of Anthropology at the University of Pennsylvania and in the Psychiatry Division of the School of Medicine, and Curator of Social Anthropology at the University Museum. He is a past president of the American Anthropological Association, the American Folklore Society, and the Society for Projective Techniques. Winner of the Viking Medal in 1955, Professor Hallowell's special interests are personality and culture and the history and culture of American Indians.

level of human existence. The position he took was a challenge to the theory of natural selection as espoused by Darwin himself. "Natural Selection," Wallace wrote, "could only have endowed the savage with a brain a little superior to that of an ape, whereas he actually possesses one but very little inferior to that of the average members of our learned societies." [2] Eiseley points out that, in 1864, Wallace "set forth the idea that with the rise of man, natural selection was ceasing to act upon the body and was coming to act almost solely upon the human intelligence. Man, he contended, was old and had attained the upright posture long before the final changes in the skull and brain which characterize our living species. Other animals had continued to change and modify under evolutionary pressures; in man, by contrast, all but mental evolution has largely ceased." [3]

Although Darwin was later accused of gross anthropomorphism by some of his critics, he did stimulate scientists to think and write about mental evolution. Wallace's views, on the other hand, were discounted because, in the end, he fell back upon a theological explanation. Romanes, a disciple of Darwin coined the term "comparative psychology," [4] and it was not long before a phylogenetic dimension had been added to the program of scientific psychology. In its early stages, however, comparative psychology had little interest for anthropologists. In reaction against anecdotalism, more rigorously controlled observations were demanded by psychologists, and lower mammals, like the rat, and insects too, became preferred laboratory subjects. The results of these observations, even though highly reliable, did not throw much light on the phylogenetic roots of human psychology. Laboratory studies of infrahuman primates like the chimpanzee, initiated by Köhler, Yerkes, and Schultz, only developed to a point where they engaged anthropological interest in the twentieth century.[5] Even today, as Nissen points out, "of the 50-odd living genera of primates, only a very few have been studied to any extent in regard to behavior: man, chimpanzee, the macaques, and cebus monkeys." [6]

It was also under the stimulus of Darwin's ideas as applied to man that historians, economists, sociologists, linguists, cultural anthropologists, and others began to apply evolutionary ideas to human institutions on a wide scale. Language, religion, art, marriage and the family, law, and economic organization were studied comparatively in order

[2] Quoted by Eiseley (1955), p. 63.

[3] Eiseley (1955), p. 63. Cf. Eiseley (1958), p. 306.

[4] His *Animal Intelligence* (1883) "is the first comparative psychology that was ever written, and its author used this term believing that comparative psychology would come to rank alongside of comparative anatomy in importance" (Boring, p. 473).

[5] See Yerkes (1943), "Epilogue: The Story of an Idea," pp. 289–301.

[6] Nissen (1955), p. 100.

to discover whether orderly developmental sequences could be established. But the fact should not be overlooked that these efforts were chiefly confined to developments in a single species of the Hominidae —*Homo sapiens.* Fossil material and archeological remains were scanty at the time and field studies of non-hominid primates living in their native state were non-existent. Besides this, the evolutionary hypothesis was closely linked with the older idea of progress as applied by social scientists and humanists. This re-enforced the reconstruction of series of unilinear stages which somewhat paralleled the concept of orthogenesis in biology. In this form theories of social and cultural evolution persisted into the early years of this century.

Since it was assumed that processes of evolution were not confined to the organic sphere alone, a corollary psychological question arose in conjunction with the attempts that were being made to establish stages of cultural evolution which had taken place in the course of man's long struggle upward from savagery to civilization. Could it be shown that in the cultures of primitive peoples there was a reflection of primitive mind? J. G. Frazer, who adhered to the recapitulation theory, was among those who explicitly linked this problem with the generic question of mental evolution. He thought that not only ethnographic data were relevant but likewise studies of patients in mental hospitals and of the ontogenetic development of the child. He said that "this comparative study of the mind of man is thus analogous to the comparative study of his body which is undertaken by anatomy and physiology." [7] But when unilinear stages of cultural evolution were rejected by most twentieth century anthropologists, the notion of "primitive mind," as applied to non-literate peoples, collapsed with them; the conclusion was drawn that culture change and development in *Homo sapiens* are not primarily linked with evolution in mentality. Outside of anthropology, the more inclusive concept of genetically determined mental evolution—insofar as it sought support in the theory of recapitulation in its original extreme form—became generally defunct with the rejection of this theory by biologists.[8] Thus the psychological dimension of evolution which to Darwin himself was an integral part of the total evolutionary process and of vital significance for our comprehension of man's place in nature fell upon evil days.[9] It is true that animal psychologists continued to investigate

[7] Frazer, p. 586. In this paper Frazer expresses a preference for "the more general name of mental anthropology," rather than "social anthropology" for the division of the subject in which he worked.

[8] See Hallowell (1954), pp. 167–70; and (1955), chap. 2; De Beer, chap. 1.

[9] Wayne Dennis (1951, p. 2) points out that "it would not be inaccurate to say that developmental psychology began with a theory—the theory of recapitulation. Child psychology, the most productive segment of developmental psychology, began shortly

some problems comparatively; but special areas of investigation, such as learning behavior in rats, emerged to the foreground, while a primary focus on evolutionary questions as such receded. Schneirla, in a review of trends in comparative psychology (1952), emphasizes the fact that "most American animal psychologists at present seem to be *really* non-evolutionary minded, in the sense that they show no special zeal to find how man differs mentally from lower animals and vice versa, but rather focus strenuously on general problems without much attention to phyletic lines." [10]

So far as anthropology is concerned, the rejection of nineteenth century unilinear theories of cultural evolution along with the notion of a demonstrable level of primitive mentality in *Homo sapiens* meant that evolution, once so inclusively conceived, was reduced in effect, to investigations in the area of physical anthropology. Physical anthropologists, moreover, concerned themselves chiefly with morphological problems, not behavior. Thus, the question arises whether, in the centenary year of the publication of the *Origin of Species,* our thinking about human evolution must remain confined to the investigation of morphological facts? I do not think so. I believe that, in the light of contemporary knowledge, it is both possible and desirable to reconsider human phylogeny in a more inclusive frame of reference than has prevailed in the immediate past, without returning to pseudo-evolutionary problems and theories of an earlier day. For, in addition to the reappraisement of problems in the area of human culture history, undertaken by other contributors to these volumes, we still are faced with the question, What were the necessary and sufficient conditions

---

after the promulgation of the theory of evolution when all scientific minds were inflamed by this great conceptual achievement." But after the movement had arrived at the "concise hypothesis" that "ontogeny recapitulates phylogeny" its decline was imminent. "The evolutionary viewpoint had seemed to open up wide unconquered vistas to child psychology. But on closer approach these beckoning plains proved to be inhabited only by unsubstantial figures and retreating will-o'-the-wisps. There was not a testable hypothesis in the entire landscape." While some psychologists of an older generation, like G. Stanley Hall and J. M. Baldwin, had adopted the theory of recapitulation and made it an integral part of their thinking, as did Frazer and Freud, this working hypothesis is not a necessary assumption of "developmental psychology." Heinz Werner (1940, p. 3) points out that the concept "is perfectly clear if this term is understood to mean a science concerned with the development of mental life and determined by a specific method, i.e., the observation of psychological phenomena from the standpoint of development." While some psychologists, when they employ this term, refer only to ontogenesis, "the mental development of the individual is, however, but one theme in genetic psychology." Related to it "is the developmental study of larger social unities, a field of interest intimately linked with anthropology and best known by the name of *ethnopsychology*. The question of the development of the human mentality, if not arbitrarily limited, must lead further to an investigation of the relation of man to animal and, in consequence, to an *animal psychology* oriented according to developmental theory."

[10] Schneirla (1952), p. 563.

that made possible, through evolutionary processes, adaptation at the level of primate existence that we find culminating in *Homo sapiens?* Simpson has pointed out that "the generally accepted modern theory of evolution is called 'synthetic' but comparative psychology has been an element not yet fully incorporated in the synthesis." [11]

What I am suggesting is that we more boldly extend the synthetic principle by probing more deeply the *anlagen* of the psychological, social, and cultural dimensions of man's existence, in order to define with more precision the relevant continuities, as well as the discontinuities, which link the behavior of *Homo sapiens* with his predecessors in the evolutionary process. The reason it seems worthwhile to consider human evolution afresh in a wider frame of reference is that now we have sources of information at our disposal which were not available in the nineteenth century, or even in the earliest decades of this century.

1. In the field of human paleontology new fossil material of importance has turned up while, on the other hand, the confusion created by the fake fossil, once known as Piltdown Man, has been resolved. Of the new material, the remains of the australopithecines are of paramount importance. Their over-all dental morphology conforms to that found in more advanced hominids such as Pithecanthropus. And besides this, although their cranial capacity is lower than that of the latter, they had achieved bipedal locomotion. Consequently these primates, associated with Villafranchian fauna of Lower Pleistocene date, are now placed among the Hominidae. Structurally and behaviorally they represent an earlier level of hominid development than the Pithecanthropus group, once considered to represent the earliest known hominid type.[12] We now know that the earliest hominids were small-brained and newly bipedal. Large brains did not announce the advent of hominid evolution, as Sir G. Elliot Smith argued in 1912 and as Sir Arthur Keith repeated with reference to Piltdown. "Recent finds of fossil men and other primates," writes Strauss, "indicate that it is the brain that was the evolutionary laggard in man's phylogeny; indeed, the studies of Tilly Edinger of the phylogeny of the horse brain suggest that this may well be a general rule in mammalian evolution." [13] Darwin himself remarked: "We must bear in mind the comparative insignificance for classification of the great development of the brain in man." [14]

In the light of this new knowledge it is now apparent that the familiar

[11] Roe and Simpson, p. 1.

[12] See Le Gros Clark, 1955, 1958, 1959; Howell, 1959; and Washburn and Howell in this volume.

[13] Strauss (1955*a*), p. 370.

[14] Quoted by Le Gros Clark (1958), p. 192.

terms "man" and "human" are colloquial terms. They are *not* equivalent to the zoological term Hominidae, or to the adjectival form, "hominid." The australopithecines were early hominids, but they were not "human" in the sense that we are human. We represent the terminal product of hominid development. It is difficult now to make use of the colloquial term "man" in discussing evolution.[15] Our temporal predecessors within the zoological family Hominidae were not all "men" in the usual lay meaning of the term. This situation is indicative of a broadening of our knowledge as well as the reality of the evolutionary process itself. In current zoological classification, the subfamily Homininae, as distinguished from the subfamilies Oreopithecinae and Australopithecinae, serves to differentiate later from earlier groups of hominids. And the term "euhominid" is coming into use to designate "men" both living and extinct, belonging to the most evolved group of the family Hominidae, the Homininae.[16]

2. Beginning with the pioneer field observations on chimpanzee by Henry Nissen, published in 1931, shortly followed by the studies of C. R. Carpenter (1934, 1940) on New and Old World monkeys and the gibbon, a new body of information on non-hominid primates began to accumulate. We now have reliable data on a few samples of the ecology and organization of primate societies,[17] supplementing behavioral observations made under laboratory conditions.[18] While hitherto both kinds of studies were made by psychologists, a few anthropologists have now begun to study non-hominid primates in the field.

3. Another source of relevant data comes from what psychoanalytic theory has been able to tell us about the structure and functioning of the human personality and from cross-cultural studies of the relation of cultural variability to personal adjustment. By directing attention to the conditions under which primary processes in human adjustment occur in *Homo sapiens,* these have raised further questions of general

[15] "The confusion of ideas," warns Le Gros Clark (1958, p. 193), "to which the loose and uncritical use of the colloquial term 'man' can lead is particularly well shown in those lists of anatomical features which are from time to time enumerated with the intention of demonstrating his uniqueness in the animal world. For example, it has been claimed that 'man' is distinguished from all other Primates by such characters as the structure of the genital organs, the prominence of the calf muscles, the red lips, the shape of the female breast, the comparative nakedness of the body, and so forth. It does not always seem to be realized that features of this kind may be no more than distinctions at the specific or generic level—characteristic of the species *Homo sapiens,* and perhaps of the whole genus *Homo*. But, of course, we have no idea whether they were also characteristic of extinct hominids such as Neanderthal man, *Pithecanthropus* and *Australopithecus*." Cf. Schultz, 1957.

[16] See Howells on classification of primates (1959), p. 137, p. 351, and Glossary.

[17] For bibliography to 1957, see Carpenter, 1958*b*.

[18] At the present time experimental research on non-hominid primates is also expanding. See, e.g., Harlow (1956), p. 273.

anthropological and psychological interest.[19] As case studies these investigations have been mainly concerned with specific differences in personality structure and functioning which can be shown to be related to cultural differences, but implicit in these data are indications that universal dynamic processes are involved which are related to the psychobiological nature of modern man as a species. Likewise, capacities are implied which must be related to generic psychological attributes of *Homo sapiens* that have deeper roots in the evolutionary process. All human individuals, through learning processes, become psychologically structured for participation in concrete sociocultural systems. On the other hand, hominids considered as an evolving group became the "creators" of culture in the generic sense.

In conventional terminology the psychological dimension of evolution has long been phrased as the evolution of "mind." But this terminology reflects the mentalistic concepts of an older period of psychology when "mind," "intelligence," and "reason" were key terms. One of the seminal contributions of the psychoanalysts—daily faced with persons who need concrete help in readjusting to their life situation—was to hypothecate a model of personality organization, conceptualized in "structural" terms. Whether we accept their particular model or not, it is one which has proved useful in clinical practice. It has likewise been fruitful in analyzing the dynamics of human behavior in sociocultural contexts of all sorts. In phylogenetic perspective, the psychoanalytic model of personality structure suggests that one of the things we must account for in human evolution is not simply a "human mind" in the abstract, but a generic type of personality organization which did not exist, perhaps, at the earliest hominid level. In terms of the psychoanalytical model, too, the "rationality" of the human mind is counterbalanced by an irrationality, linked with the constant play of biologically rooted forces which are intelligible in an evolutionary perspective. Furthermore, with the advent of a cultural mode of adaptation, the biological adaptation of human individuals becomes subordinate to their psychological adjustment.[20] They become an integral

[19] For a guide to culture and personality studies, see Hallowell, 1953, and Honigman, 1954 and 1959.

[20] "Intelligence," "reason," and other "mental traits" become specific functions of the total personality structure. Furthermore, whether described as "mind" or "personality structure," the psychological organization of the human being is just as much a function of his membership in an organized social group as it is a function of his inherited organic equipment. John Dewey emphasized this point prior to the time when culture and personality studies were initiated and also before social psychology had assumed its present form. He said (1917), "What we call 'mind' means essentially the working of certain beliefs and desires, and that these in the concrete,—in the only sense in which mind may be said to *exist,*—are functions of associated behavior varying with the structure and operation of social groups." Thus instead of being viewed as "an antecedent and ready-made thing," mind "represents a reorganization of original

part of the perpetuation of sociocultural systems to the extent that variations in personality structure and the roles which human beings are groomed to play become a necessary condition for the survival and functioning of such systems.

4. Kroeber has said that "the most significant accomplishment of anthropology in the first half of the twentieth century has been the extension and clarification of the concept of culture." [21] Without considering formal or elaborated definitions here, the essence of the concept as originally developed is that learned behavior, socially transmitted and cumulative in time, is paramount as a determinant of human behavior. The cultural systems which characterize human societies are products of social action and, at the same time, are conditioning factors in further action. It was the application, as Kroeber points out, of the culture concept by twentieth century anthropologists, among whom it assumed central importance, that "immensely advanced the growth of anthropological science."

With respect to the evolution of man, this emphasis on culture led to a somewhat paradoxical situation in the early decades of this century. While continuing to give lip service to organic evolution, although rejecting nineteenth century theories of cultural evolution, a crucial evolutionary issue was held in abeyance. Culture was taken for granted and stressed as the unique possession of *Homo sapiens* and earlier types of hominines, dating far back into the Pleistocene. The chief evidence was the association of tools with these early euhominids. In the 1920's Kroeber's paper, "Sub-human Cultural Beginnings," was practically unique. So far as *Homo sapiens* was concerned, it was assumed that all living races shared equally the necessary psychological capacities for acquiring culture. Culture in its concrete manifestations, considered as an attribute of all human societies, was abstracted and studied as such. Culture traits, complexes, and patterns became key terms. In effect, this preoccupation with culture led to a *re*-creation of the old gap between man and the other primates which, it was once thought, the adoption of an evolutionary frame of reference would serve to bridge. The repeated emphasis given to speech and culture as *unique* characteristics of man sidestepped the essence of the evolutionary problem.[22] Distinctive characteristics of the most highly

activities through their operation in a given environment. It is a formation, not a datum, a product and a cause only after it has been produced."

[21] Kroeber (1950), p. 87.

[22] In a biological frame of reference, Le Gros Clark points out (1958, p. 186), "The opposition to Darwin's thesis of the evolutionary origin of man naturally led his critics to search for anatomical characters in which the human body could be said to be 'unique,' thus providing arguments for removing man in any system of classification as far as possible from other mammals (especially the apes). In some cases, indeed, these arguments were pushed to an extreme of absurdity, which today we are apt to find rather astonishing." Reference is made to the wrangle over "hippocampus minor."

evolved primate were asserted without reference to prior capabilities, conditions, and events in the evolutionary process that made this characteristic mode of adjustment possible. For unless culture and speech be conceived as sudden and radical emergents, they must be rooted in behavioral processes which can no more be considered apart from the general framework of behavioral evolution than the distinctive structural characteristics of man can be considered apart from morphological evolution. Without the establishment of the nature of such linkages the question arises, How far has the emphasis given to distinctive attributes of man advanced our understanding of man's evolutionary position in the animal series beyond the descriptive epithets of an earlier day? One thinks of such characterizations of man as the "rational animal," the "tool-making animal," the "cooking animal," the "laughing animal," the "animal who makes pictures," or *animal symbolicus.* All these characterizations stress man's differences from other living creatures. Like the criteria of culture and speech, they emphasize discontinuity rather than the continuity, which is likewise inherent in the evolutionary process.

A statement made by Carpenter a few years ago clearly articulates an opposition to any such sharp descriptive dichotomization between man and other primates. He said he found untenable a number of assumptions that seemed acceptable to many of his colleagues. One of these was "that the phenomena known as 'mind,' language, society, culture and 'values' exist exclusively on the level of human evolution." [23] And Hebb and Thompson say that "exposure to a group of adult chimpanzees gives one the overwhelming conviction that one is dealing with an essentially human set of attitudes and motivations." [24] Thus, while cultural anthropologists have continued to render formal homage to the idea of evolution, at the same time the full range and depth of its significance has not been actively pursued. The statements of Carpenter, and Hebb and Thompson, should remind us that there remain crucial evolutionary questions which transcend the old problem of unilinear stages of cultural development and those problems which are dealt with by physical anthropologists.[25]

In phylogenetic perspective, we must ask: Did all the aspects of cul-

[23] Carpenter (1955), p. 93.

[24] Hebb and Thompson, p. 543.

[25] However, in a recent review of "some of the achievements and a few of the problems which characterize present-day physical anthropology," J. S. Weiner directs attention to behavioral evolution when he writes: "There is one large baffling topic on which our evolutionary insight still remains very meagre—the emergence of the peculiar attributes of human intelligence, temperament, and social organization." He quotes Hebb and Thompson (note 24), refers to the work of Carpenter and Zuckerman, and concludes: "It remains an unfortunate fact that of all aspects of physical anthropology this one, which carries so much of promise to the sociologist and social psychologist no less than to the human biologist, should at the present time be the most neglected of all fields of study."

ture as observed in *Homo sapiens* come into being together at an early hominid stage? Did the australopithecines *manufacture* tools, speak, pray, exercise property rights, draw, paint, and recognize moral values? Is there any relation between the expansion of the brain and cultural adaptation? Are "half-brained" hominids as capable of cultural adaptations as those with an expanded cortex? Is speech a necessary condition for the earliest phases of cultural adaptation? And is there any relation between tool-making, as contrasted with tool-using, and speech? Do non-hominid primates show any traces of what has been called culture in *Homo sapiens?* And what of the dimension of psychological structure? What kinds of psychological capacities and mechanisms underlie a cultural mode of adjustment? And what is the relation between the development of systems of social action in the primates and the emergence of cultural systems, characterized by a normative orientation?

What I have attempted in this paper are the broad outlines of a conjunctive approach to human phylogeny in which the organic, psychological, social, and cultural dimensions of the evolutionary process are taken into account with reference to the necessary and sufficient conditions that underlie a human level of existence. At the same time, I have devoted some attention to earlier opinions, in order to bring the problems needing reconsideration into sharper focus in the light of contemporary knowledge. Behavioral evolution is, perhaps, the term which best defines the framework of a conjunctive approach. Biologists, too, are now taking an increasing interest in behavioral evolution. Some years ago Nissen remarked that "one of the weakest links in the sciences dealing with evolution, the one most needed to strengthen its facts and theoretical framework is that dealing with behavior." [26] It is in behavioral perspective that we can best conceptualize the major categories of variables that must be examined with reference to the evolutionary status of *Homo sapiens.* Whether we consider hominid evolution in an ecological, a social, a psychological, or a linguistic frame of reference, behavior is the unifying center to which we must constantly return at each adaptive level. As we proceed to higher levels we must consider new integrations of determinants brought about by potentialities for behavioral adaptations that did not previously exist,

[26] Nissen (1955), p. 106. Also Lashley in his Introduction to the classical papers of the ethologists, *Instinctive Behavior,* edited by Claire H. Schiller, notes on page ix: "They have traced patterns of instinctive activity among related species and have shown that behavior may be as clear an index of phylogenetic relationship as are physical structures." In contrast with American psychologists, these zoologists have focused their attention upon instinctive rather than learned behavior, and their observations have been made in the field rather than the laboratory. The chief animals groups studied so far have been invertebrates and lower vertebrates. See also N. Tinbergen, *Social Behaviour in Animals.* New York: John Wiley & Sons, 1953; also Tinbergen's paper in our companion volume, "The Evolution of Life."

for example, the consequences of bipedal locomotion, the adoption of new food habits, the use and manufacture of tools, the expansion of the brain, the effects of a new level of psychological integration in the later hominids, and the role of speech in the symbolic mediation and coordination of social relations. In the evolutionary process, differential behavior patterns provide major clues to significant variables.

Any attack on problems of behavioral evolution, of course, involves inherent methodological difficulties. A direct observational approach at all stages is not possible. With respect to the past, we can only make inferences and deductions from non-behavioral data. But we can observe and compare the behavior of different species of living primates, with full appreciation of the fact that they represent their own specialized modes of adaptation. In the case of the hominids, archeological data provide us with both the material products of individual activity and the consequences of social interaction as expressed in traditional usage where the manufacture of tools can be established. But the archeologist, as such, is not concerned with the problem of behavioral evolution. His attention is chiefly directed to the forms, distributions, and temporal relations of objects from which the early cultures of the euhominids can be inferred. Questions of behavioral evolution, on the other hand, force us to look behind the tool and ask questions which neither the archeologist nor the physical anthropologist can answer by a direct appeal to their data. Tools as products of behavior raise questions of another order. To account for a tool-making tradition by one creature and not another we have to consider the psychobiological capacities which are a necessary condition of tool-making; intervening variables have to be inferred. Problems of this kind must be faced sooner or later, and, indeed, behavioral criteria frequently have been invoked in dealing with questions of human evolution but without sufficient discussion of all the psychological implications involved. While it is inevitable that there will be differences of opinion in the interpretation of the facts of behavioral evolution, the areas which involve dispute will be narrowed with the accumulation of new data.

The Psychocultural Dimension of Evolution

Some years ago Le Gros Clark, referring to the question of the zoological classification of the australopithecines,[27] said:

> Taxonomic difficulties of this sort, of course, are bound to arise as discoveries are made of fossils of a seemingly transitional type, and with the

[27] Le Gros Clarke (1950, p. 73): "To say that man differs from the other primates in his capacity for tool-making and language is not very useful," to which Nissen (1955, p. 102) adds "until we have identified the mechanisms and processes which produce these complex end results."

increasing perfection of the fossil record, probably the differentiation of man from ape will ultimately have to rest on a functional rather than an anatomic basis, the criterion of humanity being the ability to speak and make tools.

We must ask, then, what special capacities and conditions underlie the phenomena of speaking and tool-making? Effective use of such criteria is hardly possible without considering what these capacities and conditions may be. We cannot depend on the evidence from human paleontology and archeology alone. Insofar as speech is concerned, it is now known that reliable inferences cannot be made from brain anatomy.[28] Furthermore, it seems doubtful that speech as observed in *Homo sapiens* possesses properties as a system of communication which can be treated as a phenomenal unity in phylogenetic perspective. The question is, How far can speech actually be projected into the past? [29] Do we not have to know more than we now know about the properties of non-linguistic systems of communication at subhuman levels in order to understand the position of speech in behavioral evolution? [30]

Hockett has recently pointed out that "part of the problem of differentiating Man from the other animals is the problem of describing how human language differs from any kind of communicative behavior carried on by non-human or pre-human species. Until we have done this, we cannot know how much it means to assert that only Man has the power of speech." [31] He has approached the problem by identifying seven "key properties" of the speech of *Homo sapiens* and compared them with the available data on non-human systems of communication. Hockett discovered that there was considerable overlapping in the properties selected, although they did "not recur, as a whole set, in any known non-human communicative system." [32] This suggests that the combination of properties that characterize speech, those "design-features" which "seem to be of crucial importance in

[28] Rèvèsz (1956), p. 92.

[29] Critchley (in the present volume) notes that Keith believed that a capacity for speech could be pushed back no further than Neanderthal man, whereas L. S. Palmer, basing "his opinion upon the anatomical characteristics of the mandible," argues that perhaps the autralopithecines could speak.

[30] Schneirla (1952, p. 582) points out that an adequate comparative study of group communicative behavior "is long overdue, particularly to clarify the relationships of concepts such as 'sign,' 'signal,' and 'symbol,' as well as the criteria of 'language,' all of which appear to suffer from a heavy load of speculation and a minimum of systematic research." Cf. the historical review, "Animal Communication" by Critchley and the discussion of "Animal Languages" in Brown (p. 156 ff.). Perhaps expanding research in the area of "paralanguage" in man, as defined by Trager (1958), may provide some new leads.

[31] Hockett (1958), p. 570.

[32] *Ibid.*, p. 574.

making it possible for language to do what it does," [33] did not arise full-blown. It is argued that this assemblage of properties, considered with reference to man's lineage, "could not have emerged in just any temporal sequence. Some of them either unquestionably or with high likelihood imply the prior existence of some of the others." [34] Consequently, Hockett is led to suggest a tentative evolutionary reconstruction.

Since one of the key properties of a human system of communication is "cultural transmission," [35] a property absent in the communication systems of primates and other animals, this factor becomes highly significant chronologically and, I think, has wider implications than those developed by Hockett. The latter suggests, in effect, that, although learning and the social transmission of habits, or what he calls a "thin sort" of culture, may have existed at a very early stage in the development of the higher primates, the associated system of communication that prevailed may have operated without "cultural transmission." [36] In other words, what I prefer to call a "protocultural" stage may have been chronologically prior to speech but not, of course, to some other system of communication. The evolutionary significance of this chronology as adapted to communication lies in the fact that the conditions which permitted a protocultural stage to develop were, at the same time, among the necessary prerequisites of a communication system characterized by the total assemblage of properties considered by Hockett.

This kind of evolutionary inquiry is, of course, a far cry from earlier approaches, particularly those which began by concentrating on the problem of "primitive" languages spoken by *Homo sapiens.* These proved as fruitless as attempts to discover evidence of "primitive mind" in our species. These failures, however, may have helped to expose genuine evolutionary problems more clearly. Hockett's approach permits us to have a fresh look at speech in greater evolutionary depth. And by direct observation we know that whereas some of the great apes have been able to acquire a "thin sort" of human culture when closely associated with members of our species, they do not have the capacity to acquire and use our distinctive form of linguistic communication, even when systematically motivated.[37] There seems little

[33] Hockett, "Animal 'Languages' and Human Languages," a paper read at the annual meeting of the American Anthropological Association, Dec., 1957, but not published until 1959 (p. 32).

[34] Hockett (1958), p. 581.

[35] Hockett (1959, p. 36) says: "A behavior pattern is transmitted culturally if it is not only learned but taught, and if the teaching behavior, whatever it may be, is also learned rather than genetically transmitted." Cf. Hockett (1958), pp. 579–80.

[36] *Ibid.*, p. 36.

[37] See, e.g., Hayes (1951), chap. 8: "Teaching an Ape to Talk."

reason to doubt that, in the course of behavioral evolution, psychological capacities of crucial importance lay back of the ultimate emergence among the hominids of a characteristic system of communication. While this system shared some "design-features" with that of non-hominid primates, capacities that transcended those of other primates permitted the development and integration of novel features. These, in turn, resulted in the functional potentialities of speech as we know it in *Homo sapiens.*

Man has long been defined as the "toolmaker," yet, if tools are taken as an index of a human status, considerable preliminary analysis is required to make this criterion useful. Oakley [38] has been more precise than previous writers in his *Man, the Tool Maker,* but, nevertheless, an English biologist, Pumphrey, has remarked that " 'Subman, the Implement Maker' would have been a more accurate if less impressive title at least for the first half of his book." Pumphrey sees "no valid reason for assigning intellect to a maker of implements. . . . The web of a garden-spider and the nest of a chaffinch are highly fabricated implements," whereas genuine tools, which he thinks cannot be assigned to early members of the Hominidae, "were made in order to make something else with them." [39] Even if we define the tool concept in terms of some very general adaptive function, without further analysis it is not very useful for making distinctions in an evolutionary frame of reference. Bartholomew and Birdsell say, "In contrast to all other mammals, the larger arboreal primates are, in a sense, tool-users in their locomotion [since,] as they move through the maze of the tree tops, their use of branches anticipates the use of tools in that they routinely employ levers and angular movements," [40] which is a very broad interpretation of tool-using. These authors draw the conclusion, moreover, that "protohominids were dependent on the use of tools for survival."

There is ample evidence that both biologists and psychologists have had their own difficulties in dealing with the question What constitutes tool-using? [41] And because the phenomenon of "tool-using" is not confined to the primates alone, it is necessary to understand the varying factors that underlie what has been called tool-using in other animals, in order to interpret properly the phenomenon of tool-using in the behavioral evolution of the primates and the differential factors that made tool-*making* possible as a unique development within the hominids.

In psychological experiments with infrahominid primates, "instru-

[38] Oakley, 1950, 1951, 1954, 1956.
[39] Pumphrey (1951), pp. 27–28.
[40] Bartholomew and Birdsell, pp. 482–83.
[41] See, e.g., Thorpe, pp. 109, 332; Nissen, 1946.

mentation," as it is usually called, includes piling boxes to secure food, the manipulation of sticks to achieve a similar goal, or pole vaulting! What is interesting is that high proficiency in instrumentation under laboratory conditions appears to be a function of previous experience in related situations.[42] However, it is individual learning rather than social learning that is involved in "tool-using" of this order. Sultan's success in "making" a tool was a unique individual achievement.[43] While there would seem to be no question of the capacity of some primates to use tools as a means of achieving a desired goal when sufficiently motivated, this potentiality alone is only one of the necessary prerequisites to a more highly developed stage of tool-using. However, it seems quite likely that, under natural conditions, some rudimentary habits of tool-using in the narrower rather than the broadest sense may have been individually learned and socially transmitted in non-hominid or early hominid groups. If so, this would exemplify what I have called a *protocultural* stage.[44] Nevertheless, the conditions operative at such a stage in primate groups are not in themselves sufficient to account for the still more advanced level of *tool-making*. If the latter is invoked as a functional criterion for human status we need to do more than differentiate between tool-using and tool-making. We must ask whether tool-making presupposes a higher order of psychological structuralization and functioning than tool-using; whether it implies a social system different from that of non-hominid primates; or a different system of communication.[45] Tool-making as observed in *Homo sapiens* is a skilled act—learned in a social context where speech exists, and usually performed with

[42] Harlow (1952, p. 217), referring to Köhler's earlier construct of "insight learning," says "Insightful behavior on instrumentation problems apparently occurs only in animals that have had previous opportunity for experience in related situations."

[43] Nissen (1946, p. 562) says: "The nearest thing to the manufacture of tools in the ordinary sense seen in primates is the observation reported by Köhler of a chimpanzee fitting together two short sticks in order to make a long one. This observation has not been repeated."

[44] Hallowell, 1956.

[45] See White, 1942. In 1927, Grace A. de Laguna argued that "it is scarcely credible, even aside from the more theoretical psychological considerations, that the art of chipping stone implements could have been developed by men who had not yet learned to speak." In a later, unpublished manuscript, de Laguna has expressed her thought by saying: "*Homo faber* is *Homo cogitans.*" Cf. Révész (1956, pp. 92–93), who equates *Homo faber* with *Homo loquens*. Cf. Vallois (p. 211), who points out that tool-making, "un phenomène essentiel de l'hominisation culturelle," undoubtedly was preceded by an earlier stage of utilization . . . "qui n'impliquait encore qu'une hominisation à ses débuts. Les processus qui ont permis la fabrication doivent au contraire correspondre á une cérébralisation déjà avancée ainsique, peut-être, à uncertain usage de la parole. Une telle fabrication suppose en effect l'apparition de nouveaux centres corticaux et de nouvelles connexions sensitivo-motrices. Elle suppose l'idee d'une transmission des techniques d'un individu à un autre."

reference to a purposeful use at some *future* time.[46] Therefore, do we not have to make up our minds, when interpreting the archeological evidence, whether tool-making necessitates a sense of self-orientation in time, and, possibly, institutionalized property rights which assure continued control over the tool in the interval? When we have direct evidence of the persistence of characteristic techniques of manufacture and tool styles as well as evidence of innovation or invention (i.e., a tool-making tradition), we do have indices to a human level of cultural adaptation. But this involves far more than tool-making per se or mere social transmission.

The more perplexing evolutionary problems arise in cases where the material evidence is ambiguous. The problem is particularly difficult where the early hominids responsible for the archeological remains had a smaller cranial capacity than later hominids of the Middle Pleistocene and after. At first, the general opinion prevailed that the bipedal australopithecines of Villafranchian age were not toolmakers, although Dart maintained that, in addition to their hominoid anatomical characters, "they were human in employing skeletal parts to subserve the function of implements in the business of obtaining and preparing . . . food, in getting and dividing it." [47] With the Leakeys' recent discovery of *Zinjanthropus boisei,* however, dated as upper Villafranchian and classified as a new genus of the Australopithecinae,[48] the fact of tool-making in one genus of this group is now established. For, in this case, the discovery is unique in that the hominid remains were excavated from a living site, where they were associated with pebble tools of Oldowan type, along with the broken bones of small animals which had apparently been eaten. Consequently, as Howell says, "the new australopithecine from Olduvai Gorge represents the oldest, fully authenticated toolmaker so far known." [49]

These new empirical facts serve to sharpen an old question What is the relation between brain size and the psychological capacities for

[46] Many years ago (1928, p. 336), Kroeber noted the chimpanzee's inability "outside of posed problems to manufacture tools or lay them aside for the future." And Linton ("Appraisal," p. 266) noted the anticipatory dimension of the human tool-making situation. "This indicates," he said, "a distinct type of psychology, the realization of operation in the time stream, which no other animal shares. I think this is the point, actually, where the human mind emerges, even more than in the capacity for reorganization of experience we call 'thinking.'" Cf. Strauss (1955*b*, p. 133), who observes "that man is peculiar in the extent to which he lives in the three dimensions of time. It is this peculiarity that gives use to his remarkable degree of foresight or anticipation which is perhaps best expressed in tool-making, to use this term in its broadest sense."

[47] Dart, p. 335.

[48] Leakey, 1959 and 1960.

[49] Howell, 1960, (Comments on the Leakeys' discovery), and 1959.

cultural adaptation as we know it in *Homo sapiens?* Although no final answer can be given at present, Le Gros Clark, writing prior to the discovery referred to above, has reminded us that the range of variability in the cranial capacity of modern man is very wide (900 cc.–2300 cc.) and that "while the cranial capacity of fossil hominids can give information on the brain volume, it provides no information on the complexity of organization of the nervous tissue of which it was composed." [50] Washburn has indicated that there may be chronological questions that will have to be considered, that is, the sequential developments of tool-using, tool-making, speech, and a fully developed cultural mode of adaptation. It may be, he says,[51] that tool-using may require

> much less brain than does speech and might have started as soon as the hands were freed from locomotor functions. Oral traditions essential for complicated human society probably were not possible with less than 700 or 800 cc. of brain, and there is no likelihood that elaborate traditions of tool making are possible at lesser capacities, although simple pebble tools might well be.

This brief discussion of speech and tools as behavioral criteria of a human status has, I hope, indicated some of the preliminary problems met with in applying them. The evolutionary problem becomes even more complicated if, to begin with, we attempt to operate with the concept of culture as the criterion of a human status—that "complex whole" of Tylor's classic definition which, he said, is acquired by individuals as members of society. How can we apply such an abstract generic concept, derived from empirical observations of a very concrete nature, in any meaningful analysis of the developmental aspects of human evolution and adaptation?

Wissler tried to solve the problem by assuming the phenomenal unity of what he called a "universal pattern" of culture.[52] His solution

[50] Le Gros Clark (1959), p. 312. Cf. Oakley, 1958. Oakley (1957, p. 207) raises the question whether it is possible that "systematic tool-making arose, not gradually as most nineteenth century evolutionists led us to imagine, but suddenly and spread rapidly? . . . The earliest tools and weapons would have been improvisations with whatever lay ready to hand. Although the hominids must have begun as occasional tool-users, ultimately they were only able to survive in the face of rigorous natural selection by developing a system of communication among themselves which enabled cultural tradition to take the place of heredity. At this point systematic tool-making replaced casual tool-using, and it may be that this change-over took place in the Australopithecine stage. It would not be surprising, in view of the close correlation between culture and cerebral development, if there had been at this stage intense selection in favour of larger brains, with the result that the transition from the small-brained Australopithecus to the larger-brained Pithecanthropus took place in a comparatively short space of time."

[51] Washburn (1958), p. 432, and Table 19.5, p. 428.

[52] Referred to by Kroeber (1955, p. 198) as "that seed lightly tossed out by Wissler that has never germinated."

was reductionistic. He projected this pattern—including speech—full-fledged from the properties he conceived the "germ plasm" to possess. "The pattern for culture is just as deeply buried in the germ plasm of man as the bee pattern in the bee," he said. "The human pattern . . . is a part, if not the whole, of man's inborn behavior. . . . Man builds cultures because he cannot help it, there is a *drive* in his protoplasm that carries him forward even against his will." [53] Wissler, however, did not specify any particular genus or species of the Hominidae. He did not say whether the same universal pattern for culture was imbedded alike in the genes of Pithecanthropus and Homo, and, at the time he wrote, the problem presented by the australopithecines had not yet arisen. While it is doubtful that any simple biologistic approach to the evolutionary roots of culture can be any more fruitful than preformationistic theories in biology, at the same time, it must be recognized that Wissler was grappling with a genuine problem. It seemed clear to him that, despite the plasticity of the behavior of *Homo sapiens* and the varying traits, complexes, and patterns of different cultures, there were constant and recurrent categories of culture that transcended any particular mode of cultural adaptation.

Thirty years later Kluckhohn, discussing the question of universal categories of culture,[54] pointed out that, although in the earlier history of anthropology there were those who recognized universal categories for a decade or more before Wissler and for an even longer period subsequently,

> the attention of anthropologists throughout the world appears to have been directed overwhelmingly to the distinctiveness of each culture and to the differences in human custom as opposed to the similarities. The later, where recognized, were explained historically rather than in terms of the common nature of man and certain invariant properties in the human situation.

The point I wish to stress here is that there are inescapable psychological as well as evolutionary questions raised by "cultural universals," once such phenomena are in any way thought to be related to the nature of man and the human situation. Even if we do not accept Wissler's "universal pattern" concept as such or his reductive explanation, he was correct in viewing universals in phylogenetic perspective. It seems probable that some of them at least point directly to the functioning of basic features of a human personality structure that would appear to be a necessary condition for the

[53] Wissler (1923), pp. 264–65.
[54] Kluckhohn (1953), p. 511.

existence of many aspects of cultural adaptation. For instance, categories of Wissler's "universal pattern" were subsequently elaborated by Murdock, who itemized a long list of what he called "common denominators" of culture which occur, he says, "in every culture known to history or ethnography." [55] Among the many items he lists is *eschatology*. It is a particularly interesting item when its underlying psychological implications are considered. For concepts concerned with a future life, in order to become functionally significant, require a concept of self as being, in some sense, indestructible and persistent in future time. Consequently, a capacity for self-awareness and self-identification must be assumed as psychological universals. Furthermore, since this future existence of the self requires a locale, a level of personality organization is indicated which not only implies ego functioning but a capacity for symbolizing self in space, as well as in time.[56] In phylogenetic terms the evolutionary status of *Homo sapiens* implies common psychological potentialities. These would appear to be as necessary for the functioning of notions of eschatology as for the manufacture of tools and other forms of cultural adaptation.

In the light of our present anatomical and archeological evidence we oversimplify the problem of human evolution if we do not press beyond such general categorical correspondences as, man: speech: tools: culture. Without qualification and further analysis, we cannot associate every aspect of the kind of cultural adaptation we find in *Homo sapiens* with all members of the Hominidae, any more than we can attribute to them a common "human nature." This latter concept always has proved difficult.[57] Sometimes it has been given a purely biological content. Among anthropologists it often has received a relativistic connotation, despite lip service to the "psychic unity of mankind." [58] Spiro has given the concept a more precise meaning by asserting that "the structure and functioning of human personality constitutes man's universal human nature, psychologically viewed. Its universality is not only descriptively true; it is analytically true, as well. In the absence of human personality there could be no human culture." [59] In phyletic perspective "human nature" is, then, the consequence of an evolutionary process. However conceptualized, it cannot be attributed to the earliest hominids in any meaningful sense. In a psychological frame of reference, a human personality structure did not arise as a sudden mutation in the evolution of the hominids

[55] Murdock (1945), p. 124.
[56] Cf. Hallowell (1955), p. 100.
[57] See Bidney, 1953; Kroeber, 1955.
[58] Spiro (1954), p. 21.
[59] *Ibid.*, p. 29.

any more than a saltatory constellation of anatomical traits suddenly gave rise to "man." Howells [60] said a number of years ago:

> Heretofore we have been given to talking about 'the appearance of man'—the tyranny of terminology—as if he had suddenly been promoted from colonel to brigadier general, and had a date of rank. It is now evident that the first hominids were small-brained, newly bipedal, proto-Australopith hominoids, and that what we have always meant by 'man' represents later forms of this group with secondary adaptations in the direction of large brains and modified skeletons of the same form.

Analogically, it is equally doubtful whether we should any longer talk in terms of the "appearance of culture," as if culture, along with "man," had suddenly leaped into existence. Moreover, if the ancestral hominids were at all like the australopithecines, it seems unlikely that they could have had a system of communication that was fully the equivalent of human speech.[61] There is no positive evidence, it might also be noted, that they had fire.[62] Further discoveries and analysis, no doubt, will illuminate the nature of their tool-making, particularly with respect to the degree of tool differentiation and standardization of technique and form which prevailed. In the light of our present knowledge, we can attribute neither a fully developed cultural mode of adaptation nor a human personality structure to all the Hominidae.

Thus, instead of assuming that culture possesses a phenomenal unity from the start and trying to identify its existence in the past, it seems more fruitful to consider certain aspects of behavioral evolution that are non-cultural in nature, but which are among the indispensable conditions that made cultural adaptation possible in the later phases of the evolution of the hominids. The most important of these conditions are sociopsychological in nature. Our empirical data are derived from observation on subhominid primates in their natural

[60] Howells, 1950, and the important paper of Washburn in the same symposium. Cf. Heberer who writes: "Wir dürfen wohl sagen, dass, wie bereits Nehring (1895) vermutete und heute vielfältig werden kann . . . der Mensch 'zuerst mit den unteren Extremitäten Mensch geworden' ist. Die Erwerbung des Bipedalismus schuf die Vorbedingung für die definitive Hominisation durch Cerebralization. Ein pronogrades Wesen konnte keinen humanen Status erreichen, ebensowenig wie dies einem Brachiator möglich war" (p. 537). And on page 540, "Die Hominisation begann mit dem Eisetzen des evolutiven Trends, der zur Erwerbung des Bipedalism und zur Reduktion des Gebisses mit fortschreitendem Ersatz der Zähne durch die Hände (Instrumentalhilfen) führte. Mit diesen Erwerbungen wurde die kritische Phase erreicht, in der sich der Ubergang vom sub-humanen zur humanen Zustand volbezog." Cf. Le Gros Clark (1958), p. 196.

[61] Oakley (1954) does not think it necessary to assume that the earliest hominid tool-users, or even tool makers possessed speech. He likewise believes that a system of gestural communication preceded speech (1951). Cf. Critchley in this volume.

[62] Oakley, 1959.

habitat or under laboratory conditions, for deductions from comparative behavior are as methodologically legitimate as those from comparative anatomy.[63]

THE DIMENSION OF SOCIAL STRUCTURE

Social systems are not unique to *Homo sapiens*. And, even at this highly evolved level, "social structure" is now frequently differentiated analytically from culture or personality organization. Eggan, for example, has expressed the opinion that "the distinction between society and culture, far from complicating the procedures of analysis and comparison, has actually facilitated them." He goes on to say that "social structure and culture patterns may vary independently of one another, but both have their locus in the behavior of individuals in social groups." [64]

In approaching the sociopsychological dimension of primate evolution, a distinction of the same order is useful. Life in structured social groups is characteristic of primates and long antedated anything that can be called a cultural mode of adaptation among the more advanced hominids. Social structure can thus be treated as an independent variable. While at the highest level of primate behavioral evolution there are no organized societies without culture (or the reverse), at lower levels there were societies without culture. In phylogenetic perspective a necessary locus and an indispensable condition for a cultural system is an organized system of social action. It likewise seems reasonable to assume that systems of social action at lower primate levels require some system of communication for their operation. To characterize such a system as "language" is ambiguous and even misleading without further analysis of the "design-features" of the system. Then, too, consideration of the sensory mode of communication is required.[65] Among primates both visual and acoustic modes appear to be extremely important. Schultz [66] speaks of the intricate "silent vocabulary" of the non-hominid primate.

[63] Nissen (1955), p. 99, points out: "It might well be that if we had a record of behavior as complete as the fossil record of structures, this would yield as convincing a body of evidence for evolution as does the latter. As a matter of fact, a study of the behaviors of living species alone—together with the paleontological evidence regarding the order in which these forms appeared—provides in itself a substantial basis for postulating a process of evolution."

[64] Eggan, p. 746. Cf. Hallowell (1953), p. 600.

[65] For background material on animal communication see Scott (1958), chap. 9; Schneirla (1952), p. 582; Hebb and Thompson and Haldane.

[66] Manuscript of paper presented to the Conference on the Social Life of Early Man (1959). Carpenter (1952, p. 242) says: "Each known genus of primate has a repertoire of gestures which are employed consistently and which stimulate consistent reactions." Examples are given.

Crouching down, presenting buttocks, exposing teeth, shaking branches, pounding of chest, dancing in one place, etc. are all actions full of definite meaning. [Although] the long lists of different postures, gestures and facial movements characteristic of monkeys and apes have not yet been compiled, . . . any careful observer realizes that they represent an intricate "silent vocabulary" of great aid in social intercourse. In the perfectly adapted arboreal life of monkeys and apes the limited variety of sounds, together with the great variety of meaningful gestures and facial expressions, is fully adequate for all social life within such close contact as permits seeing and hearing these detailed means of expression.

So far as the utterance of sounds is concerned, Schultz says they "are the essence of primate life . . . ; the simian primates are by far the noisiest of all mammals." In species that have been closely investigated, like the howling monkeys of Panama and the lar gibbon, differentiated vocalizations have been shown to have functional significance in the social coordination of the individuals belonging to a group.[67] According to Schultz,

The primatologist regards language not as something radically new and exclusively human, but rather as the result of a quantitative perfection of the highly specialized development of man's central nervous control of the anatomical speech apparatus in the larynx, tongue and lips, the latter being as good in an ape as in man. . . . As soon as the early hominids had ventured into open spaces, had begun to use and even made tools and had cooperated in hunting, the total variety of all means of expression needed additions which could come only from an increase in sounds, since the practically unchanged anatomy had already been fully used for all possible gestures etc. The latter have never been lost in human evolution, but merely overshadowed by the infinitely greater variety of sounds in increasing numbers of combinations.

Oakley and others have suggested that early hominids may have depended primarily on gestures, "mainly of mouth and hands, accompanied by cries and grunts to attract attention" and that speech may have been a comparatively late development.[68] If so, a mode of communication, infrahominid in origin, would have persisted into the protocultural phase of hominid evolution. Unfortunately this interpretation must remain speculative. Yet it may be that, when the neurological basis of speech is clarified, we may be in a better position to make chronological deductions.[69] It is difficult to imagine, however, how a fully developed cultural mode of adaptation could operate

[67] See Bourlière, chap. 8 ("The Social Life of Mammals"), in which he discusses the differentiated vocalizations of the howlers, gibbons, and chimpanzee in the general framework of the accoustic signals of mammals. See also Carpenter (1952), p. 242.

[68] Oakley (1951), p. 75.

[69] See Spuhler, p. 8, and Du Bruhl.

without speech. If one of the necessary conditions for the functioning of a typically human system of communication is a speech community, an organized social system is as necessary for human language as it is for a cultural mode of adaptation. This condition was present even at the non-hominid level. So what we can discern in primate evolution is a behavioral plateau which provided the necessary context but, at first, not all the sufficient conditions for speech and culture.

It will be unnecessary here to consider the structure and functioning of infrahuman primate societies in detail. But a few general comments and interpretations may be ventured, despite the limitations of our present knowledge, for our samples of reliable observations on primate societies in their natural state are woefully small, particularly for prosimian groups.[70] Besides this, it is not yet possible to consider non-hominid primate societies systematically in the larger perspective of mammalian societies.[71] There are terminological difficulties, also. Descriptive terms like "family," "polygamy," "harem," "clan," and even "culture" and "acculturation," familiar enough when employed with reference to *Homo sapiens,* sometimes have been applied to primates at the infrahominid level. Since no systematic terminology has been developed, these labels must be used with caution, especially when evolutionary questions are at issue.

We do have considerable empirical data on what appears to be a characteristic association of organized primate groups with territories,[72] despite wide variation in the size of the group and the mating patterns that prevail in different species. There also seems to be some significant connection between arboreality and small groups and the occurrence of larger groups in open country.[73] Since the bipedal australopithecines were not forest dwellers, increase in size of organized groups, associated with an ecological adjustment to open country, may have evolutionary implications.

Variations in type of mateship, of course, have suggested the closest human analogies. Since lar gibbons, for example, live in groups which consist of one male and one female and their young, we have a close analogy to the "nuclear family" in man,[74] which likewise represents a monogamous type of mateship. Some biological writers have applied the term "family" exclusively to this kind of primate social unit, despite the fact that in anthropological writing the connotation of the

[70] For bibliography to 1957, see Carpenter, 1958*b;* Frisch, 1959 gives a brief review with bibliography of the work that has been done at the Japan Monkey Center.

[71] Bourlière (1956), p. 221.

[72] Carpenter (1958*a*), p. 242.

[73] See Table 1 ("Habitat and Group Structure in Different Species of Primate") in Chance, 1959.

[74] Murdock, 1949.

term "family" is never limited to the nuclear family. The gibbon type of mateship, in which the sexual drive of the male is low, would seem to be a limiting case in the range of social units found among the more evolved primates, and without evolutionary implications. In *Homo sapiens* we find two types of polygamous mateships, polygyny and polyandry, and social structures based on these are ordinarily called "families." Relatively rare in man in an institutionalized form, polyandrous mateships appear to be absent in infrahuman primates. On the other hand, polygynous mateships are common in both monkeys and apes. In chimpanzee and gorilla this type of mateship seems to furnish the basis of independent social groups. In some monkeys, for instance, the baboon, "harems" occur as subgroups within the larger "troops" or "bands" found in these animals. Monogamous mateships, on the other hand, do not occur in groups of larger size because females in heat mate with more than one male. Past attempts to establish any regular evolutionary sequence of mateship within *Homo sapiens* have failed, as have attempts to link any *particular* type of mateship in the infrahuman primates with early man, as Westermarck tried to do in the belief that there was evidence to show that the gorilla was monogamous. He urged that this "fact" was of significance in the study of sexual relationships and marriage in man.[75]

Perhaps it might be better to recognize that, since there are only a limited number of possibilities in mateships, it is not surprising to find them recurring at both the non-hominid and hominid levels of evolutionary development in the primates and in social units of varying size and composition. Whatever form they take, all these mateships serve the same reproductive ends. Their importance lies in this constancy in biological function rather than in any direct relation that

[75] Westermarck, however, was on the right track and must be seen in historical perspective. Hart points out (p. 108), "What had really happened to evolutionary theory between 1859 and 1891 was that, while Huxley had spent his life labouring on the genetic front to get his contemporaries to accept 'the unitary view of organic nature' and to reject the old dualistic view which saw man on one level, the rest of the animal world on another with an impassable gap eternally fixed between, the pass had been betrayed by Spencer and his followers, who, by assuming that society was one thing and biology another, had merely substituted a new dualism for the older one, and had opened up as big a gap between man and the rest of nature as had been there in pre-Darwinian days. The extraordinary thing is not that this should have happened, but that nobody seems to have been aware of what was happening until Westermarck pointed it out." The latter in his *History of Human Marriage* (p. 9) said: "If we want to find the origin of marriage, we have to strike into another path . . . which is open to him alone who regards organic nature as one continued chain, the last link of which is man. For we can no more stop within the limits of our own species, when trying to find the root of our psychical and social life, than we can understand the condition of the human race without taking into consideration that of the lower animals." Etkin argues for a monogamous protohominid social structure but on quite different grounds than did Westermarck at the turn of the century.

can be shown to the evolution of group organization. They all lie close to biologically rooted central tendencies and continuities in behavioral evolution which link *Homo sapiens* to his precursors. What we find as the common social core of all but the lowest primate groups, despite their variation, is the continuous association of adults of both sexes with their offspring during the portion of the latter's life cycle that covers the period from birth to the threshold of maturity. This core pattern of associated individuals, when considered with reference to their interrelated roles, is linked with the fact that basic functions are involved—the procreation, protection, and nurture of offspring—born singly, relatively helpless at birth, and dependent for a considerable period thereafter. Variations in mateship or size of the group may occur without affecting these functions. Besides this, the sex needs of adults and the food needs of all members of the group can be taken care of. The role of the female in relation to her young does not seem to vary widely nor the behavior of infants and juveniles. The protective role of the male in relation to infants and juveniles is similar in gibbon and howler, even though the young of the group in the latter case are not all his own offspring and the actual zoological relationship between these two species is remote. Among monkeys and apes, the adult males never provide food for juveniles or females. After weaning the juveniles always forage for themselves. Whether we call nonhominid primate groups "families," "clans," "troops," or "bands," their basic social composition can be expressed by the same general formula:

$$X \text{ males} + X \text{ females} + X \text{ infants} + X \text{ juveniles.}$$

Whatever the mating types or size of early hominid groups may have been their social composition must have conformed to this fundamental pattern. This generic type of social structure, associated with territorialism, must have persisted throughout the extremely long temporal period during which major morphological changes occurred in the species of the primate order, including those which ultimately differentiated the Hominidae from the Pongidae and later hominids from earlier ones. Underlying it, physiologically, was the type of ovarian cycle characteristic of practically all the primates. In contrast with some mammalian species in which females have only one oestrus period a year, primate females along with those of a limited number of other mammalian species, are characterized by the recurrence of successive oestrus cycles in the course of a year. The primates belong to this group or permanent polyoestrus species.[76] Breeding is not seasonal but continuous.[77] In the course of primate

[76] Bourlière (1956), p. 147.

[77] Chance (1959) points out: "Female macaques and baboons are sexually receptive for approximately 9 out of the 28 days of their reproductive cycle, so that in a

evolution, however, as Beach has pointed out,[78] some emancipation from strictly hormonal control of sexual behavior occurred, which further distinguished the higher primates from other mammalian species. Cortical control came to play an increasing role in sexual behavior, and, in hominid evolution, with the remarkable expansion of the brain, this tendency reached its culmination. Thus, the way lay open for the development in human societies of a normative orientation toward sexual behavior.

The evolutionary significance of the social organization of primate groups cannot be fully appreciated, however, without considering behavior patterns other than those directly connected with reproduction, for the structuralization of these infrahuman societies is by no means a simple function of differential roles determined by sex and age. Of central importance, particularly in groups with a terrestrial habitat, and the Old World monkeys as compared with New World species, we find inter-individual behavior influenced by an order of social ranking in the group, a dominance gradient. Males are, quite generally, dominant over females and the females associated with them may outrank other females. While it appears that in different species the "slope" of the dominance gradient varies considerably, some kind of rank order occurs. The importance of this factor in the operation of the social structure lies in the fact that it serves to reduce aggression between males, it determines priorities to mates and food, it influences the spatial disposition of individuals within the group, affects the socialization of group habits, and may determine the relations of groups adjacent to one another.

The ranking position of individuals, nevertheless, is not fully determined once and for all; an individual's role in the dominance hierarchy may change. Psychological factors such as individual experience in inter-individual relations and social learning become involved in its functioning and affect the motivation of behavior, for one of the basic conditions of the operation of infrahuman structures is a *socialization* process, as Carpenter has indicated.[79] Individuals become socially adjusted from birth through the mediation of learning processes. "Descriptions of mother-infant relations in monkeys and chimpanzee leave no doubt as to the importance of learning in the filial responses of immature primates. The infant learns to obey

---

group of monkeys where two or more adult females are present, the males will be in the presence of a sexually active female for more than half of the time. And in larger groups there will be continual sexual provocation, a situation found nowhere else in the animal kingdom except for a two-month interval during the mating season of the Pribilof seal."

[78] Beach, 1947, and Ford and Beach. Cf. Beach, 1958.

[79] Carpenter (1942), pp. 256–57.

gestures and vocal communications given by the mother and derives considerable advantage from her tuition and guidance," Beach says.[80] Indeed, modern research is showing that the primates are by no means unique among gregarious animals with respect to the importance of social learning and a dominance gradient. J. P. Scott [81] asserts:

> In animals which are capable of learning, social behavior becomes differentiated on the basis of mutual adaptation and habit formation as well as on the basis of biological differences. As shown by Ginsburg and Allee (1942) the formation of a dominance order is at least in part related to the psychological principles of learning. Once such a relationship is formed and firmly established by habit, it may be extremely difficult to upset it by altering biological factors, as shown by Beeman and Allee (1945). . . . Experiments which modify the social environment have tended to bring out the general principles of socialization. Any highly social animal that has been studied so far has behavioral mechanisms whereby, early in development, an individual forms positive social relationships with its own kind and usually with particular individuals of its kind.

With respect to the socialization factor in behavioral evolution, Collias points out: "In both insect and vertebrate societies, maintenance of cooperative relations depends to a large extent on socialization of the young. Among vertebrates, this trend reaches its climax in the primates." [82] It seems reasonable to assume, therefore, that the intimate relation between learning and social structure, so fundamental to the functioning and elaboration of cultural adaptation, was well established in the non-hominid primates prior to the anatomical changes that led to both erect posture and the expansion of the brain.

Furthermore, by direct observation of both monkeys and apes, we know that learned habits may be socially transmitted, even in the absence of speech. The most striking cases have been reported by observers who have been studying *Macaca fuscata* at the Japanese Monkey Center during the past decade. These "Japanese Apes" have been lured from their forest habitat into open feeding places, where, among other things, they have been offered new foods. Systematic observation has shown that newly acquired food habits, such as eating candies, became quite readily socialized. Imanishi points out, moreover, that young macaques acquire the candy-eating habit more quickly than adults and that some mothers learned to eat candies from

[80] Beach (1951), p. 426. See also Schneirla (1951, p. 104 ff.) in regard to ontogenetic factors influencing group organization. Cf. Paul H. Schiller, who believed there is evidence that primates have distinctive *manipulative* patterns of activity available that are not derived from experience.

[81] Scott (1956), pp. 217, 218.

[82] Collias, p. 1087.

their offspring, rather than the other way round.[83] It has likewise been observed that the spread of a new food habit may be directly related to the dominance gradient which is a central feature of their social structure. Adult females of high rank were observed to imitate the wheat-eating of a dominant male very quickly and the habit was passed on to their offspring. Females of lower rank, in a more peripheral position in the group, only later acquired the habit from their offspring who, in turn, had picked it up through association with their playmates. The rate of transmission was extremely rapid in this case, the entire process occurring within two days.[84] In another instance, a young female initiated the habit of washing sweet potatoes before eating them. This habit, having been transmitted to her playmates, as well as to her mother, was slowly transmitted to a number of groups during the next three years. The same class of phenomenon in the anthropoid apes is illustrated by nest-building in chimpanzee [85] and the transmission of the technique of working the drinking fountain at Orange Park, which chimpanzees learned from each other.[86]

The social transmission of culture has sometimes been stressed as one of its chief earmarks. But to my mind it is only one of the necessary conditions of cultural adaptation rather than a distinguishing characteristic. Social transmission is a prerequisite of culture and an earmark of a protocultural behavioral plateau. Concepts of culture that lay primary emphasis on shared and socially transmitted behavior without qualification do not enable us to make a necessary distinction of degree between different levels of behavioral evolution.[87] Voegelin has made the acute observation that, while there is a general agreement

[83] Imanishi, p. 51.

[84] Frisch, p. 589.

[85] Nissen (1955), p. 106. See also Note 90.

[86] Yerkes, p. 52. Yerkes thought the characterization of chimpanzee as "cultureless" to be "a seriously misleading statement, if not demonstrably false." He believed that "the elements or makings of cultural exhibits are present," but that "they are relatively unimpressive because unstable, fragmentary, variable, and seldom integrated into functional relations." See Munn (pp. 129–30) for references to the experimental data on observational learning. He concludes: "It is only in monkeys and apes that anything clearly approximating such observational learning can be demonstrated and even at this level the problems solved by imitation are relatively simple." Instances of spontaneous imitation on the part of Viki were: operating a spray gun, prying off the lids of cans with a screw driver, etc. (Hayes and Hayes, 1955). In the concluding paragraphs of his summary of the observations in the Japanese macaques, Frisch (p. 595) says: "It seems doubtful that definitions [of culture] which strongly emphasize such concepts as social heredity, socially acquired response-patterns, learned traditional behavior, will be able to do justice to what Julian Huxley has called the 'uniqueness of man.' To the extent to which culture is equated with learned, traditional behavior, monkeys appear to have indeed much more 'culture' than anthropologists have often thought." There is a terminological problem, too, since the author discusses some of the case material referred to above under the caption "acculturation," to which he adds quotes in the text. Imanishi employs the same term without quotes.

[87] Bidley (1953), p. 27, says: "The identification of culture with the social heritage is, to my mind, not only a misnomer but also a serious error, since it implies that the

that all culture involves learned behavior, "additional conditions are generally invoked before learned behavior is granted the status of culture," and that "if ever the converse statement were made (*that all learned behavior is culture*), it would necessarily imply that infra-human animals have culture." [88] The fact that even some animals other than primates may learn from each other,[89] that in primate groups there seems to be good evidence that social learning and socially transmitted habits do occur, and that some chimpanzees in social interaction with members of our species have acquired "culture traits," does not indicate that a full-fledged level of cultural adaptation has been reached in these species.[90] Other capacities and conditions were

---

essential feature of culture is the fact of communication and transmission, whereas I maintain that the essential feature is the combination of invention and acquisition through habituation and conditioning." Cf. the remarks of Kroeber on the use of the term "social heredity" (1948, p. 253).

[88] Voegelin, p. 370. Harlow (1951, p. 127) clearly discriminates between infra-hominids and *Homo sapiens* when he says, "In a limited sense . . . any animal living in a group and capable of facile learning must develop a *semblance of culture,* since it must have learned to be influenced in its behavior by the way of its fellows," but, at the same time he points out that "no animal other than man has a *true culture* in the sense of an organized body of knowledge passed down from generation to generation." (Italics ours.)

[89] See Hochbaum for a discussion of "tradition" in birds. Dobzhansky (1955, pp. 340–41) discusses "Rudiments of Cultural Transmission among Animals." Several of the concrete instances cited are taken from observations on bird behavior; he does not discuss this phenomena among primates. The major point he stresses, however, involves a fundamental distinction between man and other animals, relative to Bidney's point. "In animals," he says, "the individuals of one generation transmit to those of the next what they themselves learned from their parents—not more and not less. Every generation learns the same thing which its parents have learned. In only very few instances the evidence is conclusive that the learned behavior can be modified or added to and that the modifications and additions are transmitted to subsequent generations."

[90] In recent years some observers have used the term "culture" where it appears that certain habits have been socially learned and transmitted. Nissen (1955, pp. 105–6) referring to nest-building in chimpanzees writes: "There is pretty good evidence . . . that this nest-building is not instinctive, as in birds, but is, rather, transmitted by imitation or tuition from one generation to the next; it is, therefore, one of the very few items of behavior seen in these animals which may be classified as cultural." Cf. Nissen (1951*a*), p. 426. Fuller and Scott (1954, p. 29) write: "The possibility of *cultural* as well as biological inheritance can be tested by raising animals either in complete isolation or in contact with a foster species with very different habits of behavior. The results of such experiments on birds have been described extensively by Lorenz who has found evidence of both *cultural* and biological factors." (Italics ours.) I believe that *protoculture* is a term which could be used in the context of either quotation or some other word such as *tradition* (employed by ornithologists, see Hochbaum) which clearly indicates that the species referred to does not possess a system of culture that is equivalent to that of *Homo sapiens.* While Bidney (1953, p. 127) says that "all animals which are capable of learning and teaching one another by precept or example are capable of acquiring culture," he makes a distinction between "culture in general" and human or "anthropoculture" (p. 125) which is peculiar to man. He goes on to say that "this implies an evolutionary approach to the concept of culture which recognizes degrees of culture from the sub-human to the human level." Nevertheless, I find a generic concept of culture unsatisfactory for reasons I have indicated in my discussion.

required before this higher level could be realized. Indeed, neither learning nor the socialization and transmission of learned habits seems to have reached an optimum level of functioning in any non-hominid species.

Perhaps this limitation may be attributed to the absence of a psychological capacity of a higher order than was necessary for the transmission of relatively simple habits in the groups described. Learning could not acquire paramount social importance until it could function in social structures of a higher order and wider range than those represented in the infrahominid primates. In social structures of this latter type, the phenomenon of territoriality which, according to Carpenter, "reduces stress, conflict, pugnacity, and non-adaptive energy expenditure" [91] within each group by isolating it from other groups, sets up a barrier at the same time to the integration of groups and the development of social structures of a wider range and more complex order.[92] Speaking more generally, in the case of the Japanese macaques, for example, groups are almost totally isolated from each other in their natural state. It is said that "even where several groups live in contiguous territories, the inter-group relations are practically non-existent. Encounters between distinct groups are extremely rare, and even when they occur both groups keep at a safe distance from each other." [93] Offspring do not associate with parents after sexual maturity has been reached. They leave their primary group and form new ones. Individuals of two or more generations are not continuously associated in the same group during their lifetime. Consequently continuity in learned habits is strictly limited. There is no way for experience to become cumulative, either spatially or temporally, beyond the narrowest range. In order for a cultural level of adaptation to be reached, structures of a wider range were required as a necessary social setting. This further step was contingent upon the development and functioning of psychological capacities that transcend those sufficient to account for the dynamics of the narrow-range social structures described. In short, the social integration of groups larger in size, distributed more widely in space, and characterized by a greater diversity in roles required a transformation in psychological structure.[94]

[91] Carpenter (1958*a*), p. 245.

[92] Carpenter (1955), p. 98.

[93] Frisch, p. 591.

[94] We do not know what objective factors underlay the increase in the size and range of early hominid groups. Change to a carnivorous diet and hunting have been suggested. (Washburn and Avis, 1958, p. 434.) But, as Washburn says (*op. cit.*) "whether early man scavenged from the kills of the big carnivores, followed herds looking for a chance to kill, drove game, or followed a wounded animal, his range of operations must have been greatly increased over that of arboreal apes. The world view of the early human

The Biological Dimension
Neoteny and Brain Enlargement

A concomitant condition for the maximization of the sociopsychological importance of learning appears to have been the extension of the period during which the young become socialized. In the late nineteenth century John Fiske, an ardent follower of Spencer and Darwin, linked such an extension of the learning period in man directly with evolution through what he called the "prolongation of infancy." In this fact alone he thought he had discovered the essential key to man's distinction from other animals and the explanation of human psychological, familial, and cultural development. Fiske was impressed both with A. R. Wallace's account of the behavior of an infant orang raised by hand after its captured mother died [95] and by Wallace's suggestion "that natural selection, in working toward the genesis of man, began to follow a new path and make psychical changes instead of physical changes." [96] Fiske developed the thesis that the human being was born "in a very undeveloped condition, with the larger part of his faculties in potentiality rather than in actuality." [97] The period of helplessness is the period of "plasticity. . . . The creature's career is no longer exclusively determined by heredity . . . it becomes educable . . . it is no longer necessary for each generation to be exactly like that which has preceded." [98] Thus, "man's progressiveness and the length of his infancy are but two sides of the same fact"; "it is babyhood that has made man what he is." Infrahuman primates approached the point where "variation in intelligence" came to be "supremely important, so as to be seized by natural selection in preference to variations in physical constitution." But in a remote period "our half-human forefathers reached and passed this critical point, and forthwith their varied struggles began age after age to result in the preservation of bigger and better brains, while the rest

---

carnivore must have been very different from that of his vegetarian cousins. The interests of the latter could be satisfied in a small area, and other animals were of little moment, except for the few which threatened attack. But the desire for meat leads animals to know a wider range and to learn the habits of many animals. Human territorial habits and psychology are fundamentally different from those of apes and monkeys."

[95] Fiske (1909, p. 26) says it occurred to him immediately that "if there is any one thing in which the human race is signally distinguished from other mammals, it is in the enormous duration of their infancy"; a point he did not recollect ever seeing any naturalist so much as allude to. But Fiske was not quite as original as he thought. See, e.g., Lovejoy.

[96] Fiske (*op cit.*), p. 28.

[97] *Ibid.*, p. 9.

[98] *Ibid.*, p. 2.

of their bodies changed but little. . . . Zoologically the distance is small between man and the chimpanzee; psychologically it has become so great as to be immeasurable." [99]

We can see from these passages that Fiske anticipated a number of points frequently emphasized later in cultural anthropology and in evolutionary biology. But the theory he develops, while emphasizing the important role of learning in human experience and the potentialities of man for cultural development, does not account for the biological foundations of the extended period of dependency. He likewise makes "bigger and better brains" chronologically subsequent to the distinctive human condition that fired his imagination. Nor could he have anticipated the fact that later knowledge of the social organization of the non-hominid primates would fail to support his conviction that the prolongation of infancy "must have tended gradually to strengthen the relations of the children to the mother, and eventually to both parents, and thus give rise to the permanent organization of the family." For in Fiske's view, when this step was accomplished, "the Creation of Man had been achieved." [100]

While Fiske's theory, although once so widely known, is seldom referred to today, the fact should not be overlooked that the relations between the factors dealt with by him have not yet been satisfactorily resolved. Even now it is sometimes forgotten that an extended period of dependency and opportunities for social learning in man do not explain the genesis of cultural adaptation, even though these conditions may be of primary categorical importance in understanding the adjustment processes that relate an individual to his culture. While we now know more about the phylogenetic basis of what Fiske called the "prolongation of infancy," its precise psychological significance is a matter of dispute.

From comparative anatomy the fact seems well established that the larger apes, and particularly the gorilla, develop adult characteristics much earlier than does *Homo sapiens*. The latter has been called a "fetalized" animal; [101] that is, certain features that are characteristic of the fetal stages of apes persist in human adults. It is an example of a well-known evolutionary process which, generically, is usually referred to as "neoteny": fetal and/or juvenile features of an ancestral form persisting in the adult stage of descendants.[102] In man, the rate of development of some characters has been retarded. On the other hand, says De Beer,[103]

[99] *Ibid.*, p. 11.
[100] *Ibid.*, pp. 12–13.
[101] Bolk is mainly responsible for this particular term.
[102] Carter, 1951, 1953; De Beer.
[103] De Beer, pp. 75–76.

> The reproductive glands have probably not varied their rate of development, for the human ovary reaches its full size at the age of about five, and this is about the time of sexual maturity of the apes and presumably of man's ancestors. The human body is, however, not ready for the reproductive glands to function until several years later. The retardation is due to the action of hormones which play an important part in regulating the speed of development. . . . At the same time, of course, in other directions, the evolution of man has involved progressive changes of vast importance, some of which, however, might not have been possible (e.g., the development of the brain), had it not been for certain features of neoteny (e.g., the delay in the closing of the sutures of the skull).

It is the combination of various characters, considered with reference to their rate of ontogenetic development, that is peculiar to man.[104]

While such anatomical facts are well established, the psychological inferences drawn from them have varied in emphasis. Roheim maintains that the temporal disharmony between the development of what he calls the Soma and the Germa is the crucial point. Human sexuality becomes precocious because it develops at about the same rate as in other higher primates, but in our species full bodily growth is delayed. The consequence is that unconscious psychological mechanisms have come into play to repress, project, or transform sexual impulses before the individual is mature enough in other respects to engage in actual sexual activities. The Oedipus complex is universal not because it is derived from past events that have become inherited,[105] but because it "is a direct derivative of our partly premature, partly conservative (prolonged or retarded) rate of growing up." [106] "Our sexual ethics are based on juvenalization." [107] Montagu, on the other

[104] Schultz (1955), p. 53. And see his 1956 publication for an authoritative comparative treatment of the details of growth and development in various primate species. Schultz concludes that it is erroneous to emphasize retardation exclusively in man's ontogenetic development, since "ontogenetic specializations can consist of accelerations as well as retardations in man as well as in all other primates" (p. 959).

[105] Roheim (p. 424) says, "This ultra-Lamarckian point of view is untenable," i.e., Freud's Primal Horde theory.

[106] *Ibid.*, p. 424. Roheim (p. 409) says: "It is a curious fact that while man's delayed infancy is univerally admitted hardly anybody uses this fact in the sense that I do. The usual statement is that the delayed infancy makes it possible to condition human beings and that it is why psychology depends on conditioning, i.e., on culture. What culture depends on is then of course the kind of question no well behaved anthropologist should ask, because looking for origins is 'outmoded,' in fact it is nineteenth century, a truly terrible thing, a word loaded with the worst possible kind of *mana*. Quite apart, however, from this aspect of the question, how is it that nobody recognizes that in this one fact we have one of the most important keys to the understanding of human nature?"

[107] *Ibid.*, p. 413. Perhaps it should be added that while in *both* man and the anthropoids sexual organs reach maturity earlier than full body growth, in man the time difference is greater.

hand, sees in neoteny an evolutionary step whose major psychological significance is related to man's potentialities for learning.[108] He says:

> The shift from the status of ape to the status of human being was the result of neotenous mutations which produced a retention of the growth trends of the juvenile brain and its potentialities for learning into the adolescent and adult phases of development. It is clear that the nature of these potentialities for learning must also have undergone intrinsic change, for no amount of extension of the chimpanzee's capacity for learning would yield a human mind.

Besides this, account must be taken of the biological fact that in primate evolution the life span of individuals became progressively lengthened while the onset of puberty and the beginning of fertility became more and more chronologically delayed. Culminating in man, the outcome was that the interval between generations became greater. This fact, then, needs to be considered both with reference to the association of individuals in larger social groups and in relation to the need for the development of the kind of psychological structure that would permit the co-ordination of the behavior of individuals of both sexes and widely differing ages over a large time period, in order that inter-individual relations in these more complex social systems might be successfully integrated.

While it is impossible to sustain the view that fetalization is completely responsible for all of modern man's distinctive psychocultural characteristics, perhaps we may follow Sir Julian Huxley's view [109] that while

> it will not account for all the special characters we possess, notably the special enlargement of the association areas of our cortex, and the full adaptation of our feet and legs to bipedal terrestrial existence, it has certainly helped us to escape from anthropoid specialization. It is this possibility of escaping from the blind alleys of specialization into a new period of plasticity and adaptive radiation which makes the idea of paedomorphosis [fetalization, neoteny] so attractive in evolutionary theory. Both its possibilities and its limitations deserve the most careful exploration.

If so, important steps in sociopsychological evolution beyond the nonhominid or early hominid level may have been contingent upon the situational effects produced by biological factors which prolonged dependency of the young, delayed reproduction, and increased the life span in an already advanced hominid whose psychological func-

[108] Montagu (1956), p. 90. Cf. (1955), p. 22.

[109] Huxley (1954), p. 20. The unkindest cut of all has come from Cuenot (1945) who has said that man "can be considered a gorilla fetus whose development and growth have been greatly retarded."

tions were, at the same time, being greatly enhanced through the enlargement of certain areas of the brain.[110]

With respect to this particular development, there may well have been a critical transition period; however, an arbitrary Rubicon of 750 cc.[111] between the higher apes and the australopithecines on the one hand, and the early Homininae and recent man on the other, while perhaps of some crude taxonomic value, does not in itself permit significant behavioral inferences. "It is quality of brain rather than quantity, absolute or relative, that is all important," as Strauss says.[112]

Today we know considerably more than we did a generation ago about the functioning of various parts of the cortex as well as other parts of the brain. And new insights and hypotheses with evolutionary reference are coming to the fore. Washburn, referring to the diagram in Penfield and Rasmussen (1950), showing the way the body is represented on the cortex, points out that there is unequal representation but that "the areas which are largest are the ones of greatest functional importance." Thus, "when the brain increased in size, the area for hand increased vastly more than that for foot," a fact which "supports the idea that the increase in the size of the brain occurred after the use of tools, and that selection for more skillful tool-using resulted in changes in the proportions of the hand and of the parts of the brain controlling the hand." The areas concerned with speech are also large and so are the frontal lobes which have been said to be connected, in part, with foresight and planning.

Our brains are not just enlarged, but the increase in size is directly related to tool use, speech, and to increased memory and planning. The general pattern of the human brain is very similar to that of ape or monkey. Its uniqueness lies in its larger size and in the particular areas which are

[110] Bernhard Rensch, who has been investigating the effects of increased body size on the relative size of the brain and its parts, and on higher psychological functions, has advanced the hypothesis (pp. 197–98): "In man's line of descent we may at least consider the increase of the cortex, the relative increase of 'progressive,' i.e. more complicated cortex-regions, the absolute increase of the number of neurons and of dendritic ramifications, as . . . selectively advantageous factors. Thus the trend towards the human level of brain organization may be regarded as inevitable. Another important factor here is the prolongation of the juvenile phase found in many large animals. This could only occur where multiple births, and therefore intrauterinal selection for rapidity of development, had been eliminated. But once this had taken place, the prolongation of the juvenile phase was favored by selection because thereby the period of learning, that is to say the period of gaining experience and of exploration by play, is also extended. Thus the evolution of man, too, was inevitable." Cf. Rensch, 1956 and 1959.

[111] Keith. See comments by Schultz (1955), pp. 49–50. The Hayeses, however, suggest "the possibility that most of the fourfold increase in cranial capacity from anthropoid to man took place after the appearance of culture and language, and therefore after primate behavior had become essentially human" (1955, p. 116).

[112] *Appraisal,* p. 262.

enlarged. From the immediate point of view, this human brain makes culture possible. But from the long-term evolutionary point of view, it is culture which creates the human brain.[113]

In recent years, too, as a consequence of rapid advances in neuroanatomy and physiology, there has been a revival of interest in, and many discussions of, the brain mechanisms which underlie the phenomena of awareness, consciousness, attention, memory, and the functional integration of experience.[114] So far as integrative functions are concerned, the present weight of evidence appears to focus upon the influence exercised by the masses of nerve cells in the upper part of the brain stem upon the more recently evolved cortical areas. An older notion that the cortex itself was of prime significance because it was somehow the "seat of consciousness" no longer seems to make complete neurological sense. Although no unanimity of opinion has been reached, hypotheses should emerge in time which will lead to further clarification of the relations between neurological evolution, psychological functioning, and cultural adaptation. Of central importance in this complex web of relationships is the distinctive psychological focus of consciousness in *Homo sapiens*—the capacity for self-objectification which is so intimately linked with the normative orientation of all human societies.

Sociopsychological Evolution and Normative Orientation

Although we can never check developmental stages in the enlargement of the brain by direct observation of behavior, we do know what the behavioral outcome was in the most highly evolved hominid. Here, along with a greater diversification in the forms of social struc-

[113] Washburn (1959), pp. 27–29.

[114] See, e.g., Penfield and Rasmussen; Penfield and Roberts; Bremer *et al.* (eds.); Von Bonin; and Kubie. Penfield and Rasmussen write (p. 204): "It is apparent that there are important connections which conduct both ways between areas of cortex and specific nuclei of the diencephalon, and that in the process of encephalization a varying degree of autonomy has been handed over to the large cortical projections. It does not necessarily follow, however, that all function, either new or old, has been handed over in this way nor that correlation between the activities of the different cortical areas is necessarily carried out in the cortex rather than in the diencephalon. . . . Popular tradition, which seems to be largely shared by scientific men, has taken it for granted that the cortex is a sort of essential organ for the purposes of thinking and consciousness, and that final integration of neural mechanisms takes place in it. Perhaps this is only natural since there has been an extraordinary enlargement of the cortex in the human brain, and, at the same time, man seems to be endowed with intellectual functions of a new order." However, "the whole anterior frontal area, on one or both sides, may be removed without loss of consciousness. During the amputation the individual may continue to talk, unaware of the fact that he is being deprived of that area which most distinguishes his brain from that of the chimpanzee" (pp. 205–6, 226).

ture in *Homo sapiens,* we are confronted with a radical change in their underlying dynamics. At this more advanced stage a normative orientation becomes an inherent aspect of the functioning of all socio-cultural systems, since traditionally recognized standards and values are characteristic of them. Techniques are appraised as good or bad; so are the manufactured objects themselves. Property rights are regulated according to recognized standards. Knowledge and beliefs are judged true or false. Art forms and linguistic expression are brought within the sphere of normative orientation. Conduct is evaluated in relation to ethical values. All cultures are infused with appraisals that involve cognitive, appreciative, and moral values.[115]

It has been said by a biologist that the foundation of any kind of social order is dependent upon role differentiation.[116] The general principle underlying social organization at any level is that role behavior on the part of individuals is, within limits, predictable in a wide variety of situations.[117] This is what makes it possible to establish empirically characteristic patterns of behavior interaction whether in invertebrates, vertebrates, or primates, despite the fact that the relative importance of innate versus learned determinants may vary widely at different levels. Normative orientation in man implements regularities in social systems at a more complex psychological level of development through role differentiation that is mediated by socialized values and goals. While some contemporary biologists, like Darwin a century ago in his *Descent of Man,* have given particular emphasis to the moral sense of man,[118] this aspect of social adjustment is but one facet of man's normative orientation. If the total

[115] See Edward C. Tolman, in Parsons and Shils, pp. 344–46; Clyde Kluckhohn and others, "Values and Value—Orientations in the Theory of Action" (pp. 388–433). A value-orientation, whether "held by individuals or in the abstract-typical form, by groups," and varying from explicit to implicit, is defined by Kluckhohn *et al.* (p. 411) as "a generalized and organized conception, influencing behavior, of nature, of man's place in it, of man's relation to man, and of the desirable and non-desirable as they may relate to man-environment and interhuman relations."

[116] Jennings (p. 105), assuming a phylogenetic perspective and speaking of infra-human animals, said: "Only if the individuals play different functional roles is there social organization."

[117] Cf. the discussion of "role expectations" in Sarbin (p. 226 ff., and p. 255): "Persons occupy positions or statuses in interactional situations. Psychologically considered, positions are cognitive systems of role expectations, products of learning. Role expectations are bidimensional; for every role expectation of other there is a reciprocal role expectation of self. The organized actions of the person, directed towards fulfilling these role expectations, comprise the role."

[118] Dobzhansky says, "It is man's moral sense which makes him truly human" (p. 376). And Simpson, asserting that "man is a moral animal," says: "It requires no demonstration that a demand for ethical standards is deeply ingrained in human psychology. Like so many human characteristics, indeed most of them, this trait is both innate and learned. Its basic mechanism is evidently part of our biological inheritance" (p. 294).

ramifications of the normative orientation of human societies are taken into account, we have a major clue to the kind of psychological transformation that must have occurred in hominid evolution which made this level of adaptation possible and some measure of its depth and significance for an understanding of the dynamics of human systems of social action.

In their analysis of the functional prerequisites of a human society, Aberle and his associates introduce the concept of an "actor," with cognitive, affective, and goal-directed orientation, but do not discuss the psychological prerequisites of this actor. While this is irrelevant in their frame of reference, in phylogenetic perspective the capacities of the actor are crucial. For the functioning of a system of action as a normatively oriented social order requires a capacity for self-objectification, identification with one's own conduct over time, and appraisal of one's own conduct and that of others in a common framework of socially recognized and sanctioned standards of behavior.[119] Without a psychological level of organization that permits the exercise of these and other functions, moral responsibility for conduct could not exist, nor could any social structure function at the level of normative orientation. Learning remains important, of course, but it functions at a higher level of sociopsychological integration. The relations between needs, motivation, goals, and learning become more complex. The analysis of Aberle and his associates inevitably includes the "normative regulation of means," the "regulation of affective expression," and the "effective control of disruptive forms of behavior." Value systems have an ordering function in social interaction; they promote the broad behavioral expectancies which are of the essence of role differentiation in a *sociocultural* system.

Man, for example, has departed very radically from his primate forerunners in ecological development through the invention and use of technological devices of all kinds and in economic organization. A normative orientation in these spheres of activity is epitomized by the standards applied to the distribution of goods and services and to the ownership of property. One of the universal functions of all systems of property rights, which are among the common denominators of culture, is to orient individuals in human societies toward a complex set of basic values which are inherent in their day-to-day operation. This kind of value orientation is just as crucial in relation to the motivation and interpersonal relations of individuals as are the values associated with sexual behavior. Property rights are not only an in-

[119] Consequently, it is thoroughly intelligible why role theorists, more than any other group, as Sarbin points out (*op. cit.*, p. 238) "have developed and used the conception of the self as an intervening variable."

tegral part of the economic organization of any human system of social action; they likewise implement the functioning of the social order in relation to the resources of the physical environment through normative means. Discussion of "property" among infrahuman animals have centered around such phenomena as food-sharing, the defense of the nest, prey, territorial domain, and so forth. The question is: In what sense are such phenomena comparable with the socially recognized and sanctioned rights in valuable objects that characterize property in human societies? In the latter the basis of ownership is the correlative obligations others have to allow me to exercise *my* property rights. *A* owns *B* against *C,* where *C* represents all other individuals. It is an over-simplification to omit *C* and simply say *A* owns *B*.[120] Among infrahuman animals, we meet with entirely different conditions. All we observe is the utilization, or possession (in the sense of physical custody or use) of certain objects which bear some relation to the biological needs of the organism or group of organisms. We cannot properly speak of rights, obligations, and privileges in societies where there is no normative orientation. We can only refer to such abstractions when a cultural system as well as a system of social action exists. "Use-values" may exist at a protocultural stage in the primates, but they function in social systems with different properties.

Another example of normative orientation in human societies is the well-known phenomenon of incest avoidance. With its associated manifestations of shame, guilt, and anxiety, it long presented a puzzling sociopsychological problem because the underlying psychological structure was not thoroughly understood.[121] Such patterns of avoidance, with both constant and variable features, do not and could not operate at a non-hominid level where genealogical relations between individuals arc not known, where socially sanctioned value-systems are not present, and where the phenomena of self-identification and moral responsibility for conduct does not exist. Kroeber has pointed out that "the incest taboo is the complement of kin recognition." Abstraction, in turn, "involves ability to symbolize, in other words, speech." [122] Consequently, incest taboos could not arise among primates incapable of self-other orientation in a web of differentiated moral relationships. In social interaction, the individual could not be held responsible for differentiated responses to kin until the latter were explicitly classified through linguistic or other means. Although precisely the same genealogical relationships existed at a lower level of primate social organization, they could not be consciously identified

[120] For a more extended discussion see Hallowell (1955) chap. 12.

[121] Lowie, e.g., in his *Primitive Society* (1920) expressed the view that incest taboos have an instinctive basis. Later, he changed his mind (1933), p. 67.

[122] Kroeber (1942), p. 206.

and utilized as a basis of differential social interaction until the individual "actors" participating in the system developed a personality structure that permitted self-objectification and the use of symbolic means in playing sanctioned roles within a common framework of values.

Further ramifications of the basic significance of normative orientation and its psychological correlate of self-awareness in the evolution of a fully developed mode of cultural adaptation cannot be considered here. But the question can be raised whether the capacity for self-objectification was common to all the Hominidae from the beginning. Perhaps we might venture to say that, although some of the psychological *anlagen* were present at a protocultural stage, a capacity for self-objectification and role differentiation functioning in intimate relations with socially sanctioned value-systems, were sociopsychological developments that only became established in typical form long after the initial steps in hominid anatomical differentiation had taken place. One of the reasons for this, as we shall see, is that these developments were contingent upon a system of communication that was not only socially transmitted but, through symbolic mediation, gave unique and characteristic scope to the novel psychological capacities that had been developing through the expansion of the hominid brain.

Ego and Self-objectification

While it has been widely recognized that self-awareness is a characteristic phenomenon in *Homo sapiens,*[123] the psychological structure that underlies it has been seriously studied only since the rise of a more general interest in personality structure, mainly under the impact of psychoanalytic theories. The evolutionary aspects of the problem have been scarcely touched.[124] Indeed, there have been "many psychologists

[123] For example, Bidney at the outset of his *Theoretical Anthropology,* (p. 3) writes: "Man is a self-reflecting animal in that he alone has the ability to objectify himself, to stand apart from himself, as it were, and to consider the kind of being he is and what it is that he wants to do and to become. Other animals may be conscious of their affects and the objects they perceive; man alone is capable of reflection, of self-consciousness, of thinking of himself as an object." The psychologist David Katz, writing more than twenty years ago, likewise stressed what he called "objectivization" as a human differential (1937, p. 253). More recently, Rollo May has given particular emphasis to human self-awareness. "We can never see man whole," he says, "except as we see him, including ourselves, as the mammal who has a distinctive capacity for awareness of himself and his world. Herein lie the roots of man's capacity to reason and deal in symbols and abstract meaning. And herein lies also the basis for a sound view of human freedom" (1955, p. 313). Cf. May (1953), pp. 84–85. Other comparable opinions could be cited.

[124] Stanley Cobb (1957, p. 202), in discussing the papers contributed to the symposium "Brain Mechanisms and Consciousness," says: "Although some of the authors seem to confuse the concepts of 'mind' and 'consciousness,' Fessard seems to agree with

of the modern period," as Asch says, "who have spoken of the individual organism as of a congeries of capacities and tendencies without a self-character." [125] It has been pointed out, moreover, that "between 1910 and 1940, most psychologists preferred not to mention 'ego' or 'self' in their writings." [126] Nowadays, ego and self are familiar terms, although the connotation given them is not standardized. However, no one uses the ego concept in any substantive sense but rather as a psychological construct useful in conceptualizing a subsystem of the total personality, objectively approached, with reference to its development, structure, and functioning. If we wish to be rigorous, it is best to speak of a group of ego processes or functions, although this is sometimes awkward. Ego functions have a wide range; they are intimately connected with such cognitive processes as attention, perception, thinking, and judgment, because ego processes are involved in determining adjustments to the outer world in the interests of inner needs, particularly in a situation where choice or decision, and hence delay or postponement of action, is required.[127]

On the other hand, the concept of self carries a reflexive connotation: "I" can think of "me." I can discriminate myself from other objects perceptually; I can conceive of myself as an object; I can develop attitudes toward myself. Thus the self is a phenomenal datum, whereas the ego is a construct. "The self can be observed and described; the ego is deduced and postulated. The ego may be conceived in quasi-physiological terms as a sub-system of the organism. . . ." [128] Furthermore,

---

me that 'consciousness' is but one attribute of 'mind.' I would say [it is] *that part which has to do with awareness of self and of environment.* It varies in degree from moment to moment in man and from fish to man in phylogeny. It may be that invertebrates and even plants have rudimentary forms of awareness of self." It is difficult, however, to follow Cobb through to this point! Sir Julian Huxley (1956, pp. 558–59) has suggested that since "*mind* and *mental* have various undesirable connotations, it is best to drop them and to speak of awareness. Psychology in the customary sense can then be regarded as part of the general study of awareness and its evolution." This would include "the way in which new possibilities of awareness are in fact realized, and also of the limitations on their realization. . . . There are two evolutionary prerequisites for a high organization of awareness involving the incorporation of individual experience by learning: (1) a long youth period . . . ; (2) Homothermy, permitting greater uniformity and continuity of awareness. Prerequisites for the further organization of the awareness-system, to enable it to incorporate experiences from other individuals and from past generations, are (1) social life, (2) the capacity to organize awareness in the form of concepts, (3) true speech. These have permitted the evolution of the unique type of awareness-system found in man." Cf. Huxley, 1953.

[125] Asch, p. 276.

[126] Sargent, chap. 20. The publication of G. W. Allport's article in 1943 initiated a renewed interest in ego and self on the part of social psychologists in particular.

[127] Cf. Symonds, p. 4. Hartmann (1950) distinguishes ego, a psychic subsystem of the total personality with functions distinguishable from the id and superego, from self, one's own person.

[128] Mac Leod (1951), p. 234. Cf. Asch, chap. 10, "The Ego," and Symonds.

the self does not mirror the ego—the subject's capacity for self-objectification does not imply his objective knowledge of the psychodynamics of his total personality.

Considered in evolutionary perspective ego may be said to be the major "psychological organ" that structurally differentiates the most highly evolved members of the Hominidae from subhominid primates and probably other hominids of lower evolutionary rank. It lies at the core of a human personality structure as we know it in *Homo sapiens*.[129] It permits adaptation at a new behavioral level. Since, in ontogenetic development, the beginnings of ego processes can be identified in the first half-year of life, well before the acquisition of speech, we can say that, while ego development occurs in a context of social interaction, in its initial stages it is not contingent upon the prior existence of either speech or culture. The underlying capacities for ego functioning must have deeper psychobiological roots.[130] This is the area in which the evolutionary problem lies.

Heinz Hartmann has made a most illuminating suggestion as to how this problem may be approached. He says that we must not overlook important relations between animal instinct and human ego functions. His point is that "many functions, which are taken care of by instincts" in the lower animals "are in man functions of the ego." But, he says, we should not identify the nature and role of instincts in animals with "drives" in man; "the id, too, does not appear to be a simple extension of the instincts of lower animals. While the ego develops in the direction of an ever closer adjustment to reality, clinic experience shows the drives, the id-tendencies, to be far more estranged from reality than the so-called animal instincts generally are." [131] In other words,

[129] See Hall and Lindzey, e.g., who point out that "among the theorists who, in some way, make prominent use of the ego or self concept are Adler, Allport, Angyal, Cattell, Freud, Goldstein, Jung, Murphy, Murray, and Sullivan" (p. 545). Cf. Spiro (1954), pp. 27–28.

[130] While there is a considerable literature on the body-image phenomenon, the relations between body-image, ego, and self concepts are still under discussion. See Fisher and Cleveland.

[131] Hartmann (1948), p. 379 ff. With reference to ontogenesis, Hartmann has been responsible for stressing an early "undifferentiated phase," in contrast with the notion that the id is chronologically older than the ego, and the concept of a "conflict-free ego sphere" (1950). In the early undifferentiated stage of ontogenetic development there are no ego functions and no differentiation of self from the world outside. With respect to phylogenetic development, Hartmann says that while psychoanalysts do "attribute a sort of ego to animals" (1958, p. 48; no species indicated) "we cannot speak, in regard to the animal, of that kind of separation into ego and id which exists in the human adult. The very fact that the concept of instincts as it pertains to the lower animals is much more comprehensive that the concept of instinctual drives as it pertains to man prevents such a separation. It is possible, and even probable, that is just this sharper differentiation of the ego and the id—the more precise division of labor between them—in human adults which on the one hand makes for a superior, more flexible, relation to the outside world and, on the other, *increases the alienation of the id from reality*." Cf. Hartmann, 1948.

between the object seen, the memory image, and the graphic symbolization.

But there also seems to be evidence in the cave art of a related human capacity, that is, the ability to project graphically synthetic images of fabulous creatures, animal-like or human-like, which were not objects of ordinary perceptual experience. These belong, rather, to the world of creative imagination. The beast with two horns at Lascaux is the prime example of the representation of a fantastic animal.[138] Many examples of ambiguous human figures—synthesizing both human and animal characteristics—are known, and it is these figures which have proved the most difficult to interpret in the whole repertoire of cave art, since in style they do not fit the realistic tradition of the animal art.[139] The older view that these semi-human figures were the representation of actual human beings wearing masks, or the skins of animals, has been steadily losing ground. In the cave of Trois Frères, the figure originally called a "Sorcerer" by Breuil and Bégouën is now thought by them to be the representation of the "Spirit controlling the multiplication of game and hunting expeditions," [140] in other words, a god or a personage of an other-than-human class.[141] If the humanly ambiguous figures are thought of as belonging to such a class, I believe that it may be argued that we have evidence which suggests that a system of beliefs is reflected in the art, which makes this category of figures equivalent to the personages that appear in the myths of living primitive peoples. In this case the cave art would offer evidence of a level of imaginative functioning and conceptual creativity that transcended a purely naturalistic reproduction of what was perceived. It could be interpreted as revealing capacities in early representatives of *Homo sapiens* psychologically equivalent to those of living peoples

[138] See Breuil (p. 118 and Fig. 89) who writes: "By its massive body and thick legs, it resembles a bovine animal or a Rhinoceros; the very short tail is more indicative of the latter; the flanks are marked with a series of 0-shaped oval splashes; the neck and ears are ridiculously small for the body; the head with a square muzzle, is like that of a Feline; two long stiff straight shafts, each ending in a tuft, are like no known animal horns, unless, as Miss Bate suggested, those of the Pantholops of Thibet . . . This is not the only example of a composite unreal animal in Quaternary art, but it is the most spectacular." The drawing measures about 5 ft. 6 in.

[139] See Saccasyn-Della Santa for illustrations of 250 examples of these figures and a systematic classification and analysis of them.

[140] Breuil (*op. cit.*), pp. 176–77.

[141] Laming is of the opinion that "the imaginary animals and the semi-human figures are . . . incompatible with the theory of sympathetic magic" which has been applied to the animal art. Considered as a whole, she also finds untenable "the theory that they represent hunting masks or have some connection with ritual hunting dances." "Why should the sorcerers, who were probably the artists of the tribe, depict themselves on the walls of the sanctuary wearing their masks?" she asks; it seems more likely that these drawings "represent mythical beings who were perhaps connected in some way with the history of the ancestors of the group" (pp. 191–92).

studied in their full cultural context where the details of world-view and religious beliefs have been recorded.[142]

While the symbolism embodied in speech is in a different mode, since sound clusters are given a meaning-content that is unrelated to the form or qualities of the objects or events represented, it seems to me that we must assume that the same basic capacities for extrinsic symbolization are involved. Art forms are as indicative of these capacities as are speech forms. Among other things, graphic art in all its manifestations requires abstraction or else it could not function as a means of representation. In any case, it is hard to believe that the peoples of the Upper Paleolithic did not possess a vocal system of representation (although we have no direct evidence of speech) as well as a fully developed mode of cultural adaptation equivalent to the non-literate peoples of historic times. Viki and other chimpanzees, if considered as representative of an advanced level of infrahominid behavior, manifest as little capacity for graphic symbolization of an extrinsic type as for vocal symbolization.[143] By the time we reach the Upper Paleolithic, the infrahominids have been left far behind on the ladder of behavioral evolution.

Systems of extrinsic symbolization necessitate the use of material media which can function as vehicles for the communication of meanings. Abstraction and conceptualization are required since objects or events are introduced into the perceptual field as *symbols,* not in their concrete reality. Thus systems of extrinsic symbolization involve the operation of the representative principle on a more complex level than do processes of intrinsic symbolization. In the case of *Homo sapiens,* extrinsic symbolic systems, functioning through vocal, graphic, plastic, gestural, or other media, make it possible for groups of human beings to share a common world of meanings and values. A cultural mode of adaptation is unthinkable without systems of extrinsic symbolization.

From a phylogenetic point of view the capacity for individual and social adaptation through the *integral* functioning of intrinsic symbolic processes and extrinsic symbolic systems enabled an evolving hominid to enlarge and transform his world. The immediate, local, time-and-

[142] Cf. the discussion of "persons" of an other-than-human class among Ojibwa in Hallowell, 1958. Persons of this category are reified beings in the behavioral world of the Ojibwa and are equivalent to characters in their myths. Among them these narratives are true stories. Since metamorphosis is possible, a hard and fast line cannot even be drawn between the outward appearance of *human* persons and animals. Persons of the other-than-human class in particular, appear in myths and dreams in animal form. If the Ojibwa had an art similar to that of the Upper Paleolithic peoples and we had no other evidence, it can be imagined how difficult it would be to interpret the graphic representations they had made of persons of an other-than-human class.

[143] See the colored reproductions of Viki's paintings in *Life* (Dec. 3, 1951).

space-bound world of other primates, who lack the capacity for dealing effectively with objects and events outside the field of direct perception, could be transcended. Speech, through the use of personal pronouns, personal names, and kinship terms made it possible for an individual to symbolize, and thus objectify, himself in systems of social action. Self-related activities, both in the past and future, could be brought into the present and reflected upon.[144] What emerged was a personality structure in which ego processes and functions had become salient at a high level of integration—self-awareness. The inner world of private experience and the outerworld of public experience became intricately meshed through symbolic mediation. In all human societies, the self-image became, in part, a culturally constituted variable; self-orientation became integrated with other basic orientations toward the world that enabled the individual to think, feel, and act in a culturally constituted behavioral environment.[145] As a result of self-objectification human societies could function through the commonly shared value-orientations of self-conscious individuals, in contrast with the societies of non-hominid and probable early hominid primates, where ego-centered processes remained undeveloped or rudimentary. In fact, when viewed from the standpoint of this peculiarity of man, culture may be said to be an elaborated and socially transmitted system of meanings and values which, in an animal capable of self-awareness, implements a type of adaptation which makes the role of the human being intelligible to himself, both with reference to an articulated universe and to his fellow men.

The central importance of ego processes and self-awareness that we find distinctive in *Homo sapiens* can be viewed from another angle. Since self-objectification involves self-appraisal in relation to sanctioned moral conduct, we can see the social as well as the individual adaptive value of unconscious psychological processes such as repression, rationalization and other defense mechanisms. Culturally constituted moral values impose a characteristic psychological burden, since it is not always easy, at the level of self-awareness, to reconcile idiosyncratic needs with the demands imposed by the normative orientation of the self. For animals without the capacity for self-objectification no such situation can arise. As Freedman and Roe [146] write,

> Only in man is there simultaneously such a rigidity of social channeling and such a degree of potential plasticity and flexibility for the individual.

[144] Cf. Révész (1956, p. 104), "Without the verbal formulation of subjective experience and ethical standards, self-consciousness is incomplete and self-knowledge and self-control equally so. To be conscious of one's own self, to examine one's own endeavors, motives, resolves, and actions, necessarily presupposes language."

[145] See Hallowell (1955), chap. 4.

[146] Freedman and Roe, p. 461, in Roe and Simpson.

Incompatible aims and choices which are desirable but mutually exclusive are inevitable conditions of human development. This discrepancy between possibility and restriction, stimulation and interdiction, range and construction, underlies that quantitatively unique characteristic of the human being: conflict.

In *Homo sapiens,* unconscious mechanisms may be viewed as an adaptive means that permits some measure of compromise between conflicting forces. They relieve the individual of part of the burden not only forced upon him by the requirements of a morally responsible existence but by the fact that the normative orientation of any human social order permeates all aspects of living. A human level of existence requires an evolutionary price; man as a species has survived despite proneness to conflict, anxiety, and psychopathology.[147] There seems to be little question that one of the crucial areas of individual adjustment turns upon the sensitivity of the self to feelings of anxiety and guilt.

Psychoanalysts, in particular, have come more and more to recognize that psychological maladjustment centers around the structural core of the human personality. David Beres, for example, writes: [148] "There is then in man this unique structure, the ego, which in its full function allows for the expression of those qualities which distinguish the human from the animal and which, in their malfunction, give to his behavior and thought the characteristically human forms of mental illness."

Leopold Bellak [149] has recently reviewed the shift in focus that has occurred in psychoanalytic thinking:

> The novelty in psychoanalysis was originally its introduction of the unconscious in the sense of the unconsciousness of feelings, the unawareness of previously experienced events, the covert nature of motivations, and the hidden meaning of dreams and symptoms. Slowly attention focused on the forces responsible for this unconsciousness, notably repression.

A new era, however, "dedicated to the analysis not only of the unconscious but of the ego and its defences," was initiated with Anna Freud's book, *The Ego and the Mechanisms of Defense* (1936). So that now, "the pendulum has swung nearly full cycle, in that there is so much talk about ego psychology today that the forces of the unconscious are possibly already somewhat in disregard." [149]

Franz Alexander, commenting on the same shift of interest,[150] says,

[147] Cf. (*ibid.*), p. 422. Freud's interpretation is to be found in his *Civilization and its Discontents* (p. 123), "The price of progress in civilization is paid for in forfeiting happiness through the heightening of the sense of guilt."

[148] Beres, pp. 170 and 231.

[149] Bellak, pp. 25–26. Cf. Stierlin, p. 146.

[150] Alexander (p. 78 ff.). Cf. Hartmann's remarks on the synthetic or organizing functions of the ego (1948, pp. 383–84).

Mental disease represents a failure of the ego to secure gratification for subjective needs in a harmonious and reality-adjusted manner and a breakdown of the defenses by which it tries to neutralize impulses which it cannot harmonize with its internal standards and external reality. . . . The highest form of integrative function requires conscious deliberation. Everything which is excluded from consciousness is beyond the reach of the ego's highest integrative functions. . . . Psychoanalytic therapy aims at the extension of the ego's integrative scope over repressed tendencies by making them conscious.

Thus, in the terminology I have been using here, psychological functioning at a level of self-awareness is as important for rational personal adjustment as it is for the functioning of sociocultural systems. Furthermore, as Schneirla points out, it is an error stemming from an inadequate comprehension of the complex nature of a human level of existence to assume "that man's 'higher psychological processes' constitute a single agency or unity which is capable of being sloughed off" even under extreme provocation. On the contrary "socialized man even under stress of extreme organic need or persistent frustration does not regress to the 'brute level.' Rather, he shifts to some eccentric and distorted variation of his ordinary personality, which varies from his prevalent socialized make-up according to the degree of integrity and organization attained by that adjustment system." [151] This is why we find variations in the symptomatology and incidence of mental disorders in man when we consider them in relation to differences in cultural modes of adaptation. These phenomena often have been given a purely relativistic emphasis. But increasing evidence suggests that they probably can be ordered to psychodynamic principles and etiological factors that operate universally.[152] Direct comparison, moreover, between the psychopathology of the "civilized" individual mind and the "primitive mind" savors more than ever of a pseudo-evolutionary problem.

Summary

What we observe in the behavior of *Homo sapiens* is the culmination and distinctive integration of processes and capacities which require analytical discrimination and investigation in a long evolutionary perspective. The nature of man cannot be fully understood outside this framework. A cultural level of adaptation could not arise *de novo*. There were prerequisites of various kinds. Simple forms of learning, some socialization of the individual, a social structure based on role

[151] Schneirla (1949), p. 273.
[152] Benedict and Jacks; Teicher.

differentiation in organized social groups, the transmission of some group habits and perhaps tool-using, and a "non-syntactic" (Hebb) form of communication [153] may be identified as necessary but not sufficient conditions for a human level of existence. The development of these conditions in combination provided a preadaptive or protocultural stage. But all of them were raised to a new level of functional organization and inclusiveness by the psychological restructuralization that must have occurred during the evolution of the Hominidae when cerebral expansion took place subsequent to the development of bipedal locomotion and readjustment to new ecological conditions. Without this psychological factor, expressed in part by a capacity for the development of extrinsic forms of symbolization, a cultural level of adaptation could not have been reached in the first place and could not be maintained in its characteristic forms.

So far as learned behavior is concerned, its importance in relation to culture has been both exaggerated and over-simplified. "Experience will not make a man out of a monkey," as Nissen says.[154] Learning in the form of simple conditioning is found far down the animal scale.[155] Of equal importance in behavioral evolution is how much is learned and what is learned, considered in relation to the capacities and total life adjustment of the animal.[156] In anthropological writing prior to the culture and personality movement, the connection between learning and culture remained vague because it had not been considered

[153] Carpenter, (1952, p. 242) says: "The limitations of capacities for communication, especially for symbol communication seem to stop non-human primate social development at the level of limited contemporary social groupings, to preclude the development of tribal kinship and to make it impossible for them to have any except the anlagen of cultural traditions."

[154] Nissen (1955), p. 105.

[155] Harlow (1958), argues that "there is no evidence that any sharp break ever appeared in the evolutionary development of the learning process" (p. 288); at the same time, "it is quite clear that evolution has resulted in the development of animals of progressively greater potentialities for learning and for solving problems of increasing complexity" (p. 269).

[156] Hilgard is not content with the implicit, if not always explicit, generalization from comparative studies that "there are no differences, except quantitative ones, between the learning of lower mammals and man." At the human level, he says (p. 461), "There have emerged capacities for retraining, reorganizing, and foreseeing experiences which are not approached by the lower animals, including the other primates. No one has seriously proposed that animals can develop a set of ideals that regulate conduct around long-range plans, or that they can invent a mathematics to help them keep track of their enterprises. . . . Language in man is perhaps the clearest of the emergents which carries with it a forward surge in what may be learned. It seems plausible enough that other advances in the ability to learn must have come about as the nervous system evolved through successive stages below man. . . . There are probably a number of different kinds of learning which have emerged at different evolutionary periods, with the more highly evolved organisms using several of them. It is quite probable that these different kinds of learning follow different laws, and it is foolhardy to allow our desire for parsimony to cause us to overlook persisting differences."

in relation to the development of personality structure. The fact had been overlooked that the only way in which a culture can be perpetuated is through the characteristic psychological structuralization of individuals in an organized system of social action.

What seems to be significant in primate evolution is that social learning became linked with the functioning of social structure and the transmission of habits at the subhominid level. But at this level *what* was learned appears to have been quantitatively limited. In the case of the Japanese macaques the transmission of new food habits has been primarily stressed. In *Homo sapiens* we have a quantitative maximization of social learning which has led to qualitatively distinctive consequences because of the essential role that learning plays in the development of the higher levels of integrative functioning of the human personality.[157] But this is not all. There are other psychological functions manifested at this level that are of paramount importance. We find cognitive processes raised to a higher level of functioning by means of culturally constituted symbolic forms which can be manipulated creatively through reflective thought and expression. Cultural modes of adaptation, or certain aspects of them, learned and transmitted as they may be, also can be objectified, thought about, analyzed, judged, and even remodelled.

The great novelty then, in the behavioral evolution of the primates was not simply the development of a cultural mode of adaptation as such. It was, rather, the psychological restructuralization that not only made this new mode of existence possible but provided the potentialities for cultural *re*-adjustment and change. The psychological basis of culture lies not only in a capacity for highly complex forms of learning but in a capacity for transcending what is learned; a potentiality for innovation, creativity, reorganization, and change.[158]

While self, society, and culture must, of course, be conceptually differentiated for special types of analysis and investigation, they cannot be postulated as completely independent variables. Considered in phylogenetic perspective the temporal depth of their intimate connections is brought into focus. Besides this, the significance of these integral connections becomes more apparent, both with respect to the

[157] The use of terms like "acculturation" and "enculturation" as well as "superego," without qualification, in reporting observations made on the Japanese macaques seems of dubious value. (See Imanishi, 1957.) If these Old World monkeys already possess culture and have superegos, what does human evolution mean?

[158] Cf. Henry (pp. 221–22) who points out that "because his mechanism for determining personal relations lack specificity" man's unique evolutionary path is set for him "by his constant tendency to alter his modes of social adaptation. Put somewhat in value terms, man tries constantly to make a better society, i.e., one in which he can feel more comfortable. When he makes a 'mistake,' he tries to change. This is one way in which he evolves."

psychological nature of man as a product of evolution and the primary adaptive process inherent in the achievement of a sociocultural level of existence.

BIBLIOGRAPHY

ABERLE, D. F., COHEN, A. K., DAVIS, A. K., LEVY, M. J., JR., and SUTTON, F. X. 1950. "The Functional Prerequisites of a Society," *Ethics,* LX, 100–111.

ADRIAN, E. D., BREMER, F., DELAFRESNAYE, J. F., and JASPER, H. H. (eds.). 1954. *Brain Mechanisms and Consciousness.* Springfield: Charles C Thomas.

ALEXANDER, FRANZ. 1950. "The Evolution and Present Trends of Psychoanalysis," *Acta Psychol.,* VII, 126–33. Reprinted in *The Study of Personality: A Book of Readings,* HOWARD BRAND (ed.). New York: John Wiley & Sons.

ALLPORT, GORDON W. 1943. "The Ego in Contemporary Psychology," *Psychol. Rev.,* L, 451–78.

ASCH, SOLOMON E. 1952. *Social Psychology.* New York: Prentice-Hall, Inc.

An Appraisal of Anthropology Today. 1953. Edited by SOL TAX, LOREN C. EISELEY, IRVING ROUSE, and CARL F. VOEGELIN. Chicago: University of Chicago Press.

BARTHOLOMEW, G. A., JR., and BIRDSELL, J. B. 1953. "Ecology and the Protohominids," *Amer. Anthropologist,* LV, 481–98.

BEACH, FRANK A. 1947. "Evolutionary Changes in the Physiological Control of Mating Behavior in Mammals," *Psychol. Rev.,* LIV, 297–315.

———. 1951. "Instinctive Behavior: Reproductive Activities." In *Handbook of Experimental Psychology,* S. S. STEVENS (ed.). New York: John Wiley & Sons.

———. 1958. "Evolutionary Aspects of Psychoendocrinology." In *Behavior and Evolution,* ANNE ROE and G. G. SIMPSON (eds.). New Haven: Yale University Press.

BELLAK, LEOPOLD. 1956. "Psychoanalytic Theory of Personality." In *Psychology of Personality: Six Modern Approaches,* J. L. MCCARY (ed.). New York: Logos Press.

BENEDICT, PAUL K., and JACKS, IRVING. 1954. "Mental Illness in Primitive Societies," *Psychiatry,* XVII, 377–89.

BERES, DAVID. 1956. "Ego Deviation and the Concept of Schizophrenia," *The Psychoanalytic Study of the Child,* XI, 164–235.

BIDNEY, DAVID. 1953. *Theoretical Anthropology.* New York: Columbia University Press.

BOLK, L. 1926. *Das Problem der Menschwerdung.* Jena: Fischer.

BORING, EDWIN G. 1950. *A History of Experimental Psychology.* New York: Appleton-Crofts.

BOURLIÈRE, FRANÇOIS. 1952. "Classification et caractéristiques des principaux types de groupements sociaux chez les Vertébrés sauvages." In *Structure et physiologie des sociétés animales,* pp. 71–79. Paris: Centre National de la Recherche Scientifique.

———. 1956. *The Natural History of Mammals* (2d ed. rev.). New York: Alfred A. Knopf.

BREUIL, H. 1952. *Four Hundred Centuries of Cave Art.* Montignac, Dordogne: Centre d'Etudes et de Documentation Prehistoriques.

BROWN, ROGER. 1958. *Words and Things.* Glencoe, Ill.: The Free Press.

CARPENTER, C. R. 1934. "A Field Study of the Behavior and Social Relations of the Howling Monkeys," *Comp. Psychol. Mono.,* Vol. X.

———. 1940. "A Field Study in Siam of the Behavior and Social Relations of the Gibbon (*Hylobates lar.*)," *Ibid.,* Vol. XVI.

———. 1942. "Characteristics of Social Behavior in Non-human Primates," *Trans. New York Acad. Sci.,* Ser. II.

———. 1945. "Concepts and Problems of Primate Sociometry," *Sociometry,* VIII, 55–61.

———. 1952. "Social Behavior of Non-human Primates," pp. 227–45. In *Structure et physiologie des sociétés animales,* Vol. XXXIV. Paris: Centre National de la Recherche Scientifique.

———. 1954. "Tentative Generalizations on Grouping Behavior of Non-human Primates," Human Biol., XXVI, 269–76. Included in *The Non-human Primates and Human Evolution,* arr. by JAMES A. GAVAN. Detroit: Wayne State University Press, 1955.

———. 1958*a*. "Territoriality: A Review of Concepts and Problems." In *Behavior and Evolution,* ANNE ROE, and G. G. SIMPSON (eds.). New Haven: Yale University Press.

———. 1958*b*. "Soziologie und Verhalten freilebender nichtmenschlicher Primaten" in *Handbuch der Zoologie,* Bd. 8, Lief. 18: 1–32. Berlin: Walter de Gruyter.

CARTER, G. S. 1951. *Animal Evolution.* London: Sidgwick & Jackson.

———. 1953. "The Theory of Evolution and the Evolution of Man." In *Anthropology Today,* A. L. KROEBER (ed.). Chicago: University of Chicago Press.

CHANCE, M. R. A. 1959. "What Makes Monkeys Sociable," *The New Scientist* (March, 1959).

CHANCE, M. R. A., and MEAD, A. D. 1953. "Social Behavior and Primate Evolution," *Symposia of the Society for Experimental Biology,* Vol. VII.

COBB, STANLEY. 1957. "Awareness, Attention and Physiology of the Brain Stem." In *Experiments in Psychopathology,* HOCH and ZUBIN (eds.). New York: Grune & Stratton.

COLLIAS, N. E. 1950. "Social Life and the Individual Among Vertebrate Animals," *Annals N.Y. Acad. of Science,* L, 1074–92.

COUNT, EARL W. 1958. "The Biological Basis of Human Sociality," *Amer. Anthropologist,* LX, 1049–85.

CRAWFORD, MEREDITH P. 1937. "The Cooperative Solving of Problems by Young Chimpanzees," *Comp. Psychol., Mono.,* Vol. XIV, No. 2.

CRITCHLEY, MACDONALD. 1958. "Animal Communication," *Trans. of the Hunterian Society of London,* XVI (1957–58), 90–111.
CUENOT, L. 1945. "L'homme ce Neotenique," *Bull. Acad. roy. Belgique* (Brussels), Vol. XXXI.
DART, RAYMOND A. 1956. "Cultural Status of the South African Man-Apes." *Annual Report, Smithsonian Institution (1955),* 317–38. Washington, D.C.
———. 1959. *Adventures with the Missing Link.* New York: Harper & Bros.
DARWIN, CHARLES. 1871. *The Descent of Man.* London: Murray.
———. 1873. *The Expression of the Emotions in Man and Animals.* New York: Appleton. (A new edition with Preface by Margaret Mead was published by the Philosophical Library, New York, 1956.)
DE BEER, G. R. 1951. *Embryos and Ancestors.* London and New York: Oxford University Press.
DE LAGUNA, GRACE A. 1927. *Speech: Its Function and Development.* New Haven: Yale University Press.
DENNIS, WAYNE. 1951. "Developmental Theories." In *Current Trends in Psychological Theory.* Pittsburgh: University of Pittsburgh Press.
DEWEY, JOHN. 1917. "The Need for a Social Psychology," *Psychol. Rev.,* XXIV, 266–77.
DOBZHANSKY, TH. 1955. *Evolution, Genetics and Man.* New York: John Wiley & Sons.
DUBRUHL, E. LLOYD. 1958. *Evolution of the Speech Apparatus.* Springfield: Charles C Thomas.
EDINGER, TILLY. 1956. "Objects et resultats de la paleoneurologie," *Ann. paleontol.,* XLII, 97–116.
EGGAN, FRED. 1954. "Social Anthropology and the Method of Controlled Comparison," *Amer. Anthropologist,* LVI, 743–63.
EISELEY, LOREN C. 1955. "Fossil Man and Human Evolution." In *Yearbook of Anthropology, 1955,* WILLIAM L. THOMAS, JR. (ed.). New York: Wenner-Gren Foundation for Anthropological Research.
———. 1958. *Darwin's Century: Evolution and the Men Who Discovered It.* New York: Doubleday.
ETKIN, WILLIAM. 1954. "Social Behavior and the Evolution of Man's Mental Faculties," *Amer. Naturalist,* LXXXVIII, 129–42.
FISHER, SEYMOUR, and CLEVELAND, SIDNEY E. 1958. *Body Image and Personality.* New York: D. Van Nostrand.
FISKE, JOHN. 1909. *The Meaning of Infancy.* Boston: Houghton Mifflin Co. Reprinting of "The Meaning of Infancy" from *Excursions of an Evolutionist* (1884) and "The Part Played by Infancy in the Evolution of Man" from *A Century of Science, and Other Essays* (1899).
FORD, CLELLAN S., and BEACH, FRANK A. 1951. *Patterns of Sexual Behavior.* New York: Harper & Bros.
FRAZER, JAMES G. 1922. "Scope and Method of Mental and Anthropological Science," *Sci. Progress,* XVI, 580–94.
FREEDMAN, LAWRENCE Z., and ROE, ANNE. 1958. "Evolution and Human

Behavior." In *Behavior and Evolution.* ANNE ROE and G. G. SIMPSON (eds.). New Haven: Yale University Press.

FREUD, SIGMUND. 1930. *Civilization and Its Discontents.* New York: Jonathan Cape and Harrison Smith.

FRISCH, JOHN E. 1959. "Research on Primate Behavior in Japan," *Amer. Anthropologist,* LXI, 584–96.

FULLER, JOHN L., and SCOTT, JOHN PAUL. 1954. "Heredity and Learning Ability in Infrahuman Animals," *Eugenics Quart.,* I, 28–43.

GREENBERG, JOSEPH H. 1957. "Language and Evolutionary Theory." In *Essays in Linguistics.* (Viking Fund Publications in Anthropology, No. 24.) New York: Wenner-Gren Foundation for Anthropological Research.

HALDANE, J. B. S. 1955. "Animal Communication and the Origin of Human Language," *Sci. Progress,* XL, 385–401.

HALL, CALVIN S., and LINDZEY, GARDNER. 1957. *Theories of Personality.* New York: John Wiley & Sons.

HALLOWELL, A. IRVING. 1953. "Culture, Personality, and Society." In *Anthropology Today,* A. L. KROEBER (ed.). Chicago: University of Chicago Press.

———. 1955. *Culture and Experience.* Philadelphia: University of Pennsylvania Press.

———. 1956. "The Structural and Functional Dimensions of a Human Existence," *Quart. Rev. Biol.,* XXXI, 88–101.

———. 1958. "Ojibwa Metaphysics of Being and the Perception of Persons." In *Person Perception and Interpersonal Behavior,* R. TAGIURI and L. PETRULLO (eds.). Stanford, California: Stanford University Press.

HARLOW, HARRY F. 1951*a*. "Levels of Integration along the Phylogenetic Scale: Learning Aspect." In *Social Psychology at the Cross Roads,* JOHN H. ROHRER and MUZAFER SHERIF (eds.). New York: Harper & Bros.

———. 1951*b*. "Thinking." In *Theoretical Foundations of Psychology,* HARRY HELSON (ed.). New York: D. Van Nostrand.

———. 1952. "Primate Learning." In *Comparative Psychology,* CALVIN P. STONE (ed.). 3d ed. New York: Prentice-Hall, Inc.

———. 1956. "Current and Future Advances in Physiological and Comparative Psychology," *Amer. Psychologist* (June, 1956).

———. 1958. "The Evolution of Learning." In *Behavior and Evolution,* ANNE ROE and G. G. SIMPSON (eds.). New Haven: Yale University Press.

HART, C. M. H. 1938. "Social Evolution and Modern Anthropology." In *Essays in Political Economy in Honour of E. J. Urwick,* H. A. INNES (ed.). Toronto: University of Toronto Press.

HARTMANN, HEINZ. 1948. "Psychoanalytic Theory of Instinctual Drives," *Psychoanal. Quart.,* XVII, 368–88.

———. 1950. "Comments on the Psychoanalytic Theory of the Ego," *Psychoanalytic Study of the Child,* Vol. V.

———. 1958. *Ego Psychology and the Problem of Adaptation.* Trans. by DAVID RAPAPORT. (Journal of the American Psychoanalytic Association

Monograph Series, No. 1.) New York: International Universities Press.
HAYES, CATHY. 1951. *The Ape in Our House*. New York: Harper & Bros.
HAYES, KEITH J., and CATHERINE. 1955. "The Cultural Capacity of Chimpanzee." In *The Non-human Primates and Human Evolution*, arr. by JAMES A. GAVAN. Detroit: Wayne State University Press.
HEBB, D. O., and THOMPSON, W. N. 1954. "The Social Significance of Animal Studies." In *Handbook of Social Psychology*, Vol. I, GARDNER LINDZEY (ed.). Cambridge: Addison-Wesley Press.
HEBERER, GERHARD VON. 1956. "Die Fossilgeschichte der Hominoidea." In *Primatologia* (edited by H. HOFER, A. H. SCHULTZ, and D. STARK), I, 379–560.
HENRY, JULES. 1959. "Culture, Personality and Evolution," *Amer. Anthropologist*, LXI, 221–26.
HERRICK, C. JUDSON. 1956. *The Evolution of Human Nature*. Austin: University of Texas Press.
HILGARD, ERNEST R. 1956. *Theories of Learning*. 2d ed. New York: Appleton-Century-Crofts.
HOCHBAUM, H. ALBERT. 1955. *Travels and Traditions of Waterfowl*. Minneapolis: University of Minnesota Press.
HOCKETT, CHARLES F. 1958. *A Course in Modern Linguistics*. New York: Macmillan Co.
———. 1959. "Animal 'Languages' and Human Language." In *The Evolution of Man's Capacity for Culture*. J. N. SPUHLER (ed.). Detroit: Wayne State University Press (and in *Human Biol.*, February, 1959).
HOFER, H., SCHULTZ, A. H., and STARK, D. (eds.). 1956. *Primatologia: Handbuch der Primatenkunde*, Vol. I. Basel and New York: Karger.
HONIGMANN, JOHN J. 1954. *Culture and Personality*. New York: Harper & Bros.
———. 1959. "Psychocultural Studies." In *Biennial Review of Anthropology—1959*, BERNARD J. SIEGEL (ed.). Stanford: Stanford University Press.
HOWELL, F. CLARK. 1959. "The Villafranchian and Human Origins," *Science*, CXXX, 2 October.
———. 1960. Commentary on Leakey's "The Newest Link in Human Evolution," *Current Anthropology*, I, 76–77.
HOWELLS, W. W. 1950. "Origin of the Human Stock: Concluding Remarks of the Chairman." *Cold Spring Harbor Symp. Quant. Biol.*, XV, 79–86.
———. 1959. *Mankind in the Making*. Garden City: Doubleday and Co.
HUXLEY, JULIAN. 1941. *Man Stands Alone*. New York: Harper & Bros.
———. 1953. *Evolution in Action*. New York: Harper & Bros.
———. 1954. "The Evolutionary Process." In *Evolution as a Process*, J. HUXLEY, A. C. HARDY, and E. B. FORD (eds.). London: Allen & Unwin.
———. 1955. "Evolution, Cultural and Biological." In *Yearbook of Anthropology—1955*, WILLIAM L. THOMAS, JR. (ed.). New York: Wenner-Gren Foundation for Anthropological Research.

———. 1956. "Psychology in Evolutionary Perspective," *Amer. Psychologist,* XI, 558–59.

IMANISHI, KINJI. 1957. "Social Behavior in Japanese Monkeys, *Macaca fuscata,*" *Psychologia,* I, 47–54 (English).

———. 1959. "Identification: A Process of Enculturation in the Subhuman Society of *Macaca fuscata,*" *Primates,* I, 1–29 (English summary).

JENNINGS, H. S. 1942. "The Transition from the Individual to the Social Level." In *Levels of Integration in Biological and Social Systems,* ROBERT REDFIELD (ed.). ("Biological Symposia," Vol. VIII.) Lancaster, Pa.: Cattell Press.

KATZ, DAVID. 1937. *Animals and Men: Studies in Comparative Psychology.* London: Longmans, Green.

KEITH, ARTHUR. 1948. *A New Theory of Human Evolution.* London: Watts & Co.

KLUCKHOHN, CLYDE. 1953. "Universal Categories of Culture." In *Anthropology Today,* A. L. KROEBER (ed.). Chicago: University of Chicago Press.

KROEBER, A. L. 1928. "Sub-human Cultural Beginnings," *Quart. Rev. Biol.,* III, 325–42.

———. 1942. "The Societies of Primitive Man." In *Levels of Integration in Biological and Social Systems,* ROBERT REDFIELD (ed.). ("Biological Symposia," Vol. VIII.) Lancaster, Pa.: Cattell Press.

———. 1948. *Anthropology.* New York: Harcourt, Brace & Co.

———. 1950. "Anthropology," *Scientific American,* CLXXXIII, 87–94.

———. 1955. "On Human Nature," *Southwest. Jour. of Anthropology,* XI, 195–204.

KUBIE, LAURENCE S. 1953. "Some Implications for Psychoanalysis of Modern Concepts of the Organization of the Brain," *Psychoanal. Quart.,* XXII, 21–68.

LA BARRE, WESTON. 1954. *The Human Animal.* Chicago: University of Chicago Press.

LAMING, ANNETTE. 1959. *Lascaux Paintings and Engravings.* London: Penguin Books.

LASHLEY, K. S. 1949. "Persistent Problems in the Evolution of Mind," *Quart. Rev. Biol.,* XXIV, 28–42.

LEAKEY, L. S. B. 1959. "A New Fossil Skull from Olduvai," *Nature,* CLXXXIV, 491–93.

———. 1960. "The Newest Link in Human Evolution: The Discovery by L. S. B. Leakey of *Zinjanthropus boisei,*" *Current Anthropology,* I, 76.

LE GROS CLARK, W. E. 1950. *History of the Primates: An Introduction to the Study of Fossil Man.* London: British Museum.

———. 1955. *The Fossil Evidence for Human Evolution: An Introduction to the Study of Paleoanthropology.* Chicago: University of Chicago Press.

LeGros, Clark, W. E. 1958. "The Study of Man's Descent." In *A Century of Darwin,* S. A. Barnett (ed.). Cambridge: Harvard University Press.

———. 1959. "The Crucial Evidence for Human Evolution," *American Scientist,* XLIX, 299–313.

Lovejoy, A. O. 1922. "The Length of Human Infancy in Eighteenth Century Thought," *Jour. Phil.* XIX, 381–85.

Lowie, Robert H. 1920. *Primitive Society*. New York: Boni & Liveright.

———. 1933. *The Family as a Social Unit.* (Papers of the Michigan Academy of Science, Arts, and Letters.) Ann Arbor: University of Michigan Press.

MacLeod, Robert B. 1951. "The Place of Phenomenological Analysis in Social Psychological Theory." In *Social Psychology at the Cross Roads,* ed. John H. Rohrer and Muzafer Sherif. New York: Harper & Bros.

May, Rollo. 1953. *Man's Search for Himself*. New York: W. W. Norton.

———. 1955. "The Historical Meaning of Psychology as a Science and Profession," *Trans. N.Y. Acad. of Sciences,* Ser. II, XVII, 312–14.

Montagu, M. F. Ashley. 1955. "Time, Morphology, and Neoteny in The Evolution of Man," *Amer. Anthropologist,* LVII, 13–27.

———. 1956. "Neoteny and the Evolution of the Human Mind," *Explorations,* No. 6, pp. 85–90.

(Both Montagu articles reprinted in *Anthropology and Human Nature.* New York: Porter Sargent, 1957.)

Munn, Norman L. 1955. *The Evolution and Growth of Human Behavior*. Boston: Houghton Mifflin Co.

Murdock, George P. 1945. "The Common Denominator of Cultures." In *The Science of Man in the World Crisis,* Ralph Linton (ed.). New York: Columbia University Press.

———. 1949. *Social Structure*. New York: Macmillan Co.

Nissen, H. W. 1931. "A Field Study of the Chimpanzee: Observations of Chimpanzee Behavior and Environment in Western French Guinea," *Comp. Psychol. Mono.,* No. 1, p. 8.

———. 1946. "Primate Psychology." In *Encyclopedia of Psychology,* P. L. Harriman (ed.). New York: Citadel Press.

———. 1951*a*. "Phylogenetic Comparison." In *Handbook of Experimental Psychology,* S. S. Stevens (ed.). New York: John Wiley & Sons.

———. 1951*b*. "Social Behavior in Primates." In *Comparative Psychology,* C. P. Stone (ed.). New York: Prentice-Hall, Inc.

———. 1955. "Problems of Mental Evolution in the Primates." In *The Non-human Primates and Human Evolution,* arr. by James A. Gavan. Detroit: Wayne State University Press.

Oakley, Kenneth P. 1950. *Man the Tool-Maker*. London: British Museum.

———. 1951. "A Definition of Man," *Science News,* No. 20 (Penguin Books).

———. 1954. "Skill as a Human Possession." In *History of Technology,* Vol. I, C. J. SINGER *et al.* (eds.). Oxford: Oxford University Press.

———. 1956. "The Earliest Tool-Makers," *Antiquity,* XXX, 4–8.

———. 1957. "Tools Makyth Man," *Antiquity,* XXXI, 199–209. (Reprinted in *Annual Report, Smithsonian Institution, 1958,* 431–45.)

———. 1958. "Tools or Brains. Which Came First?" *Archeological News Letter* (London), VI, 48.

———. 1959. "Early Man's Use of Fire." *Conference on the Social Life of Early Man,* Wenner-Gren Foundation. (Summer, 1959.)

PARSONS, TALCOTT, and SHILS, EDWARD A. (eds.). 1951. *Toward a General Theory of Action.* Cambridge: Harvard University Press.

PENFIELD, W., and RASMUSSEN, T. 1950. *The Cerebral Cortex of Man.* New York: Macmillan Co.

PENFIELD, W., and ROBERTS, LAMAR. 1959. *Speech and Brain-Mechanisms.* Princeton: Princeton University Press.

PUMPHREY, R. J. 1951. *The Origin of Language.* Liverpool: University Press. (Reprinted in *Acta Psychol.,* IX (1953), 219–39.)

RENSCH, BERNHARD, 1954. "The Relation Between the Evolution of Central Nervous Functions and the Body Size of Animals." In *Evolution as a Process,* J. HUXLEY, A. C. HARDY, and E. B. FORD (eds.). London: Allen & Unwin.

———. 1956. "Increase of Learning Capability with Increase of Brain Size." *American Naturalist,* XC, 81–95.

———. 1959. *Homo Sapiens Vom Tier zum Halbgott.* Göttingen: Vandenhoeck und Ruprecht.

RÉVÉSZ, G. 1953–54. "Is There an Animal Language?" *Hibbert Jour.,* LXX, 141–43.

———. 1956. *The Origins and Prehistory of Language.* New York: Longmans, Green.

ROE, ANNE, and SIMPSON, GEORGE GAYLORD (eds.). 1958. *Behavior and Evolution.* New Haven: Yale University Press.

ROHEIM, G. 1950. *Psychoanalysis and Anthropology.* New York: International Universities Press.

ROMANES, G. J. 1883. *Animal Intelligence.* New York: Appleton.

———. 1888. *Mental Evolution in Man.* New York: Appleton.

SACCASYN-DELLA SANTA, E. 1947. *Les figures humaines du paléolithique supérieur Eurasiatique.* Antwerp: De Sekkel.

SAHLINS, MARSHALL D. 1959. "The Social Life of Monkeys, Apes and Primitive Man." In *The Evolution of Man's Capacity for Culture,* J. H. SPUHLER (ed.). Detroit: Wayne State University Press.

SARBIN, THEODORE R. "Role Theory." In *Handbook of Social Psychology,* Vol. I, GARDNER LINDZEY (ed.). Cambridge: Addison-Wesley Press.

SARGENT, S. STANSFELD. 1950. *Social Psychology.* New York: Ronald Press Co.

SCHILLER, CLAIRE H. (ed.). 1957. *Instinctive Behavior.* New York: International Universities Press.

SCHILLER, PAUL H. 1957. "Innate Motor Action as a Basis of Learning:

Manipulative Patterns in the Chimpanzee." In *Instinctive Behavior,* CLAIRE H. SCHILLER (ed.). New York: International Universities Press.

SCHNEIRLA, T. C. 1949. "Levels in the Psychological Capacity of Animals." In *Philosophy for the Future,* R. W. SELLARS, V. S. MCGILL, and M. FARBER (eds.). New York: Macmillan Co.

———. 1951. "The 'Levels' Concept in the Study of Social Organization of Animals." In *Social Psychology at the Cross Roads,* JOHN H. ROHRER and MUZAFER SHERIF (eds.). New York: Harper & Bros.

———. 1952. "A Consideration of Some Conceptual Trends in Comparative Psychology," *Psychol. Bull.,* XLIX, 559–97.

———. 1953. "The Concept of Levels in the Study of Social Phenomena." In *Groups in Harmony and Tension,* M. SHERIF and G. W. SHERIF (eds.). New York: Harper & Bros.

———. 1956. "Interrelationships of the 'Innate' and the 'Acquired' in Instinctive Behavior." In *L'instinct dans le comportement des animaux et de l'homme.* (Foundation Singer-Polignac.) Paris: Masson et Cie.

———. 1957. "The Concept of Development in Comparative Psychology." In *The Concept of Development,* DALE B. HARRIS (ed.). Minneapolis: University of Minnesota Press.

SCHULTZ, ADOLPH H. 1955. "Primatology in Its Relation to Anthropology." In *Yearbook of Anthropology, 1955,* WILLIAM L. THOMAS, JR. (ed.). New York: Wenner-Gren Foundation for Anthropological Research.

———. 1957. "Past and Present Views of Man's Specialization," *The Irish Journal of Medical Science* (August, 1957).

———. 1959. "Some Factors Influencing the Social Life of Primates." *Conference on the Social Life of Early Man,* Wenner-Gren Foundation. (Summer, 1959.)

SCOTT, JOHN PAUL. 1956. "The Analysis of Social Organization in Animals," *Ecology,* XXXVII, 213–21.

———. 1958. *Animal Behavior.* Chicago: University of Chicago Press.

SIMPSON, GEORGE G. 1950. *The Meaning of Evolution.* New Haven: Yale University Press.

SPIRO, MELFORD E. 1954. "Human Nature in Its Psychological Dimensions," *Amer. Anthropologist,* LVI, 19–30.

SPUHLER, J. N. 1959. "Somatic Paths to Culture." In *The Evolution of Man's Capacity for Culture,* J. N. SPUHLER (ed.). Detroit: Wayne State University Press.

STIERLIN, HELM. 1958. "Contrasting Attitudes toward the Psychoses in Europe and in the United States," *Psychiatry,* XXI, 141–47.

STONE, CALVIN P. (ed.). 1951. *Comparative Psychology.* 3d ed. New York: Prentice-Hall, Inc.

STRAUSS, WILLIAM L. J. 1955*a*. "The Great Piltdown Hoax," *Annual Report, Smithsonian Institution, 1954.* Washington, D.C.

———. 1955*b*. "Closing Remarks." In *The Non-human Primates and Human Evolution,* arr. by JAMES A. GAVAN. Detroit: Wayne State University Press.

SYMONDS, PERCIVAL M. 1951. *The Ego and the Self*. New York: Appleton.

TEICHER, MORTON I. 1954. "Three Cases of Psychoses among the Eskimo," *Jour. Ment. Sci.*, C, 527–35.

THORPE, W. H. 1956. *Learning and Instinct in Animals*. London: Methuen.

TINBERGEN, N. 1953. *Social Behaviour in Animals*. New York: John Wiley & Sons.

TRAGER, GEORGE L. 1958. "Para-Language-a First Approximation," *Studies in Linguistics,* XIII, 1–12.

VALLOIS, HENRI V. 1958. "Le problem de l'hominisation." In *Les processus de l'hominisation*. Paris: Centre National de la Recherche Scientifique.

VOEGELIN, C. F. 1951. "Culture, Language, and the Human Organism," *Southwest. Jour. Anthropol.,* VII, 357–73.

VON BONIN, GERHARDT. 1950. *Essay on the Cerebral Cortex*. Springfield: Charles C Thomas.

WASHBURN, S. L. 1950. "The Analysis of Primate Evolution with Particular Reference to the Origin of Man," *Cold Spring Harbor Symp. Quant. Biol.,* XV, 67–78.

———. 1959. "Speculations on the Interrelations of the History of Tools and Biological Evolution." In *The Evolution of Man's Capacity for Culture,* J. N. SPUHLER (ed.). Detroit: Wayne State University Press.

WASHBURN, S. L., and AVIS, VIRGINIA. 1958. "Evolution of Human Behavior." In *Behavior and Evolution,* ROE, ANNE and SIMPSON, G. G. (eds.). New Haven: Yale University Press.

WEINER, J. S. 1957. "Physical Anthropology . . . an Appraisal," *Amer. Scientist,* XLV, 79–87. (Reprinted in *Evolution and Anthropology: A Centennial Appraisal*. The Anthropological Society of Washington, 1959.)

WERNER, HEINZ. 1940. *Comparative Psychology of Mental Development*. New York: Harper & Bros.

WESTERMARCK, EDWARD. 1903. *History of Human Marriage*. London: Macmillan & Co., Ltd.

WHITE, LESLIE A. 1940. "The Origin and Nature of Speech." In *Twentieth Century English,* WILLIAM S. KNICKERBOCKER (ed.). New York: Philosophical Library.

———. 1942. "On the Use of Tools by Primates," *Jour. Comp. Psychol.,* XXXIV, 369–74.

WISSLER, CLARK. 1923. *Man and Culture*. New York: Thomas Y. Crowell.

YERKES, ROBERT M. 1943. *Chimpanzees: A Laboratory Colony*. New Haven: Yale University Press.

ZUCKERMAN, SOLLY. 1958. "L'hominisation de la famille et des groupes sociaux." In *Les Processus de l'hominisation*. Paris: Centre National de la Recherche Scientifique.

HENRY W. BROSIN

EVOLUTION AND UNDERSTANDING DISEASES OF THE MIND

Some Contributions to Psychiatry of Darwin's Theory of Evolution

It is appropriate for psychiatrists and other students of mental disorders to pay homage to the work of Charles Robert Darwin and the theory of evolution, for without this work it is difficult to imagine what the state of our discipline would be like. Medicine, of which psychiatry is a branch, depends upon the physical, biological, and social sciences for data and concepts germane to the diagnosis and treatment of disease and thus has benefited enormously from the considerable advances made by these basic sciences.

Yet, in spite of the many achievements in the biological and social sciences and the more obvious mastery over natural forces attained by the physical sciences, man is far from secure in his sojourn on earth as we enter the space and atomic age. In the eight thousand years since Neolithic man, we have learned much about the world and how to live better by utilization of material things, but we have learned relatively little about ourselves, the life of the spirit, and how to live with one another. Most of us agree with Mr. Churchill that "the empires of the future are the empires of the mind." It is our good fortune that Darwin furnished us with some of the concepts and data which make possible the systematic study and experimentation that will enable us to enter into these future empires with greater skill and thereby avoid self-destruction if we so wish. His work gave impetus which continues to inspire and serve as the framework for fresh discovery. From his work has come the insight that human activity is a determinant, as well as a product, of evolution, thereby giving man the tools by which he may become the architect of his own future.

Although all the sciences are concerned with this goal, medicine and

HENRY W. BROSIN is Professor and Chairman of the Department of Psychiatry at the University of Pittsburgh and Director of Western Psychiatric Institute. He was a Fellow of the Center for Advanced Study in Behavioral Sciences in 1956 and is currently Director of the American Board of Psychiatry and Neurology.

psychiatry often have the opportunity for a more immediate responsibility for human welfare. But they are not limited to this applied aspect of their disciplines. Because medicine and psychiatry are vitally concerned with basic processes as well as with the applied sciences useful in prevention and therapy, they have developed data and methods pertinent to the more general problems of human development and adaptation as seen in evolution. A few of such general problems are the differential survival of variants in the struggle for existence and the probable resolutions available to such evolving systems as the human personality, the family, clan, or larger socioeconomic units.

The major theme which binds the biological postulates of medicine to evolutionary theory is the adaptation of organisms and species by means of natural selection over a period of time in their struggle for survival. Social or cultural evolution is now a tenable hypothesis, even though it does not parallel in all respects the well-established theory of genetic evolution. The gradual emergence of mental and social activities from pre-existing structures unites psychology with biology and all living processes and makes possible an understanding of these functions as essentially biological systems interacting with an environment. The so-called mental disorders are probably a few among the many categories of deviants which inevitably occur in such systems and are, therefore, of intrinsic theoretical interest to all behavioral scientists. Moreover, there is the practical fact that more than one out of every twelve persons who are born today will require psychiatric help during his or her lifetime (Opler, 1956).

The fundamental concepts of psychiatry, which have been derived from evolutionary theory or strongly supported by it, are well known and have received more extensive elucidation by my distinguished colleagues (Herrick, 1956; Huxley, 1956; Persons, 1956; Simpson, 1949, 1953). I will list a few postulates which seem important for further discussion.

1. All behavior is determined by preceding events and in turn helps shape the sequences that follow. This includes those forms of seemingly irrational or deviant human behavior that have come to be known as mental disorder, criminality and delinquency, neurosis, alcoholism, accident-proneness, and many others. The basic *proposition* does *not* mean that all correlations or interrelations are always to be interpreted as causes. This postulate received much impetus from the demonstration that man, with all his strengths and weaknesses, was a part of an ordered biological world. While Darwin was developing his theory of evolution, he is also said to have fathered the disciplines of physical anthropology, ecology, comparative psychology, and be-

havior, the latter now known as "ethology" (Emerson, 1958; Huxley, 1958). Huxley says:

> His influence, of course, extended into many other fields. The evolutionary approach was adopted in linguistics, astronomy, comparative religion, geo-physics, archeology, and many other subjects; and recently the idea has begun to dawn that it is profitable and indeed necessary to regard all reality *sub specie evolutionis,* to think of the entire cosmos as a single stupendous process of evolution, though with differentiated component sectors or phases [Huxley, 1958, p. 6].

Emerson summarizes it in this way:

> Darwin's methods were to collect data from his own observations and experiments, to seek information from the publications of others, and to gather facts and ideas by direct discussion or through correspondence. With this material he arranged orderly sequences of phenomena, built tentative hypotheses and theories, and sought further pertinent facts that tended to substantiate or refute his generalizations. He described and explained the relations between diverse facts of geology, palaeontology, botany, zoology, and husbandry. He was primarily a naturalist, or what we should now call a taxonomist and an ecologist, but he drew data and concepts from many natural sciences . . . [1958, p. 2].
>
> During his life he not only gave evidence for organic evolution by means of natural selection, but he laid foundations for physical anthropology, the evolution of behavior and mind, systematic classification and order, biogeography, individual and community ecology, animal and plant breeding, and above all, he firmly established the evolutionary time dimension in all contemporary living systems . . . [p. 3]. He had no hesitation in including instinctive, emotional, and individually acquired behavior in his evolutionary theory, and must be given a prominent place in the history of comparative psychology. He used analogues from human society and cultural history to illustrate events and processes of organic evolution, and so helped to bridge the gap between the social and biological sciences that subsequently widened before it narrowed in recent times. However, Julian Huxley (1958) says that Darwin made no attempt to discuss or emphasize "the fact that evolution from the savage to the civilized state involved primarily not a biological but a cultural change" [p. 5].

2. The processes underlying behavior can be studied systematically, even though they are often complex and evanescent. One of Darwin's conceptual triumphs was that he did not exclude human psychology from the general rule, even though many other authorities preferred to do so. He made a considerable contribution to such study in his *The Descent of Man* (1871) and *The Expression of the Emotions in Man and Animals* (1872). Clinicians and social scientists, particularly those who are combining linguistic and kinesic analysis with the aid

of sound-film are reviving Darwin's use of the natural history method (Braatoy, 1954; Freed, 1958; Freud, 1938; Hoch and Zubin, 1958; McQuown, 1959; Osgood, *et al.*, 1957; Skinner, 1957).

The relation of the development of the central nervous system to behavior will be discussed by my distinguished colleagues, Doctors Magoun, Hilgard, von Muralt, and many others already have (e.g., Bailey, 1958; Bullock, 1958; Lashley, 1949; Pribram, 1958; Sperry, 1958). Numerous participants in this symposium will be primarily concerned with describing the nature of the evolution of human behavior, including that internal behavior called "thinking" or "mentation." In the main, the broad strategies outlined by J. Huxley, C. Kluckhohn, A. L. Kroeber, M. Mead, and G. G. Simpson for the origin and development of these activities are compatible with the hypotheses held by most leading writers on psychiatry even though much clarification is required at tactical levels (Berrill, 1955; Huxley, 1958; Kluckhohn, 1944, 1949, 1959; Kroeber, 1956; Mead, 1958; Simpson, 1949).

I am indebted to them and the other numerous authors cited in this essay for a comprehensible system describing the mental functions which, in many ways, support the clinical writings of Freud, E. Bleuler, and Adolf Meyer. These three writers have furnished most American psychiatrists with their basic theorems on thinking and the thinking disorders (Etkin, 1954), and I will develop some of the implications in the following sections.

3. All organisms will continue to evolve whether human intervention or co-operation occurs or not, and I will develop some of the implications of this in the following sections. Human influences are apparently decreasing the power of some of the devices of natural selection while increasing those of others through birth control or regulated lack of it; through enormously increased productivity, not paralleled through temporary higher levels of nutrition for a small part of the world's population for the present; through control of infectious diseases and many similar mechanisms discussed in other chapters of the symposium. The long-term effect of increased radiation, malnutrition, and climate control are not known, but their effect can be envisioned and studied as opportunities present themselves.

Man can consciously study and plan his own future development, both genetic and historical. This unique fact in evolutionary history needs considerable evaluation. Even though we do not know the probable outcome, we can apparently improve our survival and fulfilment value through serious effort. This may be a most important force in this ongoing battle (Huxley, 1943, 1944, 1956; Simpson, 1949).

4. Character, temperament, personality, mind, and similar systems

usually subsumed under those labels, also evolve both phylogenetically and ontogenetically. Freud (1856–1939), thoroughly trained in neuroanatomy and neurophysiology in the Darwinian era by some of the same men who gave us modern physiology, particularly Ernest Bruecke demonstrated his competence in neurology in thirty-seven original papers and monographs before he turned to a clinical problem in psychiatry (Grinstein, 1956). Freud's biographer, Jones, informs us that Freud wrote: "the theories of Darwin, which were then (1873) of topical interest, strongly attracted me, for they held out hopes of an extraordinary advance in our understanding of the world" and that this was one of the reasons he chose medicine as a profession (Brunner, 1958; Jones, 1953, p. 28). Jones also cites evidence that Freud had a course on "Biology and Darwinism" given by a zoologist, Claus, in addition to the regular courses in zoology and physiology, and that Freud was greatly influenced by Bruecke in a purely physical but dynamic physiology and in his evolutionary orientation (Jones, 1953, pp. 36, 41–42). It is not surprising that in numerous papers, especially *Three Essays on the Theory of Sexuality* (1905), Freud tried to show that every ordinary human adult must traverse various stages of biological development in order to arrive at the state of "normality" in the sense of conformity to conventional standards. Numerous clinicians and social scientists, particularly social anthropologists, have collected convincing data that human beings must learn to act like humans. This is equally true for adults and children. Xenophon, in his *Memorabilia,* tells about the farmer landowners of ancient Athens who used to go into the city to walk up and down and talk to Socrates and other philosophers. "Not that we might become popular politicians or clever lawyers, but that we might become gentlemen, in order, I mean, that we should learn how to behave rightly to our own families and dependents, how to perfect our relations with our friends and how to serve our country and our fellow-citizens" (Xenophon, 1923). This may be one of the compelling reasons for scholars to attend conferences and committee meetings.

Freud, more than anyone, demonstrated the numerous ways in which the innate forces and the acquired unconscious patterns affect all human behavior, both normal and pathological. He also speculated upon the nature of those forces which shape human thinking, specifying some of their properties. One of our most urgent tasks is better to understand the nature of the relation of the "primary process" to the "secondary process," in order better to control them (Freud, 1955). Social anthropologists are also learning the means by which this goal is attained and the numerous societal pressures that maintain the state of being human (Erikson, 1950; Haring, 1949; Kluckhohn,

1949; Kluckhohn and Murray, 1953; Thomas, 1956). In a following section evidence will be presented to show that hand in hand with better exploratory observation there have been developed more rigorous methods for gathering data about the human developmental process in natural or artificially created experimental settings.

In this section, as in others, I will not try to present in full the relations of psychiatry, Freud, and psychoanalysis to Darwin and evolutionary theory. However, I would like to express my belief that Freud and many of his followers are in the Darwinian tradition both by direct descent, as it were, and also from indirect pressures from such collateral sources as Spencer's psychological synthesis, Galton's inheritance of abilities, Lombroso's clinical studies on criminals, William James's description of the behavior of animals and men in evolutionary terms, and G. Stanley Hall's phylogenetic course on human psychology (1923):

> Freud saw man as a biological being with basic drives derived from his physiology, with behavioral propensities and limitations arising from his anatomy, all modified by his social experience during his life. Ferenczi, like Freud, utilizing a lamarckian theory of evolution, postulated an ingenious relationship between the psychosexual and personality development of man and his geological-evolutionary history. Others, including Jung, Rank, Adler, Burrow, Fromm and Kardiner, have emphasized one or another evolutionary principle in their efforts toward a unified concept of human psychology [Freedman and Roe, 1958].

Recent writers on some aspects of the subject are Sillman (1949, 1953), Ostow (1957, 1958), Székeley (1957), Róheim (1956), Murphy (1956, 1958), and Fromm (1955, 1956).

There seems to be no doubt that Darwin deserves major credit for making explicit the idea, even if he was preceded by Lucretius, that thinking in all of its ramifications, including the most abstract mathematical, ethical, or poetic constructs, is a biological activity which evolved over a period of time, perhaps one to two hundred million years, from the very simple forms which are the early substitutes for motor activity following a felt need or inner experience of a drive to eat, for example (C. R. Darwin, 1874; Spector, 1959). This view apparently was common during the later nineteenth century so that it is not surprising that Freud made thinking as a problem-solving activity the cornerstone of his model for "mind" in the epoch-making *The Interpretation of Dreams* (1900). Americans familiar with the writings of John Dewey since his essays on *The Influence of Darwin on Philosophy* (1910) and with G. H. Mead in *Mind, Self, and Society* (1934) will not find this novel or strange. Mead anticipated in many ways the themes now dominant in the literature on evolution of man's cultural development in time and his ability and responsibility to fashion his own world if he is to survive.

Recent work by Rose corroborates the experiments of Poetzl (1917) and Fisher (1956, 1958). Rose (1955) suggests that the scanning and screening of incoming perceptions common in much thinking begins at the retina and that his concept of perception is similar to Schilder's concept of the process of thought. Schilder said: "Every thought is the product of a process of development . . . every thought recapitulates the phylogenesis and ontogeneses of thinking" (1951). Further light on the scanning process as it relates to perception has been thrown by Kragh (1955), as quoted by Faergeman:

> An experimentally isolated perception is found to recapitulate in totally condensed temporal form the phylogenesis as well as the individual development of the subject. By use of tachistoscope and after images he obtains pre-states of perception, then investigates the content of the perceptual prestages by use of TAT, live drawings and other similar stimuli and finds that they exactly parallel ontogenesis [1956].

One powerful outgrowth of the evolutionary thesis is that new properties, functions, and activities can evolve from old structures. The mammalian bat's wing and the whale's fin illustrate the changes in quality of function which can occur with some alteration in form. The nasopharyngeal structures originally used for breathing and eating became adapted for subtle speech. These examples do not strain our credulity as much as the transformation evident in the mental abilities for abstraction and creation of an Isaac Newton in comparison with some of our prehistoric ancestors, such as Cro-Magnon man, or even with some of our preliterate contemporaries in Africa or South America. Yet Darwin himself pointed out that these apparent differences were only in quantity, not in quality:

> Nevertheless, the difference in mind between man and the higher animals, great as it is, certainly is one of degree and not of kind. We have seen that the senses and intuitions, the various emotions and faculties, such as love, memory, attention, curiosity, imitation, reason &c., of which man boasts, may be found in an incipient, or even sometimes in a well-developed condition, in the lower animals. They are also capable of some inherited improvement, as we see in the domestic dog compared with the wolf or jackal. If it could be proved that certain high mental powers, such as the formation of general concepts, self-consciousness, &c., where absolutely peculiar to man, which seems extremely doubtful, it is not improbable that these qualities are merely the incidental results of other highly-advanced intellectual faculties; and these again mainly the result of the continued use of a perfect language. At what age does the new-born infant possess the power of abstraction, or become self-conscious, and reflect on its own

existence? We cannot answer; nor can we answer in regard to the ascending organic scale. The half-art, half-instinct of language still bears the stamp of its gradual evolution. The ennobling belief in God is not universal with man; and the belief in spiritual agencies naturally follows from other mental powers. The moral sense perhaps affords the best and highest distinction between man and the lower animals; but I need say nothing on this hand, as I have so lately endeavoured to shew that the social instincts;—the prime principle of man's moral constitution—with the aid of active intellectual powers and the effects of habit, naturally lead to the golden rule. "As ye would that men should do to you, do ye to them likewise"; and this lies at the foundation of morality.

In the next chapter I shall make some few remarks on the probable steps and means by which the several mental and moral faculties of man have been gradually evolved. That such evolution is at least possible, ought not to be denied, for we daily see these faculties developing in every infant; and we may trace a perfect gradation from the mind of an utter idiot, lower than that of an animal low in the scale, to the mind of a Newton [1874, pp. 494–95].

The biological view of thinking is a unifying concept precluding mind-body dichotomies. Although this is one of the psychiatrists' most important working postulates, it is noteworthy that, in spite of preponderant philosophic agreement, there is much operational difficulty in medicine and psychology about this problem (Hayek, 1952; Huxley, 1941). The philosophic position of "emergent evolution" which has grown out of Darwin's theories has probably been the most popular and only adequate positive alternative to reductionism and one which has powerfully influenced psychiatric thinking.

The theory of *emergent evolution*—momentarily at least—came closer than any other proposed integrative principles to providing a synthesizing nucleus for modern philosophy. This theory of levels, as it was sometimes called, in one form or another had the vigorous support of such outstanding thinkers as Wilhelm Wundt, Lester Ward, C. Lloyd Morgan, Samuel Alexander, Jan Smuts, Alfred North Whitehead, G. P. Conger, C. D. Broad, Roy Wood Sellars, and many others [Reiser, 1958, p. 42].

5. The concepts of instincts as basic motivation forces for behavior is prominent both in evolutionary theory and in most psychiatric theories. Darwin found such global concepts as instincts and emotions useful, as did Freud, so that today many clinicians use them in spite of their shortcomings. However, with the rise of a new field called "ethology," of which Darwin may be said to be the father, associated with the work of K. Lorenz, N. Tinbergen, W. Thorpe, E. Hess, and others, we now have comparative studies which help to throw light on some of the mechanisms of behavior previously called "instinctive"

(Huxley, 1958). The homing flight of pigeons, the migrations of the central European warbler, the ability of bats to fly in the dark, the mating patterns of monogamous animals like the wolf and the fabulous Canada goose, the organized family patterns of large ape colonies with the concept of "territoriality" as affecting the fighting behavior, were formerly thought to be instinctive action. Now all these are seen to depend variously on special hitherto unrecognized sense organs or the ability to recognize complex signal systems and to live in a world of coordinated communication systems. The older concept that the animal seems to be born with these systems intact is now mitigated by the knowledge provided by controlled experiments that some of these systems are acquired very early and quite permanently in the process known as "imprinting" (Fletcher, 1957; Hess, 1958; Lorenz, 1958; Róheim, 1956; Schiller, 1957; Scott, 1958; Thorpe, 1956). However, this does not eliminate the need for a concept of instinct if this is carefully defined.

6. The fact that many individual differences are based on gene differences is stressed both in evolutionary theory and by psychiatry. Although evolution deals largely with species, their origins and destinies, it also calls attention to the importance of deviants of all types, for here is often found the manner in which new species are formed and maintained. No attempt will be made to recount the history of the development of these ideas following the publication of the *Origin of Species* in 1859, since this requires the time and talent of a specialist, but the recognition of the importance of biological variations and individual differences is one of the cornerstones of modern psychiatric theory. It is true that psychiatrists differ regarding the reciprocal influence and importance of the genetic inheritance and the environmental influences. Few would insist on an absolute belief in the overriding importance of the genetic inheritance in all mental disorders, whatever they may be, except where the evidence is strongly in favor of such a phenomenon, as in phenylpyruvic acid oligophrenia. This is especially true from the time when human genetics has come to a new maturity, a period which enables qualified scientists to examine the evidence in the light of appropriate criteria (Neel and Schull, 1954; Srb and Owen, 1955; Waddington, 1958). Neel and Schull list thirty-four human disorders, most of them uncommon or unimportant, in which it may be possible to recognize a carrier state. Only a dozen of these have a high reliability of recognition of the carrier state, and the probable mode of inheritance is often complex and, in the case of diabetes mellitus, uncertain. In such important diseases as hypertension, schizophrenia, and epilepsy the evidence is far from completely convincing (1954). The largest single group of serious mental dis-

eases—the schizophrenias—probably have diverse origins, including powerful genetic components, as shown by Kallmann (1959). Contrary to much public opinion, many leading psychiatrists who study psychotic reactions as primarily a complex set of human interactions believe also that genetic roots are vital factors in the formation of these serious disorders (Benjamin, 1958). The devotion of most psychiatrists to psychotherapy in its many forms, from individual psychoanalysis to group and occupational therapy, is motivated by the practical necessity of finding better ways to help sick people get well and is not necessarily a denial of genetic or other organic factors which are not understood and not amenable at present to direct intervention. It would be desirable for us to have a balanced program of investigation into the manifold causes of mental disorders, but this must be brought about through informed understanding in the scientific community and not enforced through the emotional prejudices of groups dispensing funds or the narrowly conceived administrative act of an official. No doubt, self-regulating social and professional devices will grow so efficient that, over a period of two or three centuries, inequities will be corrected, but, in the meantime, some genuine benefits will be lost by restricted research.

Freud stressed the differences in the "constitution" of children as one reason why absolute rules cannot be given for child rearing and stressed the need for finding that proper balance between measures which helps the child control his impulses so that he can become a civilized person, and yet permits him sufficient freedom to grow into manhood without crippling frustrations (Freud, 1933). Several quotations may clarify the position of many psychiatrists and give us some background for discussing future development in this area:

If we can find an optimum of education which will carry out its task ideally, then we may hope to abolish one of the factors in the etiology of neurotic illness, viz., the influence of accidental infantile traumas. The other factor, the power of a refractory instinctual constitution, can never be got rid of by education [1933, p. 204].

The expectation that we shall be able to cure all neurotic symptoms is, I suspect, derived from the lay belief that neuroses are entirely superfluous things which have no right whatever to exist. As a matter of fact they are serious, constitutionally determined affections, which are seldom restricted to a few outbreaks, but make themselves felt as a rule over long periods of life, or even throughout its entire extent. Our analytic experience that we can influence them to a far-reaching degree, if we can get hold of the historical precipitating causes and the incidental accessory factors, has made us neglect the constitutional factor in our therapeutic practice. . . . Some specific tendency, some particular instinctual component, is too strong in comparison with the counter-forces that we can mobilize against it.

This is quite generally so in the case of the psychoses. We understand them in so far as we know quite well where we ought to apply the levers, but they are not able to lift the weight. In this connection we may hope that in the future our knowledge of the action of hormones . . . will provide us with a means of coping successfully with the quantitative factors involved in these diseases [*ibid.*, pp. 210–11].

It is noteworthy that H. S. Sullivan, who is usually cited as a leading exponent of the environmentalist position for the causation of the major psychoses, estimated that over 20 per cent of schizophrenias are organic in nature:

In brief, I have come to the opinion that there are two unrelated syndromes confused under the rubric of *dementia praecox,* or—as it is often used synonymously—schizophrenia. One syndrome is the congeries of signs and symptoms pertaining to an organic, degenerative disease usually of insidious development. These patients are finally discovered to be psychotic, although no one can say how long the state has been developing. Their outlook is very poor—even, I surmise, under the treatments by partial decortication which now enjoy such vogue. I am content that this syndrome be called *dementia praecox.*

The other syndrome is the one about which I am offering some data. It is primarily a disorder of living, not of the organic substrate [1947, p. 73].

This view is an old one held by many clinicians, but, unfortunately, neither physical nor psychological tests, including Rorschach, enable us to distinguish these groups with any certainty, and consequently their existence remains an attractive but unproved hypothesis. It is worthwhile noting that after clinicians made more serious efforts to use psychological methods of treatment of the major psychoses, the older pessimism, as expressed by most central European psychiatrists and by Freud, was found to be unwarranted. Without overlooking the genetic and organic components which inevitably play their part in any disease process, the Past fifty years have shown us that we must master some better methods of studying the fate of an organism in its environment, because that interaction is the locus of much of the disorder. The more intelligently psychotherapy has been used, the more effective has it been found, in many cases. The summary of this advance up to 1950, by Manfred Bleuler, is a judicial appraisal:

It seems as if the coming years will be predominantly dedicated to the investigation of those older concepts of schizophrenia which have seen in schizophrenia, primarily if not entirely, an individual disturbance of the adaptation to the difficulties of life. It will be a great experience to find out, in a few years or decades, in which way and to what extent the hopes based on these working hypotheses will have been confirmed [1955, p. 76].

In another place, Freud says that he believes ego-instincts to be involved only secondarily to the pathogenic functions of the libido. He also recognizes the possibility, which must be shown by future experiments, that in severe psychoses these ego-instincts may themselves be primarily deranged (1935, p. 372).

These opinions by leading writers do not constitute evidence, but they illustrate that the clinical study of deviants, in man's long struggle to evolve, centers about the nature-nurture axis. More recent work with chemicals such as mescaline LSD-25 (lysergic acid-diethylamide) and Taraxein opens up new experimental means whereby we can study the effects of possible chemical mediators (Council for International Organizations of Medical Sciences, 1954; Spiegel, 1950–58).

The study of mother-child separation and the so-called sensory deprivation support the hypothesis that postnatal learning may be crucial for some mental disorders (Bowlby, 1951; Group for the Advancement of Psychiatry, 1956*a, b,* 1957; Solomon *et al.*, 1957).

7. The study of sex is important in both evolution and psychiatry. It may be straining a point to say that Darwin's intensive studies in sex as a mechanism in natural selection give respectability and justification to persons who wish to study the human aspects of this tabooed subject. However, today, in spite of our vaunted freedoms we are often not much more tolerant than in the early days of Havelock Ellis and Freud in the nineteenth century—as witness the difficulties encountered by Alfred Kinsey—but we are more tolerant of animal investigation in this area.

8. Populations are in a complex state of balance with their environment. This is beautifully described in the well-known closing paragraph of the *Origin of Species*. It has taken psychiatrists many years to have a better grasp of similar concepts in relation to family and group dynamics. The concept that such groups have a tendency to maintain a stable equilibrium for the system as a whole and that subclasses are definable systems with properties which can be tested has grown in popularity after K. Lewin (1935) and W. Köhler (1938) developed it in the framework of experimental psychology. The foregoing concepts are being used in the study of the family, in industrial and military groups, in hospital populations and other social units studied by anthropologists and sociologists (Benedek, 1949; Bond, 1952; Caudill, 1958; Caudill and Stainbrook, 1954; Greenblatt *et al.*, 1957; Grinker and Spiegel, 1945; Menninger, 1948; Miller and Swanson, 1958; Pederson-Krag, 1955; Stainbrook, 1955; Stanton and Schwartz, 1954; von Mering and King, 1957).

From the foregoing evidence, albeit brief, it is apparent that the basic problems of psychiatry are those of all other biological and social sciences. Psychiatry may have new data, methods, and concepts to contribute to the general fund of information because psychiatry deals with deviants over a wide spectrum of behavior. The evidence is overwhelming that the working concepts which make up evolutionary theory contribute heavily to an understanding of mental disorders. The leading hypotheses of psychiatry, including the focus upon the failures of adaptation to the ordinary difficulties in early living, may become useful to students of general and human biology and the social sciences because they highlight processes in these complex systems which may not be easily visible in the usual developmental schemes.

It is commonplace that what the biologist may term "deviant behavior" the lawyer may call a "crime," the clergyman a "sin," and the psychiatrist a "mental disorder." From the viewpoint of customary behavior, such irregular patterns may be considered behavior in reaction to stress, which may furnish scientists with advantageous experimental material. Several analogies have been used to make the point that the internal dynamics of an organism are sometimes best studied under conditions of stress, or even past the point of tolerance. For example, the lines of stress in a crystal may not be visible to inspection by the naked eye but are inferrable when the crystal suddenly shatters. On the proving grounds or test facilities, a motorcar or airplane is put through its paces far past all ordinary performance requirements in order to determine the probable areas of initial failure.

If students of human evolution wish to study the ways in which genetic, constitutional, and environmental forces influence human beings, they will find the data from psychiatry of more than passing interest, both for analysis of processes and for methods of amelioration. While the home, religion, government, and education are the most obvious institutions for acculturation, tremendous changes may possibly be effected by means of the application of sound principles of public health and mental health.

Diagnostic Categories, Normality, and Individuality

There is so little general agreement among psychiatrists regarding the definition and delineation of the various diagnostic categories of mental disorders that a parenthetical statement may clarify a question which will undoubtedly trouble many readers. This defect in taxonomy is a serious drawback which will hamper many investigations depending upon precise description of such quasi-entities as dementia praecox, schizophrenia, or manic-depressive psychoses. Nissen (1958) proposes a metataxonomy, which is a categorization of possible categorizations, as a helpful initial analysis of this problem. Simpson (1958) comments upon some of the difficulties involved in Nissen's six functional categories and also upon other bases for classification, such as courtship and mating behavior, social behavior, territoriality, and description by levels such as first-order or molecular behavior (example: locomotion, Colbert) and second-order or compound behavior (example: courtship, Hinde and Tinbergen). Simpson (1958) suggests that the third level of compounding or complexity beyond simple movements are those involved from social to cultural behavior, examples of which are found in the writings of J. S. Huxley, H. F. Freedman, A. Roe, and M. Mead. In spite of the limitations described above, significant work is being done, and each new idea or method helps define the problem of proper categorization a bit more sharply. More investigators are becoming accustomed to working in such significant functions as perception, memory, learning, and hallucinations, which are not dependent upon disease categories. As more information about the formal character of disease processes is discovered, new definitions which are operationally useful will be created (Crookshank, 1949; Riese, 1953; White, 1926).

Perhaps, more important, we are beginning to understand how better to study behavior with the view of placing demonstrable "causes" in their proper place in a wide spectrum and at their proper level of organization. Examples of widely separated clinical disorders are the cases of phenylpyruvic acid oligophrenia at one pole and the emotional vomiting under social stress at the other. In between these extremes we must establish flexible criteria appropriate to the subculture with which we are dealing for definitions of health and for disease.

If "disease" is considered as a relative failure in adaptation to existing life-circumstances, including gaseous, fluid, and mineral intake and output, nutrition, family and job activities, aims, hopes, fears, and talents, we can begin to describe more concretely the physical, psychosocial, and sociocultural interactions in the cultural setting. Significant relations and eventually, we hope, causal relations will be revealed by the study of interactions at the various levels of development. We are better equipped at the moment to study the adaptive failures in walking, talking, and driving a car than we are in the internal disharmonies within the body and the various aspects of the mental functions. It is encouraging to know that, by means of improved specification of the multiple methods involved in intensive psychotherapy, we are now receiving new data which can be ordered in more meaningful ways. We expect that prediction and other methods of validation may become more certain in dealing with these new data and that, in consequence, we can expect much more useful interchange with other sciences.

Disease will be understood as an aspect of continuing processes both in a biological organism and in the ongoing culture of which the disease is an intimate part. This process of disordered adaptation may be more or less reversible under appropriate conditions. It may make itself known in general signs and symptoms, such as apathy, indifference, fatigue, irritability. More specific patterns, such as obsessions, phobias, specific work inhibitations, miserliness, profligacy or accident-proneness, may be the presenting complaint. Disease is not an entity but the name given to adaptive failures of already existing and operating processes which are themselves always in an unstable state of change, although the unit of time may occasionally seem prolonged to a single observer with calendar urgencies of his own. Even delusions and hallucinations in disoriented psychotic patients are only expressions of conflicts and defective solutions by means of media available to the patients from their animal heritage (Brosin, 1952; Rennie, 1957).

The need for multidimensional and multidisciplinary studies has been a well-publicized requirement among psychiatrists. Clinicians Adolf Meyer and W. A. White, in the past generation, consistently taught their students the crucial nature of this need. Unfortunately, they lacked the data which are now available to make such enterprises more feasible. The specifications for the study of a unit of behavior by Margaret Mead are probably more precise than those of three decades ago:

> We can get some picture of how change occurs only when each individual is fully specified in his genetic and experiential peculiarity, when the culture which these individuals share, the set of social relations within which they participate, the wider situation of which they are a part—a world system of trade or politics, a chain of volcanoes which periodically erupt, the cycle of hurricanes or droughts within which they must survive, the other systems with which they are in competition—are simultaneously included within the definition of the unit to be investigated . . . [1958, p. 496].
>
> The procedures for the microstudy of evolutionary change at the point of initiation must meet the same requirements: the study *in situ* of a group of fully identified living individuals, of known culture, in a known network, within a known ecological system, specified in inorganic, biological, and social terms [p. 500].

All students of the behaviorial sciences are familiar with the difficulties inherent in cultural or statistical concepts of normality, so that we need not belabor them here (Wegrocki, 1953). Some new insights about defining normality in human biological systems are outlined by the biochemist, Roger J. Williams (1956). Williams, as well as Conway Zirkle and Frederick A. Hayek, discuss the associated problems of individuality from the viewpoint of genetics and economics with relation to future developments. Our ignorance about both psychological and cultural processes and our lack of control over them is shown (Hayek, 1958; Williams, 1958; Zirkle, 1958).

Other authors in this symposium will discuss these topics and the resultant questions of whether we can control our destiny via proper governance of our human relations at all levels. In this paper the reciprocal relations between deviant behavior and related freedoms for the best development of individuality, as contrasted with ordered predictable behavior which may be dehumanizing and stultifying are emphasized. These are the more general forms of the problems facing a physician who wants to help the person who is in distress because of his failure to adapt to his environment. If we are thorough Darwinians, committed to the belief that mind is also evolving because all its components are unstable systems in constant flux, we will be able to face some of our problems more imaginatively than heretofore.

Current Investigations Pertinent to Psychiatry

If we agree to view a man as a highly individualized complex, but the evolving product of his genetic potential interacting with his environment from the moment of conception to his inevitable dissolution, we have a unifying principal from which to study almost all diseases of men, including the so-called mental, as a part of the total biological-physical world. Man learns to become human in the best sense through the ordered acquisition of new properties and activities by means of the interaction between the innate and acquired forces acting in him. The new levels of activity are also ordered events, with relations which can be described and defined and, in some cases, even predicted and controlled. At all stages of the evolving process, some disorders or deviations may occur which may be called by mankind (depending upon circumstances) either "disease," "sin," or "crime." If we accept the hypothesis that all these corrupted processes are resultants of the natural forces made familiar to us by the physical, biological, and social sciences, we can more easily understand that most, if not all, diseases are the disorders of being a member of a species in a physical-psychosocial matrix (Group for the Advancement of Psychiatry, 1957; Hinkle and Wolff, 1957, 1958; Levy, 1958).

These disorders may be grouped into several large categories. One category would be those genetic and congenital anomalies occurring in the conceptual and prenatal phases. These complex processes are grouped together for convenience only. With more understanding of chemical genetics and embryology, we may make tremendous strides toward improving controls over their varied operations. Apart from the usual developmental anomalies resulting from unknown sources, we may improve the probabilities that a fetus will become a healthy human being by controlling nutritional, toxic, infectious, and even stress factors. Just as we are becoming more aware of man's extreme dependence upon his environment after birth, so we are also becoming sensitized to the environmental factors affecting intrauterine development (Herrick, 1956; Hooker, 1952).

Another category in that group of disorders inherent in the biological organization of man are those attendant upon his inevitable aging. The metabolic and structural changes in our sense organs and other physiological systems are now known to be processes which can be understood by natural science methods. In the same sense we may view most infections and deficiency diseases and tumors as natural events in the history of man, for if our progenitors had not survived these inner and outer onslaughts for many thousands of years, mankind would not have survived. Immunological resistance to disease may have been a more important factor in human survival before man began to control this aspect of his environment than his intricate nervous system. Epidemics have been powerful forces in determining the destinies of man, and even today such diseases as malaria influence the development of a culture.

The groups of derangements which are currently of interest to many scientists, in addition to the specialists who have responsibility for their care, are those associated with man as a human being. From chemistry, physiology, and communication theory, not to mention other frontiers, we learn of man's delicate sensitivity to his internal and external environment, both the psychosocial and the physical (Cannon, 1932; Hinkle and Wolff, 1957; McQuown, 1959). With increased insights into the interrelations between the influence of external and internal stress in the human nervous system, particularly the autonomic system and the hormones, we are now slowly learning that many disturbances, both acute and chronic, are inherent in man's relations to his fellow man. Freud called attention, forcibly, to the fact that many of the operations influencing behavior were out of awareness and might directly affect body function. It is here that medicine can provide valuable concepts and data to other disciplines which might not regard these pathological manifestations of body malfunction as expressions of psychosocial dysfunction. Because of the contributions from the field of medicine, we are now better able to recognize and study somewhat deviant behavior at various levels in otherwise well-regulated organisms.

We can see only dimly into the future because the parameters of activity are too numerous and the chains of causes and effects often too long for convincing demonstration at this time. It is necessary to be on constant guard against assigning causative roles to components of a system which may be merely concomitant or secondary to the major factors. However, we do have enough evidence in such diseases as peptic ulcer, ulcerative colitis, and bronchial asthma, to name only a few of many, to be fairly certain that ideas and emotions are organic activities which are indissolubly integrated as activities of the entire organism and, in time, influence the body profoundly. Schiller's observation "Es ist der Geist der sich ein Körper baut" ("It is the spirit which builds itself a body") is more than a poetic philosophical statement. Future experimentation and clinical research will give us better insights about these examples of disturbed behavior ("stress diseases") as well as about the major psychoses, neuroses, and character disorders which also fit into the Darwinian pattern of evolution.

The Need for Better Theory

In view of the multiplicity of factors influencing the formation of an individual's behavior and, consequently, the deviant behavior inherent in the system, it is neither necessary, desirable, nor possible for me to attempt coverage of all pertinent investigations in the problems of mental disorders. In a sense, most of the papers in this symposium are relevant to this subject. All of us are dealing with aspects of behavior relevant to human behavior and are employing the Darwinian concept of man and his "mind" as ceaselessly evolving processes in a complex system of interrelated organizations. While we do not have a common vocabulary or consensually validated hypotheses to help us translate the data from one field of investigation or level of organization to another, it may be possible even now to stake out a few lampposts to help us in the search for unifying theories. Perhaps the directions proposed by current biology and the behavioral sciences will help us devise methods to avoid being mired in the old dilemmas of mind versus body, determinism versus freedom, "wholism" and emergent evolution versus reductionism. The strategy for avoiding such meaningless questions (P. W. Bridgman, 1927) is to consider that science is moving toward the methods in which it studies organism rather than limiting itself to particles. The primary units in the various levels of organization are events, and the focus is upon the relations between events. The organism is an unstable system capable of change at any time and in many directions (Bridgman, 1927; Whitehead, 1925, 1929). These views seem compatible with those of many others in the new fields of chemical genetics, functional anatomy, neurophysiology, and biochemistry, as well as in anthropology, ethology, communication theory, experimental psychology, and related sciences, and would provide a framework for more convenient translation of disparate ideas. I am aware that even this simplified statement is controversial, and I look to the future for improved elucidation of these complex and often subtle problems (Born, 1949, 1951; Gillespie, 1958; Schrödinger, 1957; Woodger, 1956).

It seems highly probable that the idea of complementarity, introduced by Niels Bohr and vigorously supported by J. Robert Oppenheimer, is a valuable concept for reconciling abstruse differences (Bohr, 1958; Grene, 1958; Oppenheimer, 1953, 1955):

> You know that when a student of physics makes his first acquaintance with the theory of atomic structure and of quanta, he must come to understand the rather deep and subtle notion which has turned out to be the clue to unraveling that whole domain of physical experience. This is the notion of complementarity, which recognizes that the various ways of talking about physical experience may each have validity, and may each be necessary for the adequate description of the physical world, and may yet stand in a mutually exclusive relationship to each other, so that to a situation to which one applies, there may be no consistent possibility of applying the other. Teachers very often try to find illustrations, familiar from experience, for relationships of this kind; and one of the most apt is the exclusive relationship between the practicing of an art and the description of that practice. Both are a part of civilized life. But an analysis of what we do and the doing of it—these are hard to bed in the same bed [1955, pp. 82–83].

A brief statement about the limitations and dangers of the use of the complementarity principle in arriving at pseudo-solutions to physical and philosophical problems through avoidance of the law of contradiction has been made by Northrop (1958). The tremendous freedoms in conceptualization learned by physicists from their long experience with trying to understand puzzling experimental data enables Heisenberg (1958) to give us a valuable suggestion:

> These new results had first of all to be considered as a serious warning against the somewhat forced application of scientific concepts in domains where they did not belong. The application of the concepts of classical physics, e.g., in chemistry, had been a mistake. Therefore, one will nowadays be less inclined to assume that the concepts of physics, even those

of quantum theory, can certainly be applied everywhere in biology or other sciences [p. 199].

Because numerous distinguished essays in this symposium are directly concerned with the evolution of man as a social being, I will not attempt to restate the major relevant propositions. Suffice it to say that it seems improbable that man as we now know him will alter much genetically in the foreseeable future. Perhaps other essayists will show ways in which *Homo sapiens* might permit marked deviants to live and develop, but it seems unlikely. Aldous Huxley has given us a graphic picture in his notable volumes, *Brave New World* and *Brave New World Revisited,* of some of the possibilities and also of some of the human defenses against such mechanization. In view of the early stage of development of this subject and the consensus of experts that marked genetic deviants will not easily be tolerated, it will be most profitable to turn to the relevance of psychiatry and medicine to historical and cultural change, which operates much more quickly and decisively for most human individuals and societies (Chance and Head, 1953; Herrick, 1956; A. L. Huxley, 1946, 1958; J. S. Huxley, 1941, 1943, 1944, 1953, 1956, 1958; McKinley, 1956; Mead, 1958; Montagu, 1955; Simpson, 1949; Smith, 1958; Whyte, 1948). The acceleration of the rate of change in human attitudes and social relationships, including those in government, law, education, religion, and industry, during the past four hundred years, has convinced many writers that, although progress in human improvement is not inevitable, it remains a definite possibility if we are sufficiently skilful to devise methods for saving ourselves (Baillie, 1951; Bury, 1955). Similarly, many authorities in various fields are giving much more precise definition to the old dictum "Human nature cannot be changed," for historians as well as biologists and physicians are becoming aware that this statement is approximately true only for the genetically determined potentials for individuals as they develop in appropriate environments (Curti, 1956; Simpson, 1949). By "human nature" most of us ordinarily mean the conglomeration of prejudices, habits, attitudes, and beliefs with which we apparently meet our problems in life. I say "apparently" because, in many instances, we do not make our judgments merely because we belong to the Blue clan rather than the Gray, but rather in response to subliminal and minimal cues which are meaningful in the environment in the areas of vocal modifiers, gestures, movements, facies, posture, odor, or habit patterns. Linguists and kinesicists are beginning to recognize signal systems which may account for some behavior patterns in the activities between persons of different sex and age or between different social and ethnic groups (McQuown, 1959). The insight that each individual is bombarded with thousands of sensory stimuli continuously, both internal and external, is not new, since they are stated quite clearly by Freud in *The Psychopathology of Every Day Life* and were developed by numerous followers. They are present in the brilliant pioneering work of Edward Sapir and Leonard Bloomfield and were dramatically related to neurophysiology, information and communication theory, and computer construction by such members of the Macy Conferences on Cybernetics (1946–51) as Von Neumann, McCulloch, Pitts, Wiener, Bateson, and others (see Cherry, 1957; Conference on Cybernetics, 1950–55; Jeffress, 1951; Ruesch and Bateson, 1951; Wiener, 1948).

More refined and detailed studies by anthropologists and linguists such as Bateson, Birdwhistell, Hockett, and McQuown present adequate support for the common observation that every person must in some way avoid attending to the entire cacophony of messages entering his system from thousands of end organs (McQuown, 1959). It would be highly instructive to know how the organism "learns" such selective inattention or differentiated inhibition because in experimental psychology, linguistics, and psychiatry there are many examples, such as a mother's ability to sleep in a noisy city apartment oblivious to all but her baby's cry. Most academic persons have ample reason to believe the adage that we hear only what we wish to hear or that only the prepared mind can perceive new data or relations (L. Pasteur).

Because we believe that cultures evolve and that the individuals in them learn the signal systems by which life becomes possible, we are led to the hypothesis that individuals and the culture they live in are interacting systems whose properties and relations can be pragmatically described. The ready availability of film and sound-tape records which can be studied in minute detail ($\frac{1}{50}$ second per frame for film, and about $\frac{1}{3}$ second for tape) over as much time as is necessary by multiple analysts using various techniques enables scientists to study human communication systems as they have not been studied before. It is now possible to produce permanent records of ongoing human transactions in the home which apparently do not disturb the basic processes involved. With due clinical caution to protect the subjects against noxious social consequences of undesirable publicity, these records do not ordinarily represent a violation of privacy. The highly intricate signal systems employed by family members in the ordinary course of their routine tasks of living will provide the scientists of the future with enormous data for increasing our understanding of some of the means by which we become and remain in an evolving process which we term "human." In order to make relatively reliable interpretations of sound-film raw material, it is essential that data from comparative populations be available. Some progress is being made in this direction by the current preliminary studies of Birdwhistell on children of various populations—American, Balinese, and Iatmul—utilizing the films made by M. Mead and G. Bateson. The logical position taken by Darwin that an increase in those qualities which promote the welfare of the group would give the group greater survival value than would be possessed by groups whose members acted principally for their individual safety—a theme supported by Kropotkin, J. Huxley, and others—receives considerable corroboration from the linguistic-kinesic analysis of human interaction. We are only beginning to gather specific evidence on how mutually interdependent we really are (T. H. and J. S. Huxley, 1947; Kropotkin, 1939).

It may not be amiss here to reiterate a plea which fortunately is now being heard more frequently on behalf of natural history methods utilizing the observation of intact organisms in a "natural" habitat. The comparative and correlational techniques associated with this method can furnish us with valuable exploratory leads until such time as the more quantitative methods of the physical sciences can be used more appropriately (Cronbach, 1957; Fine, 1958–59; Hilgard, 1958, 1959; Oppenheimer, 1953, 1958; Rioch, 1958; Scott, 1958; Thorpe, 1956; Tinbergen, 1951).

> The second purpose that the statistical method has been called upon to accomplish is the prediction of the individual case from a precise knowledge of the group or mass. . . . We have seen that this kind of statistical method gives only a somewhat sterile kind of knowledge so far as concerns individuals, namely, a knowledge of betting odds. . . . Its place in the methodology of science is not an independent one. By it alone one cannot discover new truths about phenomena. But it is a highly important adjunct to other modes of research [Pearl, 1940; quoted by Berkson, 1958].

With increasing experience, it seems probable that investigators in the field of human behavior will gradually become more comfortable with the use of natural history methods when these are indicated by the nature of the material and will plan both research strategy and tactics in terms of methods and designs uniquely fitted to deal with the systems or processes at hand, instead of attempting prematurely to use quantitative techniques in areas which do not lend themselves to such treatment. Perhaps one of the best legacies Darwin left us is the record of his success with the natural history method made specific by relatively limited but good experimentation, thus reminding us of the numerous genuine advances made in clinical medicine since Hippocrates by this method. Darwin's example may help restore some dignity and self-respect to those investigators who cannot bring themselves to use quantitative methods irrelevantly or trivially.

The Evidence from Ethology on Instincts and Early Behavior Patterns

An enormous amount of learning is consummated during the first two years to enable a child to master the essentials of his language and the myriad social and physical relations implicit in this mastery. From recent past experience it would appear that many decades of adequate teaching derived from experimentation will be required to influence the public to use this concept, because implicit in this concept is the idea that immediate responsibility in the parent-child relationship is involved (Greenberg, 1957; McQuown, 1959). Multiple independent observations seem to agree that much of a person's basic character is determined by the ages of four to six years. Future growth and development of personality, as well as of physique, are inevitable under favorable circumstances, but apparently within the framework of the *mechanisms built into* the person in the initial phases. The ethologists have helped enormously by furnishing evidence in small animals that early "imprinting" may result in highly stable behavior system (Beach, 1958; Bullock, 1958; Caspari, 1958; Hess, 1958; Lorenz, 1958; Pribram, 1958; Roe and Simpson, 1958; Schiller, 1957; Sperry, 1958). Jere Wilson, associated with the Hess group in Chicago, is now studying the smiling response in infants as one method of investigating the possibility of relating human behavior to imprinting (Hess, 1958). The new experimental data from small animals make possible a more systematic experimental approach to some of the vexing problems formerly subsumed under the name of "instinct" or "drive" because these problems were not testable. Societal organization in animals, including the impressive courtship behavior and subsequent monogamy of some species, and the complex operations involved in "territoriality" provide a wealth of opportunity for learning how to study the genesis and maintenance of systems other than the human, which yet resemble the human in various ways (Carpenter, 1958; Hinde and Tinbergen, 1958; Huxley and Huxley, 1947).

CLINICAL STUDIES ON EARLY INFANT BEHAVIOR

Bowlby (1951) has reviewed the literature up to 1952 and prepared an impressive array of clinical evidence from numerous workers. His evidence shows that many disturbed children and chronic delinquents who did not seem to have feelings for anyone were found to have grossly disturbed relationships with their mothers in their early years. He was struck by the tendency of these children to steal; hence he compared a group of forty-four thieves with a control group, similar in number, age, and sex, who were also emotionally disturbed but did not steal. He found that among the thieves there were fourteen "affectionless characters," while the control group had none; second, he noted that seventeen of the thieves had suffered complete and prolonged separation from their mothers or foster-mothers for six months or more during their first five years of life, while only two of the controls had suffered similar separations. This and similar studies are dramatic but subject to criticism because (1) they are retrospective clinical observations on a selected group of children who developed adversely and they fail to account for those who may have had the same experience but have developed normally; (2) retrospective studies may reflect distorted memories, particularly in irresponsible persons who want to enlist sympathy by describing overt hardships in childhood (Group for the Advancement of Psychiatry, 1958). Fortunately, direct studies are most numerous, and some of them are documented on film. Space does not permit a review of the most carefully designed and executed of these studies, but the results are uniform. In almost all cases, maternal deprivation results in retardation of a child's development physically, intellectually, and socially. Evidence of physical and mental illness may also appear (Bowlby, 1951; Levy, 1943). These experiments gain in interest as it becomes possible to specify with greater accuracy both the intensity and the duration of the deprivation under study. The concept of critical periods for maximal or optimal development in the human is a highly significant step forward, even though the critical periods are not so well defined as the imprinting period in birds or other species (Levine and Lewis, 1959; Scott, 1958). The central importance of these provocative studies is clear for preventive medicine and for all students of the behavioral sciences, since, by using more refined observational and comparative techniques, it seems possible to ask concrete experimental questions that are answerable with available methods. Presumably, very few scientists would be willing to do the obvious crucial experiments on human beings in a properly equipped and isolated island laboratory where numerous rigid controls could be rigorously enforced. However, the press has reported that the Chinese Communist government is establishing communes in which mother-child separation of various degrees is being effected, along with the dissolution of conventional family life (Shackford, 1958). Perhaps the scientists of Communist China will be able to furnish the world with more conclusive data in the future. It is entirely possible that an appropriately governed commune system may generate a population of relatively compliant, apathetic citizens, similar to the children described by Bowlby, Anna Freud, Spitz, and others and made so vivid by George Orwell in *1984*

(Bowlby, 1951; Freud, A., 1948; Freud and Burlingham, 1943, 1944; Orwell, 1949; Spitz, 1945, 1946, 1949; Spitz and Wolf, 1946*a, b*). In contrast to the plans of the Red Chinese, it is noteworthy that Soviet Russia has ceased enforcement of earlier attempts (1917–35) at communes where family life and child-mother separation was planned. No doubt more studies will be made of this phenomenon, wherever it is found, as scientists and responsible government officials become aware of its importance. With more confirmation of data on this problem, our own physicians and hospitals will encourage better mother-child relations in the first years of life, and judges and social agencies responsible for the destiny of children will use verified information in practical ways. It is not too much to hope that, insofar as some physical and mental disorders are the resultant of vulnerabilities or disabilities built into the human organism early in his life, they will be gradually corrected by more intelligent child-rearing practices.

Mention should be made of many studies of the development of thought made on infants since John B. Watson began his behavioristic studies in 1910; but, since this topic will be covered in other essays, I will only mention the long-term studies of Piaget in which various properties of thought can be distinguished at different age levels. For example, a five-year-old child in a given environment is said to deal only with concrete operations of classes and relations, whereas an adolescent can transcend the concrete by using algebraic models for his problem-solving (Inhelder, 1957; Inhelder and Piaget, 1958; Piaget, 1957).

The impetus given to studying the continuity in development of thought by these and other methods may be of considerable value to students of human evolution. The work of Kragh (1955) and of Schilder (1951) which was mentioned in the first section is noteworthy here.

THE CURRENT PRAGMATIC CLINICAL APPROACH

Now that we are developing better methods for examining more closely that which is innate because of genetic influence and that which is acquired by experience, the belief that psychoses and neuroses (deviants) are either innate or acquired becomes more amenable to meaningful systematic investigation. New data, both about the influence of enzyme systems or noxious secretions and the acquired behavior of the person and about his significant action patterns will permit both prediction and control. In view of the high stability of some patterns of behavior acquired early, it is now more feasible to think more flexibly about both normal and pathological action patterns.

While many activities of the body may be innate, and undoubtedly are, we are on the very threshold of discovering that some deeply rooted activities may be acquired by transmission through the family group.

The nature of the transmission of tradition, as well as of language, how we should walk, hold our bodies, cross our legs, blink our eyes, wrinkle our foreheads, attack our enemies, choose our mates, or reward our friends, is not well understood. Apparently, much that is most important to us in later life is acquired before we have words to discuss the acquiring operation. An industrious application by numerous workers utilizing and comparing individuals of all ages and stations from the same and different cultures, particularly in the family setting, will help us better to devise methods for detecting the relations between the innate and the learned. As we focus with microscopic care upon the split-second signals in the area of vocal modifiers and kinesics (body motion) with the aid of the new cameras and sound-recording machines upon the child-rearing process, a new world of data is revealed (McQuown, 1959).

Biologists who have no difficulty in appreciating the delicacy of the developmental process from conception to birth, with its myriad opportunities for chemical or physical error in this most complex process, will have little difficulty in seeing that, when the neonate enters upon his new environment, he not only must continue earlier processes of growth and development as in his osseous and hematopoietic systems but must learn new techniques such as breathing and eating, in order to survive. The neonate has an incomplete sensory organization and must master the signals whereby he can command his supplies, including food, warmth, and human association, to be brought to him. Once we see in microscopic detail the nature of even a few of the operations of this most impressive of all transformations, we can appreciate the wealth of potential errors in development. I do not know the probability that a fertilized human ovum will come to normal maturity. It has always seemed to me that the odds were considerably against a normal birth, but this may be a naïve view. Certainly, we can be equally astonished at the fact that so many persons grow up as well integrated as they do. From the biological point of view and sheer probability calculation, we must ask anew how it is that most men and women reach maturity with recognizable human features, attitudes, and activities, in view of the extraordinary likelihood that maladaptations could occur. This may be largely explicable by means of Waddington's work on epigenetic canalization (Waddington, 1958).

Ordinarily, as citizens, most of us are too preoccupied with our own affairs to wonder much about deviants, unless they present us with unusual dramatic, moral, or aesthetic problems. It is apparent that to gain basic data about the development of human nature and about one way in which human beings can all learn to live together in a mutually co-operative (one-world) civilized society, we can probably do no better than to study how normal organization is achieved and to study carefully the failures who present themselves as patients, prisoners, or persons otherwise unsuited for conventional living. For those deviants whom we label highly talented or geniuses, there should be a special study, in order to make possible the conditions under which such persons can occur and be maintained in a climate of heightened creativity. If our civilization could methodically improve the creativity of our best people by even a small fraction and enable more of these highly talented, now without reasonable opportunity, to function productively, we would obviously be in a better position to meet future threats and proceed to future achievements.

An important question which deserves careful study for future planning is the relative capacity for change in character, personality, or temperament after childhood. In view of current ignorance about how genes bring about innate characteristics which may determine that a person will die from vascular disease at age sixty rather than ninety or why many schizophrenic attacks seem to begin in late puberty, while others are delayed, it is not easy to hazard a guess about changes in overt behavior at any age. Theoretically we are committed to the possibility that such alterations due to genetic influences are possible (Jackson, 1959; Kallmann, 1953). Kallmann has presented evidence for this hypothesis in his twin studies (1953; see also Snyder, 1959). On the other hand, common experience, as well as psychiatric case studies and the therapeutic work of educators and the clergy, reveal that persons do alter their motivations and characteristic methods of relating to other persons and things, apparently under the sole influences of a psychosocial relationship with another person. The transformation of Jean Valjean after the bishop gave him the candlesticks is a well-known literary example which can be duplicated from his practice by any successful physician. Psychiatrists are just beginning to appreciate and document the fact that often people can make people sick and people can make people healthy again. This does not deny the powerful influences of the genes or that emotions are mediated physiologically in a most complex manner in the body. While this body machinery may be the fault, as it is in some instances of mental diseases, more often these disorders seem to be related to faulty human relationships (Bleuler, 1955). Even genetic defects, which cannot be corrected by direct methods, are often amenable to some form of ameliorative or supportive therapy. Psychiatrists are now less likely to believe what Jennings calls "the fallacy that showing a characteristic to be hereditary proves that it is not alterable by environment" (Hogben, 1938). It should be remembered that the converse is equally a fallacy.

EVIDENCE AGAINST SIMPLIFIED EARLY LEARNING THEORIES AS EXPLANATION OF PATHOGENIC FACTORS

Mention has been made of Kallmann's studies utilizing the method of comparative twin observation in numerous diseases. His observations on the manic-depressive and schizophrenic psychoses are probably of most interest to psychiatrists, although his conclusions on psychoneurotic and delinquent behavior, deficient resistance to stress and infection, and pathological sex variants will also be of interest to psychiatrists and behavioral scientists for their intrinsic value, as well as for their medical prognostic values. Because the subject is a complex one, it requires lengthy exposition of numerous categories of information. No attempt will be made here to summarize Kallmann's excellent clinical studies which support the hypothesis that genetic factors are highly important in these disorders (Darlington, 1958*a, b;* Kallmann, 1953, 1959; Lewis, 1958; Razran, 1959). Kallmann's work and that of others in human genetics deserves careful study by all clinicians, because the latter are apt to overlook the possibilities inherent in this work. At least eight of our medical schools have established departments of medical genetics; we may look forward to more data for integration with other methods of study.

Goldhamer and Marshall have written a stimulating report on a time series extending back to 1840 of comparative admission rates of psychotic patients in two hospitals (Goldhamer and Marshall, 1953). They examine carefully by statistical techniques the evidence for any change in admission rates and fail to find any change when all corrections have been made. This is an impressive conclusion, negating the popular thesis that hospitalizable psychoses of early middle life are on the increase over a prolonged period. However, the authors are careful to point out that these data do not give us information about the psychoneurotic and character reactions, organic disorders, and psychoses in those over sixty years of age, or the short-term psychotic reactions. They also emphasize that, although the homogeneity of nineteenth- and twentieth-century rates for this limited category of psychoses from both rural and urban sources may be interpreted as evidence of a genetic determination or constitutional predisposition, they do not believe that it is *direct* evidence for such a view (Greenberg, 1957, p. 96). These data do raise the question whether the changes in the socioeconomic structure of our society during the past century has caused increased stress, resulting in increased hospitalized psychoses. It is both amusing and salutary to read the excellent statements by psychiatrists in 1855 ascribing an increase in mental disorder to the increased competitive stress of living (Goldhamer and Marshall, 1953, p. 95). The possibility does therefore exist that child-rearing practices and family patterns have altered less than has been detected during this period; this may account in part for the stability of the rates. Psychiatrists do not know the crucial components of personality growth and development and therefore cannot detect clearly either the safeguards or the hazards in this biological process. Goldhamer and Marshall have helped place the problem in better perspective, because they are not satisfied with oversimplified explanations for their complex data.

Data from military experience is often interesting when viewed in proper context. It is neither easy nor safe, however, to generalize about its significance for civilian populations, because the numerous variables make it unsafe for comparative purposes. One set of observations concerning American soldiers in Korea in the years 1951–55 will be mentioned, because it may have some future follow-up value. The observed groups had a highly stable rate of diagnosed psychosis, whatever the nature of their duty. Whether in active combat, in advance or in retreat, or in the rear echelons, it seems that the admission and discharge rate for gross, overt incapacitating psychosis did not change much. Of course, the total mental health picture of soldiers did vary, as shown by changes in admission rates for such categories as "combat fatigue," psychoneurosis, panic reactions, and behavior disorders which are sensitive indicators of the "stress" present in any given sector. If it is recalled that all American soldiers were screened for mental disorder at the time of induction and that screening was in effect continuously during basic and divisional training, and even in some cases at the port of embarkation, the data become more provocative in spite of the patent gross statistical difficulties. One of the most troublesome difficulties in utilizing these data for comparative purposes is that the criteria for the diagnosis of a psychosis varied both medically and administratively. When dealing in large numbers, however, where many variables may be corrected, it is striking that the disability rate for psychoses should be 2.24 admissions per 1,000 average strength per year for the Korean campaign; 2.69 per 1,000 per year for World War II; and 2.97 per 1,000 per year for World War I. During these three war periods the admission rates for psychoneuroses were: World War I, 8.00; World War II, 25.52; Korean conflict, 8.65; but these fluctuations are probably due to artificial factors and cannot be used for comparison. In a review of admission rates for psychosis for the total U.S. Army per 1,000 per year from 1920 to 1953, including our years of depression and inflation, Col. A. J. Glass found small variations. Changes in composition of the population involved, changes in policies and procedures, and common stresses such as combat service and economic changes have not affected army psychosis rates in any identifiable way (Glass, 1957; Group for the Advancement of Psychiatry, 1960; Whitehorn, 1956). Similar stabilities, I feel, will be found in the armies of other countries where the data permit comparison, although accurate and comparable data have not been available.

The brilliant studies by Ford and Beach which relate various patterns of sexual behavior to levels of cultural development and the phylogenetic scale illustrate future possibilities (Ford and Beach, 1952). The possibility of clinical interference with genetic development, called "The Third Stage in Genetics" by D. Michie (1958) is another. A new world of chemical operations is opened, following the discovery of sexual reproduction in a strain of *Escherichia coli* by Lederburg and Tatum and manipulations by Lederburg and his colleagues of phage containing chromosomal material, desoxyribose nucleic acid, generally known as "DNA." Michie reports recent work by J. Benoit, P. Leroy, C. Vendrely and R. Vendrely in which DNA was extracted from Khaki Campbell ducks and injected into ducklings of the Pekin breed:

> The majority of the treated birds, *and of their offspring,* developed a constellation of characters (pigmentation of beak, morphology of feathers, shape of head, size and conformation of body) apparently derived from the donor breed. If the authors have correctly interpreted their observations, there had occurred, to use Darwin's phrase, a "formation of hybrids between distinct . . . varieties without the intervention of the sexual organs" [1958, pp. 83–84].

Jean Rostand discusses the possibilities of DNA injections in humans and anticipates a time "when each human infant could receive a standard DNA that would confer the most desirable physical and intellectual characteristics. Such children will not be the offspring of a particular couple, but of the entire species" (1958, p. 47). Rostand also discusses the legal and ethical aspects of parthenogenesis, artificial insemination, surgical transplantation of ova, nuclei of ova, and embryos, and surgical grafting. No doubt, future generations will find ways to deal fairly with these questions.

In a different category, but enormously important both now and in the future, are the reports concerning a recently discovered group of compounds known as "19-nor-steroids." Results show that these compounds have remarkable properties of blocking ovulation. They are now being tested as oral contraceptives in field studies in the West Indies (Pincus, 1959).

H. J. Muller's book, *Out of the Night: A Biologist's View of the Future* (1953), has been cited as a route to practical eugenics via artificial insemination. There is, however, little evidence that this method is being or will be used in the near future. Bentley Glass (1959) has suggested that someday defective human genes may be mixed in the laboratory with healthy genetic material in order to produce normal individuals or that the sex of children may be preordered by a device that electrically separates the two types of sperm.

The evidence for prenatal influence by means of the mother's adrenal-cortisone state suggests hope for control which will help guarantee healthy, well-integrated babies at birth. However, this method is probably not feasible, because the narrowing of the range of susceptibilities probably decreases the potential range of creativity (Hoch and Zubin, 1955).

The investigations of H. S. Liddell (1953), D. M. Levy (1953), J. P. Scott, P. F. D. Seitz (1958), and H. F. Harlow (1958) are only a few of many in which selective breeding and/or very early training of animals (other than man) reveal the feasibility of specifying the nature of the interactions in the neonate and infant and even some of the delayed consequences. Harlow specifically makes a good case, saying that there is no evidence that any *sharp* break ever appeared in the evolutionary development of the learning process (1958*a, b;* see also Fuller and Scott, 1954).

These examples of man's ability to understand and influence human behavior at basic levels lend real force to the contentions of many scientists that man need not necessarily accept a deteriorating course in the evolutionary process. Waddington (1959), in commenting about some current theorizing about evolution, stresses the effect of behavior upon environment:

> The relation of the behavior of an animal to the evolutionary process is not solely that of a product; behavior is also one of the factors which determines the magnitude and type of evolutionary pressure to which the animal will be subjected. It is at the same time a producer of evolutionary change as well as a resultant of it, since it is the animal's behavior which to a considerable extent determines the nature of the environment to which it will submit itself and the character of the selective forces with which it will consent to wrestle [p. 204].

Current Contributions by Social Scientists to the Study of Mental Disorders

The recognition by social scientists that mental disorders may be, to some degree, a resultant of socioeconomic and psychological forces has induced some of them to make this field their major concern. These investigators, using essentially the natural history or correlational methods employed by Hippocrates and Darwin, are trying to find out more precisely how environment alters behavior. Implicit in some of this work is the concept of man and his society as living organizations which are evolving in systematic patterns. There is increasing evidence from anthropological, ecological, psychological, and epidemiological studies that environmental forces exert a significant influence upon human behavior, including pressure upon some deviant behavior in otherwise normal persons, as well as upon those who are identified as incapacitated by mental disorders.

Some dramatic examples that environmental forces are active in shaping conduct are the following: the predominance of cardiorespiratory disorders as manifestations of "battle fatigue" during World War I in contrast to the preponderance of gastrointestinal syndromes in World War II; the varying incidence of conversion hysterias with educational levels and situations; the marked changes in the incidence of such diseases as peptic ulcer, diabetes mellitus, exophthalmic goiter in men as compared to women; the notable decrease in soiling patients in mental hospitals even before the ataraxic drugs appeared; and the decrease in paranoid syndromes in patients with cerebral syphilis as newer drugs and heat treatments became available in the 1930's. These are but a few among many similar suggestive clinical studies (Diethelm, 1953; Dunbar, 1954; Halliday, 1948; Menninger, 1948; Mittelman and Wolff, 1942; Oppenheimer, 1956; Stouffer *et al.*, 1950).

Of interest is the opinion that the social sciences, particularly economics, political science, and sociology, are now, or have been, strongly influenced by the theory of evolution. I defer to the opinion of experts and will mention the strongly negative opinion of D. G. MacRae to avoid careless optimism. MacRae (1956) says:

> What, in all probability, sociology most needs at the moment is not either a Newton or a Darwin, but a Linnaeus to elaborate a really workable classification of social structures and of the range and variety of institutional patterns and sequences [p. 311].

Faris (1956) in the United States voices similar opinions. I do not believe the situation to be as barren as painted by MacRae or Faris in social anthropology, social psychology, or psychiatry, but I submit his question for general discussion. Since the social science area of investigation is very new, methods need to be developed which will satisfy the usual canons of science, even though handicapped by the fact that many basic questions in genetics, physiology, chemistry, anthropology, medicine, and psychology cannot be answered at this time. Moreover, a major discovery in any one of these areas will considerably alter the current outlook.

Although the subject matter, methods, attitudes, and goals of social psychiatry are not well defined, it is being recognized as a field distinct from the conventional learning and practice in psychiatry in which the individual and his relations to others is the major focus. While overlapping occurs from several areas, the newer field can usually be identified. Its focus is upon the composition and dynamics of large populations, preventive measures and treatment methods applicable to large groups, and the various techniques for studying and working in these areas. The textbook by Lemkau (1955), monographs by Halliday (1948), Jahoda (1958), Opler (1956), and several symposia give the main outlines of the current methods and projects of social scientists and physicians who, with varying backgrounds, are investigating the biological deviants known as the mentally ill in our society (Erikson, 1950, 1958; Halliday, 1948; Hollingshead and Redlich, 1958; Jahoda, 1958; Kruse, 1957; Leighton *et al.*, 1957; Lemkau, 1955; Lewis, 1953; Linton, 1956; Rose, 1955; Opler, 1956; *Symposium on Preventive and Social Psychiatry,* 1958; U.S. National Advisory Mental Health Council, 1955; Wittkower and Fried, 1959).

Émile Durkheim (1858–1917) pioneered in these fields, particularly by his application of statistical methods to anthropological and sociological material, which is well exemplified in his famous studies on suicide (1897). Following this brilliant leader, a number of others have attempted to relate sociocultural factors with personality patterns and with behavior, including the more limited goal of relating cultural pressures to those deviations called "psychopathology." The scope of interest of workers in social psychiatry is quite broad. The data accumulated by these workers on the prevalence and incidence of various disorders by the demographers or epidemiologists are probably of general interest. Some of the best known of the foreign and domestic studies have been reviewed by Opler (1956). In general, using various methods of interviewing and testing and by results from questionnaires, investigators try to relate age, sex, occupational, ethnic, socioeconomic, cultural, and related factors to the presence or absence of one or more manifestations of mental, emotional, or social disorders. Sometimes the physical manifestations often associated with emotional stress are included, such as asthma, migraine, peptic ulcer, colitis, rheumatoid arthritis, and similar diseases, or delinquency, alcoholism, accident-proneness, domestic discord, and divorce (Caudill, 1953).

To give one example, the widely quoted study of Faris and Dunham (1939) purported to show that the slums of Chicago were a breeding ground of psychoses, even though this seemed unlikely from ordinary clinical observation and from gross inspection of the data (Faris, 1956). Hollingshead and Redlich, of New Haven, in another popular study (1958) have described the prevalence of mental disorders as associated with an economic class (Hollingshead and Redlich, 1958). On the face of it, these are valuable studies, but Opler (1956) points out that they contain weaknesses which need analysis and correction before the major thesis can be accepted without qualification. Gerard and Houston in the Worcester studies found that areas with high rates of schizophrenics could be accounted for by highly mobile, unattached individuals who drifted into these areas. Many of them were probably ill before arrival (1953). M. S. Schwartz (1946) has found that schizophrenics tend to go lower in the occupational scale prior to hospitalization. Opler says (1956):

> . . . If the additional evidence of Gerard, Houston, and Schwartz is accepted, both the prevalence and gross ecological data of Faris and Dunham, and the prevalence data of New Haven are open to serious question . . . [p. 94].
>
> While the New Haven study of Hollingshead and Redlich aims at more than allusion to "successful interaction," "organized agencies of society," and "serious gaps" in the interactive process, it deals only with prevalence, the number of mentally ill persons as of a given stated day, with no consequent separation of incidence and disease duration. Thus the high prevalence rates in certain socioeconomic strata may reflect an unmitigated course of illness in some persons plus the number of those becoming ill, or seeking treatment as of a given date [*ibid.*].

Perhaps this brief résumé of the two well-known studies in social psychiatry will illustrate some of the technical problems to be mastered. New data and methods are now available which may open the door to sorely needed new approaches. The cautions by Professor Aubrey Lewis concerning inevitable psychiatric bias and against premature frontal attacks upon the problems of complex populations deserve careful study (1953).

In everyone's mind must be the controlled experiment in which a small colony would be set up, preferably on an agreeable but small island, where social and physical constraints could be imposed with a minimum of effort, to study, with the best technical design available, human infants, particularly twins. The primary goal would be to detect native abilities as differentiated from the incredibly swift learning that appears to occur in the first 20–30 months after birth. The unfortunate fate of the Stockard Farm must be kept in mind to insure continuity (1941). As previously mentioned in the section on linguistic-kinesic studies, tangible advances will probably be made on the functioning of communication systems in families, with more expert use of sound-film. The studies may then provide the data and the impetus for more direct experimental work with early infant learning.

Planning for the Future

I have referred to my indebtedness to J. Huxley, G. G. Simpson, and A. Huxley and others regarding the possibilities for planning a better life for all children throughout the world. This is a difficult task because conditions for which we are planning are changing at an increasingly accelerated rate, as propounded by Henry Adams and many others at the beginning of this century. Major advances anywhere along the frontiers of science, particularly the free use of nuclear energy, biological control in genetics, and increased understanding and control of unconscious processes, will alter the program. Vickers (1958, 1959) has brilliantly analyzed the limitations of planning health programs. DuBos in *Mirage of Health* (1959) argues that "complete freedom from disease and from struggle is almost incompatible with the process of living," and it seems likely there will be some deviants and maturation failures in any conceivable biological-cultural system.

Ordinarily, we think of the goal of education as centering for the most part around how best to develop the freedom for self-development and creativity compatible with law and justice. The virtues of self-reliance, self-discipline, and self-motivation will probably be high in the minds of most parents and educators as desirable character traits. However, the discourse might go somewhat further in asking what kind of society we are presumably training them for. The traits mentioned may be much more important and take distinctive forms in a relatively crude, pioneering culture than in a comparatively stabilized, refined one. We might specify that training for optimal flexibility and adaptability should be a high priority in our educational goals, because the prospect of rapid change seems inevitable for generations to come. If by conscious planning, we set about to inculcate such virtues because they appear to have high survival value, how do we go about it in such a way that the cost in human values is not too high? Perhaps these questions seem too remote to be of value at this time. As a rough exercise in planning, it is possible to envision future societies which have been described for us and try to project the best preparatory education for them.

One example would be a culture characterized by overpopulation (C. Darwin), overorganization (A. Huxley and G. Orwell), and resultant semistarvation, limited opportunities for self-realization, and increased constraints upon personal liberties (C. G. Darwin, 1953; A. L. Huxley, 1946, 1958; Orwell, 1949).

Another example would be a culture of abundance in which leisure ("the mother of discoveries"), ample opportunities for self-cultivation, and personal freedom exist. Marcuse (1955), in *Eros and Civilization*, has made a good initial statement of the possibilities in the second case. He utilizes known Freudian hypotheses and data to argue that it would be feasible to train children for a society of plenty, in spite of the skepticism of Freud and many other psychoanalysts about the limits of altering the personality. Marcuse's position is logical and stimulates speculation about what may be possible. In fact, J. Huxley saw, in 1943, the possibilities when he discussed the development of personality (T. H. and J. S. Huxley, 1947).

Most authorities agree that man can take steps to improve his survival values. As in dealing with other problems, we need to put our best people to work on the problems of the future, as suggested by M. Mead (1957). Possibly our best single defense against the noxious forces is self-awareness in its most profound sense, as suggested by Simpson (1949) and by Roe (1959). It seems probable that we have now some of the elements upon which to build a comprehensive science of man, or social psychology in the broadest sense, which Bertrand Russell referred to as "the science to save us from science" (1957). Such a science will include material from the observations on the activities of normal and sick people made by physicians and psychiatrists.

Conclusions

Medicine, including psychiatry, has been strongly influenced by Darwin's evolutionary theory, and many current advances are directly related to it. Most students of mental disease and mental health accept the hypothesis that genetic evolution and psychosocial or cultural evolution are both necessary for understanding mental disease as among the various types of behavioral deviants. With better evidence about man's activities from the basic sciences, particularly genetics, psychology, and the behavioral sciences, including history, the prospects for improved prevention and control of mental disorders are inevitably increased. We shall have to encourage more study of the less conscious elements in man's thinking and acting and definitely assign the task of preparing for future problems in human relations to qualified persons who can devote all their energies to it, just as we now have scientists studying the physical resources of the future. The information gained among sick or distressed people will be of value to the study of normal human behavior and may give us additional leads for enhancing the well springs of creativity and well-being.

The hypothesis that all human activity must be studied as complex reciprocal interactions between internal and external systems is gaining much support as a result of the new advances in genetics, ecology, ethology, and the human behavioral sciences. With increased conceptual and experimental skills, gradually, solutions will be found to (1) the "problem of the observer" and the need to specify the interaction between the observer and the system he is observing, with appropriate emphasis upon the method utilized to study the system as well as upon data obtained; (2) the barriers to studying component systems in larger total systems; (3) the possible resolution of apparently contradictory data by means of the complementarity principle; and (4) the concepts of feed-back and related theorems from communication theory as applicable to human activities.

Human activity as a determinant as well as a product of evolution is becoming a more tenable hypothesis. That man may truly become the architect of his own fortunes, if he is sufficiently willing, is more than a faint hope.

References

Bailey, P. 1958. "Evolution and Disease of the Brain," *Perspectives Biol. and Med.*, II, 62–74.

Baillie, J. 1951. *The Belief in Progress.* New York: Charles Scribner's Sons.

Barnett, S. A. (ed.). 1958. *A Century of Darwin.* Cambridge: Harvard University Press.

Beach, F. A. 1958. "Evolutionary Aspects of Psychoendocrinology." In Anne Roe and G. G. Simpson (eds.), *Behavior and Evolution,* pp. 81–102. New Haven: Yale University Press.

Benedek, T. 1949. "The Emotional Structure of the Family." In R. N. Anshen (ed.), *The Family: Its Function and Destiny,* pp. 202–25. New York: Harper & Bros.

Benjamin, J. D. 1958. "Some Considerations in Biological Research in Schizophrenia," *Psychosom. Med.,* XX, 427–45.

Berkson, J. 1959. "The Statistical Investigation of Smoking and Cancer of the Lung," *Proc. Mayo Clin.,* XXXIV, 206–24.

Berrill, N. J. 1955. *Man's Emerging Mind: Man's Progress through*

Time—Trees, Ice, Flood, Atoms, and the Universe. New York: Dodd, Mead.

BLEULER, M. 1955. "Researches and Changes in Concepts in the Study of Schizophrenia, 1941–1950," *Bull. Isaac Ray Med. Libr.*, III, 1–132.

BLOOMFIELD, L. 1939. *Linguistic Aspects of Science.* Chicago: University of Chicago Press.

BOHR, N. 1958. "Unity of Knowledge." In his *Atomic Physics and Human Knowledge,* pp. 67–82. New York: John Wiley & Sons.

BOND, D. D. 1952. *Love and Fear of Flying.* New York: International Universities Press.

BORN, M. 1949. *Natural Philosophy of Cause and Change.* Oxford: Clarendon Press.

———. 1951. *The Restless Universe.* 2d ed., rev. New York: Dover Publications.

BOWLBY, J. 1951. *Maternal Care and Mental Health.* Geneva: World Health Organization.

BRAATOY, T. F. 1954. *Fundamentals of Psychoanalytic Technique.* New York: John Wiley & Sons.

BRIDGMAN, P. W. 1927. *The Logic of Modern Physics.* New York: Macmillan Co.

BROSIN, H. W. 1952. "Contributions of Psychoanalysis to the Study of the Psychoses." In F. ALEXANDER and H. ROSS (eds.), *Dynamic Psychiatry,* pp. 285–306. Chicago: University of Chicago Press.

BRUNER, J. S. 1958. "The Freudian Conception of Man and the Continuity of Nature," *Daedelus,* LXXXVII, 77–84.

BULLOCK, T. H. 1958. "Evolution of Neurophysiological Mechanisms." In ANNE ROE and G. G. SIMPSON (eds.), *Behavior and Evolution,* pp. 165–77. New Haven: Yale University Press.

BURY, J. B. 1955. *The Idea of Progress: An Inquiry into Its Origin and Growth.* New York: Dover Publications.

CANNON, W. B. 1932. *The Wisdom of the Body.* New York: W. W. Norton & Co.

CARPENTER, C. R. 1958. "Territoriality: A Review of Concepts and Problems." In ANNE ROE and G. G. SIMPSON (eds.), *Behavior and Evolution,* pp. 224–50. New Haven: Yale University Press.

CASPARI, E. 1958. "Genetic Basis of Behavior." In ANNE ROE and G. G. SIMPSON (eds.), *Behavior and Evolution,* pp. 103–27. New Haven: Yale University Press.

CAUDILL, W. 1953. "Applied Anthropology in Medicine." In A. L. KROEBER (ed.), *Anthropology Today,* pp. 771–806. Chicago: University of Chicago Press.

———. 1958. *The Psychiatric Hospital as a Small Society.* Published for Commonwealth Fund by Harvard University Press.

CAUDILL, W., and STAINBROOK, E. 1954. "Theoretical Symposium on Contributions of Interdisciplinary Research to Psychiatric Theory," *Psychiatry,* XVII, 27–40.

CHANCE, M. R. A., and MEAD, A. P. 1953. "Social Behavior and Primate

Evolution." In SOCIETY FOR EXPERIMENTAL BIOLOGY, *Symposium No. 7: Evolution,* pp. 395–439. New York: Academic Press.

CHERRY, C. 1957. *On Human Communication.* Cambridge: Technology Press of Massachusetts Institute of Technology.

COLBERT, E. H. 1958. "Morphology and Behavior." In ANNE ROE and G. G. SIMPSON (eds.), *Behavior and Evolution,* pp. 27–47. New Haven: Yale University Press.

CONFERENCE ON CYBERNETICS. 1950–55. *Transactions, 6th–10th Conference* (1949–53), Vols. I–V. Edited by HEINZ VON FOERSTER. New York: Josiah Macy, Jr., Foundation.

COUNCIL FOR INTERNATIONAL ORGANIZATIONS OF MEDICAL SCIENCES. 1954. *Brain Mechanisms and Consciousness: A Symposium.* Consulting editors: E. D. ADRIAN, F. BREMER, H. H. JASPER; editor for the council: J. F. DELAFRESNAYE. Springfield, Ill.: Charles C Thomas.

CRONBACH, L. J. 1957. "The Two Disciplines of Scientific Psychology," *Amer. Psychologist,* XII, 671–84.

CROOKSHANK, F. G. 1949. "The Importance of a Theory of Signs and a Critique of Language in the Study of Medicine." In C. K. OGDEN and I. A. RICHARDS, *The Meaning of Meaning,* pp. 337–55. 10th ed. London: Routledge & Kegan Paul.

CURTI, M. 1956. *The American Paradox: The Conflict of Thought and Action.* New Brunswick, N.J.: Rutgers University Press.

DARLINGTON, C. D. 1958*a*. "Control of Evolution in Man," *Nature,* CLXXXII, 14–17.

———. 1958*b*. "Cousin Marriages," *Triangle,* (Basel), III, 277–80.

DARWIN, SIR C. G. 1953. *The Next Million Years.* Garden City, N.Y.: Doubleday.

DARWIN, C. R. 1873. *The Expression of the Emotions in Man and Animals.* New York: Appleton.

———. 1874. *Descent of Man and Selection in Relation to Sex.* New York: A. L. Burt.

———. 1951. *Origin of Species.* New York: Philosophical Library, 1951. Reprint of first edition. 1859.

DE CHARDIN, PIERRE. 1959. *The Phenomenon of Man.* New York: Harper & Bros.

DEWEY, J. 1910. *The Influence of Darwin on Philosophy and Other Essays in Contemporary Thought.* New York: Henry Holt & Co.

DIETHELM, O. 1953. "Changing Psychopathology." In *Annual Report of the Payne Whitney Psychiatric Clinic.*

DUBOS, R. J. 1959. *Mirage of Health: Utopias, Progress, and Biological Change.* New York: Harper & Bros.

DUNBAR, HELEN F. 1954. *Emotions and Bodily Changes: A Survey of Literature on Psychosomatic Interrelationships, 1910–1953.* 4th ed. New York: Columbia University Press.

DURKHEIM, É. 1951. *Suicide.* Glencoe, Ill.: Free Press. Originally published in France in 1897.

EMERSON, A. E. 1958. "The Impact of Darwin on Biology." (Preliminary draft of an article and lecture presented at Southern Illinois University, October 1, 1958.)

ERIKSON, E. H. 1950. *Childhood and Society*. New York: W. W. Norton & Co.

———. 1958. *Young Man Luther: A Study in Psychoanalysis and History*. New York: W. W. Norton & Co.

ETKIN, W. 1954. "Social Behavior and the Evolution of Man's Mental Faculties," *Amer. Naturalist*, LXXXVIII, 129–42.

FAERGEMAN, P. M. 1956. Review of *The Actual-Genetic Model of Perception-Personality*, by U. KRAGH, *Psychoanalyt. Quart.*, XXV, 597–601.

FARIS, R. E. L. 1956. "Evolution and American Sociology." In S. PERSONS (ed.), *Evolutionary Thought in America*, pp. 160–80. New York: G. BRAZILLER.

FARIS, R. E. L., and DUNHAM, H. W. 1939. *Mental Disorders in Urban Areas*. Chicago: University of Chicago Press.

FINE, R. 1958–59. "The Logic of Psychology," *Psychoanal. and Psychoanalyt. Rev.*, XLV, 15–41.

FISHER, C. 1954. "Dreams and Perception: The Role of Preconscious and Primary Modes of Perception in Dream Formation," *Jour. Amer. Psychoanalyt. Assoc.*, III, 389–445.

———. 1956. "Dreams, Images, and Perception: A Study of Unconscious-preconscious Relationships," *ibid.*, IV, 5–48.

———. 1957. "A Study of the Preliminary Stages of the Construction of Dreams and Images," *ibid.*, V, 5–60.

FLETCHER, R. 1957. *Instinct in Man*. New York: International Universities Press.

FORD, C. S., and BEACH, F. A. 1952. *Patterns of Sexual Behavior*. New York: Harper & Bros.

FREED, H. 1958. "Psychoanalysis." in E. A. SPIEGEL, *Progress in Neurology and Psychiatry*, XIII, 366–83. New York: Grune & Stratton.

FREEDMAN, L. Z., and ROE, ANNE. 1958. Evolution and Human Behavior." In ANNE ROE and G. G. SIMPSON (eds.), *Behavior and Evolution*, pp. 455–79. New Haven: Yale University Press.

FREUD, ANNA. 1948. *The Ego and the Mechanisms of Defence*. London: Hogarth Press.

FREUD, ANNA, and BURLINGHAM, DOROTHY. 1943. *War and Children*. New York: Medical War Books.

———. 1944. *Infants without Families; The Case for and against Residential Nurseries*. New York: International Universities Press.

FREUD, S. 1933. *New Introductory Lectures on Psychoanalysis*. New York: W. W. Norton & Co. To be published in *Standard Edition*, Vol XXII.

———. 1935. *A General Introduction to Psycho-Analysis*. New York: Liveright Pub. Corp., 1935. To be published in *Standard Edition*, Vols. XV, XVI.

———. 1938. "Psychopathology of Everyday Life." In *Basic Writings of*

Sigmund Freud, pp. 35–178. New York: Modern Library. To be published in *Standard Edition,* Vol. VI.

———. 1949. *Three Essays on the Theory of Sexuality.* London: Imago Publishing Co. To be published in *Standard Edition,* VII, 123–243.

———. 1955. *The Interpretation of Dreams.* New York: Basic Books. To be published in *Standard Edition,* Vol. V.

———. "The Unconscious." *Standard Edition,* XIV, 159–215; *Collected Papers,* IV, 98–136.

FROMM, E. 1955. *The Sane Society.* New York: Rinehart.

———. 1956. *The Art of Loving.* New York: Harper & Bros.

FULLER, J. L., and SCOTT, J. P. 1954. "Heredity and Learning Ability in Infrahuman Mammals," *Eugenics Quart.,* I, 29–43.

GERARD, D. L., and HOUSTON, L. G. 1953. "Family Setting and the Social Ecology of Schizophrenia," *Psychiat. Quart.,* XXVII, 90–101.

GILLESPIE, C. C. 1958. "Lamarck and Darwin in the History of Science," *Amer. Scientist,* XLVI, 388–409.

GLASS, A. J. 1957. "Observations upon the Epidemiology of Mental Illness in Troops during Warfare. In *Symposium on Preventive and Social Psychiatry . . . ,* pp. 185–98. Washington, D.C.: Walter Reed Army Institute of Research.

GLASS, B. 1959. Lecture at Michigan State University reported in *Time,* LXXIII, 20, January 19, 1959.

GOLDHAMER, H., and MARSHALL, A. W. 1953. *Psychosis and Civilization.* Glencoe, Ill.: Free Press.

GREENBERG, J. H. 1957. *Essays in Linguistics.* Chicago: University of Chicago Press.

GREENBLATT, M., *et al.* 1957. *The Patient and the Mental Hospital.* Glencoe, Ill.: Free Press.

GRENE, MARJORIE. 1958. "Two Evolutionary Theories," *Brit. Jour. Phil. Sci.,* IX, 110–27, 185–93.

GRINKER, R. R., and SPIEGEL, J. P. 1945. *Men under Stress.* Philadelphia: Blakiston Co.

GRINSTEIN, A. 1956. *The Index of Psychoanalytic Writings,* I, 578–82. New York: International Universities Press.

GROUP FOR THE ADVANCEMENT OF PSYCHIATRY: 1956*a. Illustrative Strategies for Research on Psychopathology in Mental Health.* (Symposium No. 2.)

———. 1956*b. Factors Used to Increase the Susceptibility of Individuals to Forceful Indoctrination: Observations and Experiments.* (Symposium No. 3.)

———. 1957. *Methods of Forceful Indoctrination: Observations and Interviews.* (Symposium No. 4.)

———. 1958. "Some Observations on Controls in Psychiatric Research." (Circular Letter No. 283.) (Also published as Report No. 42, 1959, Committee on Research.)

———. 1960. "Committee on Preventive Psychiatry in the Armed Forces,

1959." (To be published. Admission chart with comparative data.)

HALLIDAY, J. L. 1948. *Psychosocial Medicine; A Study of the Sick Society.* New York: W. W. Norton & Co.

HARING, D. G. 1949. *Personal Character and Cultural Milieu.* Rev. ed. Syracuse, N.Y.: Syracuse University Press.

HARLOW, H. F. 1958*a*. "The Evolution of Learning." In ANNE ROE and G. G. SIMPSON (eds.), *Behavior and Evolution,* pp. 269–90. New Haven: Yale University Press.

———. 1958*b*. "The Nature of Love," *Amer. Psychologist,* XIII, 673–85.

HAYEK, F. A. 1952. *The Sensory Order.* Chicago: University of Chicago Press.

———. 1958. "The Creative Powers of a Free Civilization." In F. MORLEY (ed.), *Essays on Individuality,* pp. 183 ff. Philadelphia: University of Pennsylvania Press.

HEISENBERG, W. 1958. *Physics and Philosophy.* New York: Harper & Bros.

HERRICK, C. J. 1956. *The Evolution of Human Nature.* Austin: University of Texas Press.

HESS, E. H. 1958. " 'Imprinting' in Animals," *Scient. American,* CXCVIII, 81–90.

HILGARD, E. R. 1958. "Intervening Variables, Hypothetical Constructs, Parameters, and Constants," *Amer. Jour. Psychol.,* LXXI, 238–46.

———. 1959. "Evolution, Mathematics, and Psychology," *Contemp. Psychol.,* IV, 142.

HINDE, R. A., and TINBERGEN, N. 1958. "The Comparative Study of Species-specific Behavior." In ANNE ROE and G. G. SIMPSON (eds.), *Behavior and Evolution,* pp. 251–68. New Haven: Yale University Press.

HINKLE, L. E., JR., and WOLFF, H. G. 1957. "Health and the Social Environment; Experimental Investigations." In A. H. LEIGHTON, *et al.* (eds.), *Explorations in Social Psychiatry,* pp. 105–37. New York: Basic Books.

———. 1958. "Ecological Investigations of the Relation between Illness, Life Experiences and the Social Environment," *Ann. Int. Med.,* XLIX, 1373–88.

HOCH, P. H., and ZUBIN, J. 1955. *Psychopathology of Childhood.* New York: Grune & Stratton.

———. 1958. *Psychopathology of Communication.* New York: Grune & Stratton.

HOGBEN, L. T. 1938. *Science for the Citizen.* New York: Knopf.

HOLLINGSHEAD, A., and REDLICH, F. C. 1958. *Social Class and Mental Illness: A Community Study.* New York: John Wiley & Sons.

HOOKER, D. 1952. *The Prenatal Origin of Behavior.* Lawrence: University of Kansas Press.

HUXLEY, A. L. 1946. *Brave New World.* New York: Harper & Bros.

———. 1958. *Brave New World Revisited.* New York: Harper & Bros.

HUXLEY, J. S. 1941. *Man Stands Alone.* New York: Harper & Bros.

English edition title: *The Uniqueness of Man.* London: Chatto & Windus, 1941.

———. 1943. *Evolution: The Modern Synthesis.* New York: Harper & Bros.

———. 1944. *Man in the Modern World.* New York: New American Library.

———. 1953. *Evolution in Action.* New York: Harper & Bros.

———. 1955. "Evolution and Genetics." In J. R. NEWMAN (ed.), *What Is Science?* pp. 256–89. New York: Simon & Schuster.

———. 1956. "Evolution, Cultural and Biological." In W. L. THOMAS, JR. (ed.), *Current Anthropology,* pp. 3–25. Chicago: University of Chicago Press.

———. 1958*a*. "Cultural Process and Evolution." In ANNE ROE and G. G. SIMPSON (eds.), *Behavior and Evolution,* pp. 437–54. New Haven: Yale University Press.

———. 1958*b*. "Darwin and the Idea of Evolution." In *A Book That Shook the World: Anniversary Essays on Charles Darwin's Origin of Species,* pp. 1–12. Pittsburgh: University of Pittsburgh Press.

———. 1958*c*. "Emergence of Darwinism," *Jour. Linnean Soc., Zoölogy,* Vol. XLIV, No. 295, and *Botany,* Vol. LVI, No. 365.

———. 1959. In *The Phenomenon of Man* by PIERRE DE CHARDIN. New York: Harper and Bros.

HUXLEY, T. H., and HUXLEY, J. S. 1947. *Touchstone for Ethics, 1893–1943.* New York: Harper & Bros. English edition title: *Evolution and Ethics, 1893–1943.* London: Pilot Press, 1947.

INHELDER, B. 1957. "Developmental Psychology." In *Annual Review of Psychology,* VIII, 139–62. Palo Alto, Calif.: Annual Reviews.

INHELDER, B., and PIAGET, J. 1958. *The Growth of Logical Thinking from Childhood to Adolescence.* New York: Basic Books.

JAHODA, MARIE. 1958. *Current Concepts of Positive Mental Health.* New York: Basic Books.

JEFFRESS, L. A. (ed.). *Cerebral Mechanisms in Behavior.* New York: John Wiley & Sons.

JONES, E. 1953. *The Life and Work of Sigmund Freud,* Vol. I. New York: Basic Books.

KALLMANN, F. J. 1953. *Heredity in Health and Mental Disorder: Principles of Psychiatric Genetics in the Light of Comparative Twin Studies.* New York: W. W. Norton & Co.

———. 1959. "Heredity and Eugenics: Review of Psychiatric Progress, 1958," *Amer. Jour. Psychiat.,* CXV, 586–89.

KEMPF, E. J. 1958. "Basic Biodynamics," *Ann. New York Acad. Sci.,* LXXIII, 869–910.

KLUCKHOHN, C. 1944. "The Influence of Psychiatry on Anthropology in America during the Past One Hundred Years." In AMERICAN PSYCHIATRIC ASSOCIATION, *One Hundred Years of American Psychiatry,* pp. 589–617. New York: Columbia University Press.

KLUCKHOHN, C. 1949. *Mirror for Man.* New York: Whittlesey House.

———. 1956. "The Impact of Freud on Anthropology," *Bull. New York Acad. Med.,* XXXII, 903–07.

———. 1959. "The Role of Evolutionary Thought in Anthropology." In B. J. MEGGARS (ed.), *Evolution and Anthropology,* pp. 144–57. Washington, D.C.: Anthropological Society of Washington.

KLUCKHOHN, C., and MURRAY, H. A. (eds.). 1953. *Personality in Nature, Society and Culture.* 2d ed. New York: Knopf.

KÖHLER, W. 1938. *The Place of Value in a World of Facts.* New York: Liveright Publishing Co.

KRAGH, U. 1955. *The Actual-genetic Model of Perception-Personality.* Copenhagen: Ejnar Munksgaard.

KROEBER, A. L. (ed.). 1953. *Anthropology Today.* Chicago: University of Chicago Press.

———. 1956. "History of Anthropological Thought." In W. L. THOMAS (ed.), *Current Anthropology,* pp. 293–311. Chicago: University of Chicago Press.

KROPOTKIN, P. A. 1955. *Mutual Aid: A Factor of Evolution.* New York: Penguin Books, 1939; Boston: Extending Horizons Books, 1955. Combined with T. H. HUXLEY'S *Struggle for Existence.*

KRUSE, H. D. (ed.). 1957. *Integrating the Approaches to Mental Disease.* New York: Hoeber-Harper.

LASHLEY, K. S. 1949. *"Persistent Problems in the Evolution of Mind," Quart. Rev. Biol.,* XXIV, 28–42.

LEIGHTON, A. H., *et al.* (eds.). 1957. *Explorations in Social Psychiatry.* New York: Basic Books.

LEMKAU, P. V. 1955. *Mental Hygiene in Public Health.* 2d ed. New York: McGraw-Hill Book Co.

LEVINE, S., and LEWIS, G. W. 1959. "Critical Period for Effects of Infantile Experience on Maturation of Stress Response," *Science,* CXXIX, 42–43.

LEVY, D. M. 1943. *Maternal Overprotection.* New York: Columbia University Press.

———. 1953. "Observational Psychiatry: The Early Development of Independent and Oppositional Behavior." In R. R. GRINKER (ed.), *Mid-Century Psychiatry,* pp. 113–21. Springfield, Ill.: Charles C Thomas.

———. 1958. *Behavioral Analysis.* Springfield, Ill.: Charles C Thomas.

LEWIN, K. 1935. *A Dynamic Theory of Personality.* New York: McGraw-Hill Book Co.

LEWIS, A. 1953. "Points of Research into the Interaction between the Individual and the Culture." In J. M. TANNER (ed.), *Prospects in Psychiatric Research,* pp. 51–55. Oxford: Blackwell Scientific Publications.

———. 1958. "Fertility and Mental Illness," *Eugenics Rev.,* L, 91–106.

LIDDELL, H. S. 1953. "The Biology of Wishes and Worries." In R. R. GRINKER, (ed.), *Mid-Century Psychiatry,* pp. 104–12. Springfield, Ill.: Charles C Thomas.

LINTON, R. 1956. *Culture and Mental Disorders.* Springfield, Ill.: Charles C Thomas.

LORENZ, K. Z. 1958. "The Evolution of Behavior," *Scient. American,* CXCIX, 67–78.

McCULLOCH, W. S. 1951. "Why the Mind Is in the Head." In L. A. JEFFRESS (ed.), *Cerebral Mechanisms in Behavior,* pp. 42–111. New York: John Wiley & Sons.

McKINLEY, G. M. 1956. *Evolution: The Ages and Tomorrow.* New York: Ronald Press Co.

McQUOWN, N. A. (ed.). 1959. *The Natural History of an Interview.* New York: Grune & Stratton.

MACRAE, D. G. 1958. "Darwinism and the Social Sciences." In S. A. BARNETT (ed.), *A Century of Darwin,* pp. 296–312. Cambridge: Harvard University Press.

MARCUSE, H. 1955. *Eros and Civilization.* Boston: Beacon Press.

MEAD, G. H. 1934. *Mind, Self, and Society from the Standpoint of a Social Behaviorist.* Chicago: University of Chicago Press.

MEAD, MARGARET. 1957. "Towards More Vivid Utopias," *Science,* CXXVI, 957–61.

———. 1958. "Cultural Determinants of Behavior." In ANNE ROE and G. G. SIMPSON (eds.), *Behavior and Evolution,* pp. 480–503. New Haven: Yale University Press.

MENNINGER, W. C. 1948. *Psychiatry in a Troubled World.* New York: Macmillan Co.

MEYER, A. 1950–51. *The Collected Papers of Adolf Meyer.* 4 vols. Baltimore: Johns Hopkins Press.

MICHIE, D. 1958. "The Third Stage in Genetics." In S. A. BARNETT (ed.), *A Century of Darwin,* pp. 56–84. Cambridge: Harvard University Press.

MILLER, D. R., and SWANSON, G. E. 1958. *The Changing American Parent.* New York: John Wiley & Sons.

MITTELMAN, B., and WOLFF, H. G. 1942. "Emotions and Gastroduodenal Function: Experimental Studies on Patients with Gastritis, Duodenitis, and Peptic Ulcer," *Psychosom. Med.,* IV, 5–61.

MONTAGU, A. 1955. *The Direction of Human Development: Biological and Social Bases.* New York: Harper & Bros.

MULLER, H. J. 1935. *Out of the Night: A Biologist's View of the Future.* New York: Vanguard Press.

MURPHY, G. 1956. "Toward a Dynamic Trace-Theory," *Bull. Menninger Clin.,* XX, 124–34.

———. 1958. *Human Potentialities.* New York: Basic Books.

NEEL, J. V., and SCHULL, W. J. 1954. *Human Heredity.* Chicago: University of Chicago Press.

NISSEN, H. W. 1958. "Axes of Behavioral Comparisons." In ANNE ROE and G. G. SIMPSON (eds.), *Evolution and Behavior,* pp. 183–205. New Haven: Yale University Press.

NORTHROP, F. S. C. 1958. Introduction. In W. HEISENBERG, *Physics and*

Philosophy: The Revolution in Modern Science, pp. 1–26. New York: Simon & Schuster.

OPLER, M. K. 1956. *Culture, Psychiatry, and Human Values.* Springfield, Ill.: Charles C Thomas.

OPPENHEIMER, J. R. 1953. *Science and the Common Understanding,* pp. 68–82. New York: Simon & Schuster.

———. 1955. "Physics in the Contemporary World." In J. R. OPPENHEIMER, *The Open Mind,* pp. 81–102. New York: Simon & Schuster.

———. 1956. "Analogy in Science," *Amer. Psychologist,* XI, 127–35; also in *Centennial Rev.,* II, 351–73, 1958.

ORWELL, G. 1949. *Nineteen Eighty-four.* New York: Harcourt, Brace & Co.; also "Signet Books," No. 798 (1950).

OSGOOD, C. E., *et al.* 1957. *The Measurement of Meaning.* Urbana: University of Illinois Press.

OSTOW, M. 1957. "The Erotic Instincts—a Contribution to the Study of Instincts," *Internat. Jour. Psycho-Analysis,* XXXVIII, 304–24.

———. 1958. "The Death Instincts—a Contribution to the Study of Instincts," *ibid.,* XXXIX, 5–16.

PEARL, R. 1940. *Introduction to Medical Biometry and Statistics,* pp. 1–19. 3d ed. Philadelphia: W. B. Saunders Co.

PEDERSON-KRAG, GERALDINE. 1955. *Personality Factors in Work and Employment.* New York: Funk & Wagnalls.

PERSONS, S. (ed.). 1956. *Evolutionary Thought in America.* New York: George Braziller.

PIAGET, J. 1957. *Logic and Psychology.* New York: Basic Books.

PINCUS, G. 1959. Announcement printed in *Bull. Amer. Acad. Arts and Sci.,* XII, 1.

POETZL, O. 1917. "Experimentell erregte Traumbilder in ihren Beziehungen zum indirekten Sehen," *Zeitschr. Neurol. Psychiat.,* XXXVII, 278–349.

POLYAK, S. L. 1957. *The Vertebrate Visual System.* Edited by HEINRICH KLÜVER. Chicago: University of Chicago Press.

PRIBRAM, K. 1958. "Comparative Neurology and the Evolution of Behavior." In ANNE ROE and G. G. SIMPSON (eds.), *Behavior and Evolution,* pp. 140–64. New Haven: Yale University Press.

RAZRAN, G. 1959. "Soviet Psychology and Psychophysiology," *Behavior Sci.,* IV, 35–48.

REISER, O. L. 1958. "The Concept of Evolution in Philosophy." In *A Book That Shook the World: Anniversary Essays on Charles Darwin's Origin of Species,* pp. 38–47. Pittsburgh: University of Pittsburgh Press.

RENNIE, T. A. C. 1957. "The Psychosocial Position." In H. D. KRUSE (ed.), *Integrating the Approaches to Mental Disease,* pp. 147–54. New York: Hoeber-Harper.

RIESE, W. 1953. *The Conception of Disease.* New York: Philosophical Library.

RIOCH, D. McK. 1958. "Multidisciplinary Methods in Psychiatric Research," *Amer. Jour. Orthopsychiat.,* XXVIII, 467–82.

ROE, ANNE. 1959. "Man's Forgotten Weapon," *Amer. Psychologist,* XIV, 261–66.

ROE, ANNE, and SIMPSON, G. G. (eds.). 1958. *Behavior and Evolution.* New Haven: Yale University Press. Particularly articles in Part II on "The Physical Basis of Behavior" by BEACH, CASPARI, SPERRY, PRIBRAM, and BULLOCK.

RÓHEIM, G. 1956. "The Individual, the Group, and Mankind," *Psychoanalyt. Quart.,* XXV, 1–10.

ROSE, A. M. (ed.). 1955. *Mental Health and Mental Disorder: A Sociological Approach.* New York: W. W. Norton & Co.

ROSE, G. T. 1959. "Scanning and Screening in an Acute Aggressive Episode," paper read at Philadelphia, May, 1959, American Psychoanalytic Association Meeting.

ROSTAND, J. 1958. "La Biologie et le droit." In *Science fausse and fausses sciences.* Paris: Gallimaid. Quoted from *Time,* June 2, 1958, p. 47.

RUESCH, J., and BATESON, G. 1951. *Communication: The Social Matrix of Psychiatry.* New York: W. W. Norton & Co.

RUSSELL, B. 1951. "The Science To Save Us from Science." In M. GARDNER (ed.), *Great Essays in Science,* pp. 389–402. New York: Pocket Library.

SAPIR, E. 1949. *Selected Writings in Language, Culture, and Personality.* Edited by D. G. MANDELBAUM. Berkeley: University of California Press.

SCHILDER, P. F. 1951. "On the Development of Thoughts." In D. RAPAPORT (ed.), *Organization and Pathology of Thought,* pp. 497–518. New York: Columbia University Press.

SCHILLER, CLAIRE H. (ed.). 1957. *Instinctive Behavior: The Development of a Modern Concept.* New York: International Universities Press.

SCHRÖDINGER, E. 1957. *Science, Theory, and Man.* New York: Dover Publications. Formerly published under the title *Science and the Human Temperament.*

SCHWARTZ, M. S. 1946. "The Economic and Spatial Mobility of Paranoid Schizophrenics and Manic Depressives." Unpublished Master's thesis, University of Chicago.

SCOTT, J. P. 1958*a*. *Animal Behavior.* Chicago: University of Chicago Press.

———. 1958*b*. "Critical Periods in the Development of Social Behavior in Puppies," *Psychosom. Med.,* XX, 42–54.

SEITZ, P. F. D. 1958. "The Maternal Instinct in Animal Subjects," *Psychosom. Med.,* XX, 215–26.

SHACKFORD, R. H. 1958. "Chin Gang Empire," p. 46A–1. *The Pittsburgh Press.*

SILLMAN, L. R. 1949. "Monotheism and the Sense of Reality," *Internat. Jour. Psycho-Analysis,* XXX, 124–32.

———. 1953. "The Genesis of Man," *ibid.,* XXXIV, 146–52.

SIMPSON, G. G. 1949. *The Meaning of Evolution.* New Haven: Yale University Press.

SIMPSON, G. G. 1953. *The Major Features of Evolution.* New York: Columbia University Press.

———. 1958. "Behavior and Evolution." In ANNE ROE and G. G. SIMPSON (eds.), *Behavior and Evolution,* pp. 507–35. New Haven: Yale University Press.

SKINNER, B. F. 1957. *Verbal Behavior.* New York: Appleton-Century-Crofts.

SMITH, J. M. 1958. *The Theory of Evolution.* New York: Penguin Books.

SNYDER, L. H. 1959. "Fifty Years of Medical Genetics," *Science,* CXXIX, 7–13.

SOLOMON, P., *et al.* 1957. "Sensory Deprivation: A Review," *Amer. Jour. Psychiat.,* CXIV, 357–63.

SPECTOR, B. 1959. "Darwin—Down House—Dawn," *New England Jour. Med.,* CCLX, 119–24.

SPERRY, R. W. 1958. "Developmental Basis of Behavior." In ANNE ROE and G. G. SIMPSON (eds.), *Behavior and Evolution,* pp. 128–39. New Haven: Yale University Press.

SPIEGEL, E. A. (ed.). 1950–58. *Progress in Neurology and Psychiatry: An Annual Review,* Vols. V–XIII. New York: Grune & Stratton.

SPITZ, R. A. 1945. "Hospitalism: An Inquiry into the Genesis of Psychiatric Conditions in Early Childhood," *Psychoanalyt. Study of the Child,* I, 53–74.

———. 1946. "Hospitalism: A Follow-Up Report on Investigation Described in Volume I, 1945," *ibid.,* II, 113–17.

———. 1949. "The Role of Ecological Factors in Emotional Development in Infancy," *Child Development,* XX, 145–55.

SPITZ, R. A., and WOLF, KATHERINE, M. 1946*a.* "Anaclitic Depression: An Inquiry into the Genesis of Psychiatric Conditions in Early Childhood," *Psychoanalyt. Study of the Child,* II, 313–42.

———. 1946*b.* "The Smiling Response: A Contribution to the Ontogenesis of Social Relations," *Genet. Psychol. Mono.,* XXXIV, 57–125.

SRB, A. M., and OWEN, R. D. 1955. *General Genetics.* San Francisco: Freeman.

STAINBROOK, E. 1955. "The Hospital as a Therapeutic Community," *Neuropsychiatry,* III, 69–87.

STANTON, A. H., and SCHWARTZ, M. S. 1954. *The Mental Hospital.* New York: Basic Books.

STOCKARD, C. R. 1941. "The Genetic Endocrine Basis of Differences in Form and Behavior," *Amer. Anat. Mem.,* No. 19.

STOUFFER, S. A., *et al.* 1950. *The American Soldier,* Vol. IV: *Measurement and Prediction.* Princeton: Princeton University Press.

SULLIVAN, H. S. 1947. *Conceptions of Modern Psychiatry.* Washington, D.C.: William Alanson White Psychiatric Foundation.

Symposium on Preventive and Social Psychiatry. 1958. Sponsored jointly by the Walter Reed Army Institute of Research, Walter Reed Army Medical Center, and the National Research Council, April 15–17, 1957. Washington, D.C.: Walter Reed Army Institute of Research.

SZÉKELEY, L. 1957. "On the Origin of Man and the Latency Period," *Internat. Jour. Psycho-Analysis,* XXXVIII, 98–104.

THOMAS, W. L. (ed.). 1956. *Current Anthropology.* Chicago: University of Chicago Press.

THORPE, W. H. 1956*a*. *Learning and Instincts in Animals.* London: Methuen.

———. 1956*b*. "Some Implications of the Study of Animal Behavior," *Adv. Sci.,* XIII, 42–55.

TINBERGEN, N. 1951. *The Study of Instinct.* Oxford: Clarendon Press.

U.S. NATIONAL ADVISORY MENTAL HEALTH COUNCIL, COMMUNITY SERVICES COMMITTEE. 1955. *Evaluation in Mental Health: A Review of the Problem of Evaluating Mental Health Activities.* ("U.S. Public Health Service Publications," No. 413.) Washington, D.C.: U.S. Dept., Health, Education, and Welfare, Public Health Service, National Institutes of Health, National Institute of Mental Health.

VICKERS, G. 1958. "What Sets the Goals of Public Health?" *New England Jour. Med.,* CCLVIII, 589–96; also in *Lancet,* I, 599–607.

———. 1959. *The Undirected Society.* Toronto: University of Toronto Press.

VON MERING, O., and KING, S. H. 1957. *Remotivating the Mental Patient.* New York: Russell Sage Foundation.

VON NEUMANN, J. 1958. *The Computer and the Brain.* New Haven: Yale University Press.

WADDINGTON, C. H. 1958. *The Strategy of the Genes.* London: Allen & Unwin.

———. 1959. Review of *Behavior and Evolution,* edited by ANNE ROE and G. G. SIMPSON, *Science,* CXXIX, 203–4.

WATSON, J. B. 1919. *Psychology from the Standpoint of a Behaviorist.* Philadelphia: J. B. Lippincott Co.

WEGROCKI, H. J. 1953. "A Critique of Cultural and Statistical Concepts of Abnormality." In C. KLUCKHOHN and H. A. MURRAY (eds.), *Personality in Nature, Society, and Culture,* pp. 691–701. 2d ed. New York: Knopf.

WHITE, W. A. 1926. *The Meaning of Disease.* Baltimore: Williams & Wilkins.

WHITEHEAD, A. N. 1925. *Science and the Modern World.* New York: Macmillan Co.

———. 1929. *Process and Reality.* Cambridge: Cambridge University Press.

WHITEHORN, J. C. 1956. "Stress and Emotional Health," *Am. Jour. Psychiat.,* CXII, 773–81.

WHYTE, L. L. 1948. *The Next Development in Man.* New York: Henry Holt & Co. Also available in "Mentor Books," No. 50 (1950).

WIENER, N. 1948. *Cybernetics: Or the Control and Communication in the Animal and the Machine.* New York: John Wiley & Sons.

———. 1950. *The Human Use of Human Beings: Cybernetics and Society.* Boston: Houghton Mifflin Co.

WILLIAMS, R. J. 1956. *Biochemical Individuality: The Basis for the Genetotrophic Concept*. New York: John Wiley & Sons.

WILLIAMS, R. J. 1958. "Individuality and Its Significance in Human Life." In F. MORLEY (ed.), *Essays on Individuality,* pp. 125 ff. Philadelphia: University of Pennsylvania Press.

WITTKOWER, E. D., and FRIED, J. 1959. *Newsletter: Transcultural Research in Mental Health Problems, No. 4, June, 1958*. Montreal, Canada: Department of Psychiatry and the Department of Sociology and Anthropology, McGill University.

WOODGER, J. H. 1956. *Physics, Psychology, and Medicine*. Cambridge: Cambridge University Press.

XENOPHON. 1923. *Memorabilia—Oeconomicus,* Sec. 11, par. 48, of *Memorabilia,* p. 35. English translation by E. C. MARCHANT. London: Heinemann.

ZIRKLE, C. 1958. "Some Biological Aspects of Individualism." In F. MORLEY (ed.), *Essays on Individuality,* pp. 37 ff. Philadelphia: University of Pennsylvania Press.

HERMANN J. MULLER

THE GUIDANCE OF HUMAN EVOLUTION

Even though natural selection has been the great guiding principle that has brought us and all other higher organisms to their present estate, every responsible student of evolution knows that natural selection is too opportunistic and shortsighted to be trusted to give an advantageous long-term result for any single group of organisms. Mankind constitutes one of those relatively rare, fabulously lucky lines whose ancestors did happen to win out—else we would not be here—while the incalculably vast majority of species sooner or later vanished—that is, there are no living descendants now. Of all the species existing at any one time, only a relatively few ever function as conveyors of germ plasm that is to continue indefinitely, but most of these few branch and rebranch to more than compensate for the far greater number that are lost. Do we have reasons for believing that our species belongs in that very limited category that is to continue into the geologically distant future?

In examining this question we may first note that man is virtually excluded from ever again splitting into diversified species on this earth, so long as his technological culture remains. For that culture has the effect of shrinking the earth and removing ever more effectively the barriers to migration and interbreeding. Moreover, besides lacking the multiple chances for success which multiple speciation confers, our single species is undergoing, genetically, something analogous to an increase in entropy within itself. For its diverse sublines—hitherto numerous, partly isolated, and to some extent subject to ultimate competition with one another—are increasingly dissipating their separate

HERMANN JOSEPH MULLER, professor in the Department of Zoology, Indiana University, is perhaps best known for his experimental studies of mutations induced by X-rays on the chromosomes of *Drosophila,* work for which he was awarded the Nobel prize in physiology and medicine in 1946.

He began his genetics research at Columbia, studying under Thomas Hunt Morgan, also a Nobel prize winner (1933). Professor Muller later conducted studies at the Moscow Institute of Genetics and at the famous Institute of Animal Genetics in Edinburgh. He is co-author of the text *The Mechanism of Mendalian Heredity* (with T. H. Morgan, A. H. Sturtevant, and C. B. Bridges).

individualities by merging genetic combinations, so that ever less opportunity is afforded for the intraspecies selection among many small groups that has been so potent an evolutionary force. Finally, the remaining intragroup selective processes are becoming subject to modification in their direction of operation through the influence of social processes which, left to themselves, tend to preserve and in some ways even to aid the multiplication of characteristics that are disserviceable to the welfare of the group as a whole—that is, of the species. For all these reasons it seems to follow that the one final remaining line of man will, if he retains or amplifies his technological culture, meet with biological extinction long before the earth grows too hot or too cold to support him.

The question arises here, May not this very culture that man has made effect some further alterations in the working of the principles of selection or add features to them that will, after all, permit man's indefinite survival as a civilized being? The answer seems clear. Cultural interference can bring about the survival of man and his culture only if it makes consummate use of man's most distinctive characteristic, his foresight, so as *consciously* to evade the otherwise inevitable decline.

Genetic Benefits Resulting From Past Cultural Evolution

Before discussing what such purposeful action would imply, we may first acknowledge that cultural factors, operating *without* man's realization of the evolutionary effects they would ultimately produce—that is, without long-range foresight—have in fact exerted major influences on human evolution in the past. And most, although not all, of the changes wrought thereby were of kinds that we would nowadays classify as good.

Prime examples are man's facility in using tools, permitting better manipulation of the environment, and his facility in communicating, mainly through speech. Tools and speech themselves are, of course, cultural developments, improved through many generations of extragenically transmitted experience. The possession of these aids to living, even in their more primitive forms, gave increasing scope for the exercise of the faculties that produced them and thereby strengthened and sharpened the selection that elaborated further the genetic bases of these faculties. Involved here, primarily, were mental abilities and proclivities of diverse kinds, including the very drives to engage in such activities. For example, the invention of numerical terms and methods of measuring afforded more abundant means whereby mathematical aptitudes could be utilized advantageously and thus enabled selection to be more effective in developing the biological basis of these aptitudes; the latter process, in turn, paved the way for still more cultural advance in these directions. Thus cultural and biological evolution were mutually reinforcing in their effects on intellectual development.

Similar considerations apply to other psychological aspects of culture that were made practicable and advantageous as a result of cultural advance, such as the wider gregariousness and the impulses to comradeliness and co-operation within larger groups than the family. Susceptibility to group experiences of the type called "religious" is a part of this genetic pattern that has developed in response to culture, one of the major advantages in this case being the group solidarity that it fosters. At the same time, the cultural development of hunting and its derivative activity, warfare, served as the foundation for the intensified selection of predispositions to combativeness, xenophobia, and related impulses, which made intergroup antagonism an active complement to intragroup cohesion. And so, despite the radically different methods that underlie the cultural process of acceptance of communicated lessons derived from experience and the genetic process of multiplication of genes arising by mutation, one can no longer disentangle the influence of these two interwoven sets of factors in changing both human culture and the human genotype.

When we turn to visible morphological features, the story is in principle the same. Among the human morphological traits that have probably developed in considerable measure as a result of selective conditions that were promoted by cultural features, we must, of course, set in the first place the enlargement of the brain and features that were directly subsidiary to increased cranial size, such as the widening of the pelvis. Another obvious morphological development that was promoted by cultural advance was the recession of the jaws and the converse relative prominence of the chin and perhaps also of the nose. For these changes were surely selective consequences of cooking, cutting, scraping, clubbing, and stabbing, by the use of the fire, knives, scrapers, weapons, etc., as substitutes (by means of culture) for the operations of biting and mangling. Likewise, the cultural development of coverings for the body made it advantageous for the body hair to be genetically reduced, since temperature regulation became more refined thereby, external parasites could be brought under better control, and cleanliness in general was enhanced.

Specific features of given cultures have also led to distinctive bodily changes in given subgroups—such as those in adaptation to climates where men were enabled to live only after cultures suitable for these climates were developed. A well-known case in point is that of the narrower eyes, shorter intestines, and other adjustments to arctic and almost purely carnivorous ways of life in groups which have long complied with such conditions. Another obvious case is that of the light pigmentation of groups which, by reason of the climate they live in and types of clothing, shelter, and food they have adopted, receive a minimum of ultraviolet on their skin and at the same time relatively little vitamin D in their diet. Moreover, long-term cultural emphasis on given features of bodily form, such as steatopygy among the Hottentots and Bushmen of South Africa or a retreating forehead among some Amerindians, has evidently favored a reproductive selection that intensified the genetic factors for these traits. To what extent long-continued fashions in behavior limited to given groups have correspondingly exerted effective selection for given psychological propensities is a question to which only speculative answers can at present be given. But there can be little doubt that such a widespread practice, common to many cultures, as living by routine labor must have resulted in some selection, even though largely unconscious, of the proclivities conducive to it.

Genetic Inadequacies of Cultural Evolution

The examples of physical deformities (as we would think of them) that have been consciously selected within certain groups show that the influence of culture on selection has not always been in advantageous and wholesome directions. Numerous are the instances in which cultures have gone off on tangents by inertias of their own, as in the squandering of people's efforts on useless work among ancient Egyptians and Sumerians and modern Tibetans, the sacrificial excesses of the Aztecs, and (though in lesser degree) the puritanical excesses of the forefathers of some of us. In each such case there must have been repercussions on the workings of genetic selection, and it is evident that the resulting influences were likely to be detrimental to the advancement of some genetic traits most desirable for human welfare. In early times such tendencies must have been held in check to a considerable extent by intergroup selection. The groups were small and numerous, and, other things being equal, those with harmful cultural developments must have tended to dwindle in competition with the sounder ones. With the merging of many small into few large groups, intergroup selection, whether of culture or of genes, loses its force, and, with the formation of the all-inclusive, virtually panmictic society, it practically ceases to exist. Thereafter there remains no mechanism for the rectification of harmful practices except the conscious recognition of them for what they are and resultant socially adopted reform based on that recognition.

Certainly, as Darwin pointed out, the family, long before the arrival of man, must have afforded the primary unit for intergroup selection whereby the genetic basis of altruistic proclivities became developed. Then, as the groups came to include a number of families, they were still small enough and numerous enough to allow effective selection for the traits that predisposed their members for the wider co-operation and altruism here in order. Even if the individual sacrificed himself for the small group, he tended to foster the multiplication, through the others of that group, of genes like those that had predisposed him to this behavior. With the formation of towns and large civil units in general, this kind of influence on selection tended to disappear.

It is doubtless true that, even today, co-operation *within* the family results in some positive selection in favor of genes conducive to intrafamilial aid. In the setting of large-scale communities, the operation of these same genes can to some extent be adapted through cultural practices for the purposes of mutual aid in these larger communities. The very success of such adaptation, however, spells a corresponding decline in the selection. Moreover, so far as those proclivities are concerned that would tend to broaden the basis of the co-operation, by making a man really feel toward men in general as toward his brothers, there is no longer an automatic mechanism for enhancing their genetic basis. At the same time the characteristically human tendency to feel antagonistic toward those outside one's "circle" keeps seeking an outlet and is not effectively selected against. This situation has left men very imperfectly constructed to live by the utopian precept "Love one another." Yet any other basis of behavior for a form of living that depends on our modern global technologies must result in even greater disharmonies, whereby these technologies themselves become turned against their users.

We are now at a juncture where we must move forward or backward. No matter what the steps may be—whether gradually constructive or catastrophic—by which the major group divisions of our present transitional political setup become superseded by an effective world-embracing organization, it is evident that there is no longer any ultimate alternative to such union except the retrogression of modern culture—that is, the virtual abandonment of the benefits of science. The latter contingency would be one that few of those aware of what it would involve would willingly choose if they could make the decision. In effect, it would mean the failure of civilization and the frustration of the rationality which, limited though it is, has been the most distinctive and creative characteristic of humanity. Assuming, however, that men do win out in their efforts to achieve a universal community, this question must be faced: What effects will the universalization and enhancement of modern culture have on the genetic constitution of man and thereby on his conditions of life?

So far as the adaptive differentiations of present human geographical groups are concerned, it is evident that their inevitable swamping-out, through merger aided by relaxed selection, should give no grounds for serious concern. For the lack of these specialized features can readily be compensated for, in the given situations, by such artificial contrivances as specially constructed clothing, shelters, means of transport, heating, cooling, special food supplies, optical equipment, medicaments, insecticides, germicides, vaccines, and so forth. Thus, far greater stretches become open to comparatively dense habitation than formerly, without the requirement of distinctive genetic endowment.

One sometimes hears the assertion that similar considerations apply to all inherited variations in man: that is, with the progress of technology and more especially of medical, general biological, and biochemical methods, it will become increasingly possible to compensate for all kinds of genetic abnormalities and deviations and thus to raise each individual virtually to the optimum level in all respects. From this self-styled "progressive" view, eugenics is an old-fashioned and reactionary notion destined to be discarded. In the official Communist version of this view, the improvements and corrections brought about in the individual's makeup by manipulation of his physical and psychological environment, including careful training, become automatically incorporated into his inheritance so that men will become better and better equipped genetically with each generation as a result of their improved environment. It would be out of place for us here to state the overwhelming case against this naïve Lamarckian doctrine, so long discredited by the fundamental principles established through decades of genetic research. In the West it is commoner for those with strong environmentalist leanings no longer to espouse Lamarckism but simply to maintain that human hereditary defects, perhaps with relatively rare exceptions, will be rendered unimportant phenotypically by means of the consummate skills of future technology, even though they will probably tend to persist as gene differences. In addition, some who wish to appear up-to-date are bold enough to declare that even defective genes may in time be repaired or replaced by ultrafine substitutions of their nucleotide constituents.

Of course, it is indisputable that, as man's control over matter advances, more and more of his bodily structure and functioning can be amended and even replaced by artificial means. Thus, even as primitive man found his body hair largely dispensable and replaced it by coverings, future man will require less and less heat regulation, antibody production, natural hormone generation, digestive juice secretion, and so forth. He can use scooters for locomotion and computers for calculation. And he can finally do without himself!

Primitive man could replace his natural faculties by artifices only in certain limited aspects of his living, as in protection against weather, wild animals, and certain vicissitudes in food supply. The techniques needed for these purposes were not too burdensome, and they actually gave him much better means of meeting the given requirements than did his inherited specializations. Hence it was these biological endowments rather than his cultural devices which ultimately became too burdensome to be maintained. Certainly in the case of modern and future man this situation will apply in many more areas of living. But to acknowledge this fact is not the same as to say that there would, in general, be a net gain in substituting for our genetic endowment everything that would now or in the future be devised to take its place. Where should we draw the line?

A given natural endowment is better lost than retained if all the following three conditions hold. First, the artificial substitute should be more effective and dependable than the natural endowment. Second, the net burden to the community involved in maintenance and operation should, for a given return, be less for the artificial substitute than for the natural endowment. Third, the maintenance of the natural endowment as a supplement to the man-made contrivance should be more trouble than it is worth before its lapse can be considered justifiable.

Today, of course, no attempt is made to assess these balances when procedures are instituted which, in helping individuals, may contribute to the relaxation of selection in given directions. It is regarded as ethical to employ every available artificial aid to enable an individual to reproduce or to enable him to live and thereby reproduce, even when his reproduction would be likely to perpetuate the genetic condition that had occasioned the given difficulty. So far as the immediately treated generation is concerned, mutual aid of this kind is unquestionably a social obligation. Its over-all cost is very small in comparison with its benefits to the community in well-being, general efficiency, harmonious interrelations, sense of security, and enjoyment of life. The real issue is not whether society should in this way help the individuals themselves to live better, as if that were where the matter stopped. It is whether the acts of society should be so ordered as actually to facilitate the perpetuation of defective genetic equipment into later generations. Should we give with one hand while taking away with the other?

Let us be clear about the genetic processes here involved. Like a gas that tends to expand in all directions as a result of the random movements of its molecules, the genetic material is continually undergoing mutations in innumerable directions in a chaotic manner, and the vast majority of these mutations are, of course, detrimental in their effects on the organism, tending to disorganize and de-adapt it, both in regard to the general fitness of the organism as a whole for living and reproducing and in regard to the development, structure, and functioning of any given part or characteristic of the organism. It is only selection which, by holding in check the multiplication of the deleterious mutants after they have arisen and, conversely, by promoting the multiplication of the very rare superior types, allows the mutation pressure to result in increased fitness of the organism for the given conditions, even as the walls of a chamber containing a gas allow the pressure of the gas to shape it to fit its container. Relax the selection in any direction, and mutation pressure will cause disorganization and de-differentiation in that direction up to whatever limit is set, much as any recession or loosening of one or more walls of the chamber is followed by a corresponding diffusion of the contained gas.

Results of the Continuation of Present Practices

On the average, the counterpressure of selection, consisting in the elimination of individuals with excess detrimental genes, almost exactly equals the pressure of mutation in producing these genes. There is evidence from more than one direction that, in man, at least one person in five, or 20 per cent, carries a detrimental gene which arose in the immediately preceding generation and that, therefore, this same proportion—one in five—is, typically, prevented by genetic defects from surviving to maturity or (if surviving) from reproducing. This equilibrium holds only when a population is living under conditions that have long prevailed. Modern techniques are so efficacious that, used to the full, they might today (as judged by recent statistics on deaths and births) be able to save for life and for a virtually normal rate of reproduction some nine-tenths of the otherwise genetically doomed 20 per cent. Assuming this to be the case, there would in the next generation be 18 per cent who carried along those defects that would have failed to be transmitted in the primitive or equilibrium population, plus another 20 per cent (partly overlapping the 18 per cent) who had the most recently arisen defects. At this rate, if the effectiveness of the techniques did not diminish as their job grew, there would, after about eight generations, or 240 years, be an accumulation of about 100 "genetic deaths" (scattered over many future generations) per 100 persons then living, in addition to the regular "load of mutations" that any population would ordinarily carry. It can be estimated (on the supposition that human mutation rates are like those in mice) that this amount of increase in the load is about the same as would be brought about by an acute exposure of all the parents of one generation to 200 r of gamma radiation, a situation similar to that at Hiroshima, or by a chronic, low-dose-rate exposure of each of the eight generations to 100 r.

This result sounds worse than it is because most of the mutant genes would cause only a slight amount of damage in their usual heterozygous condition. Since the regular load of perhaps some scores of significantly detrimental genes per individual—nearly all in heterozygous condition but each a potential cause of some far-future genetic death—gives rise to a total risk of genetic extinction, as measured under primitive conditions, of something like (but probably exceeding) 20 per cent for any *given* individual, it is evident that the addition of just one more detrimental gene to his load would usually increase his own risk of extinction by only some tenths of 1 per cent. Thus the process of genetic decline would be exceedingly slow, thanks to the innumerable factors of safety that are built into our systems. The decline, in fact, would for a very long time consist mainly in a reduction in these factors of safety.

Let us next suppose that this sparing of genetic deaths by the aid of technology were to continue indefinitely at the assumed rate, a rate at which a genetic defect, on the average, subjects a person to only a tenth as much risk as it would if he were living under primitive conditions. Eventually, after some tens of thousands of years, a new equilibrium would be reached at which the load of mutations would be about ten times as large as at present. Thus as many extinctions as mutations would again be occurring. If we are to keep to our previously chosen figure for mutations, there would be one extinction for every five individuals, or 20 per cent. The frequency of genetic deaths would therewith return to the level which it had in primitive times and would be far above that now prevailing, in spite of all technological efforts. At the same time, the average individual of that time, carrying ten times today's genetic load, would, if tested under primitive conditions, be found to be no longer subject to a risk of extinction of only 20 per cent, but to one of 200 per cent. This means that he would carry twice as much defect as would suffice to eliminate him. Man would thereby have become entirely dependent on the techniques of his higher civilization. Yet, even with these techniques, he would be subject to as high an incidence of genetic misfortunes as had afflicted him in primitive times. That is, his weaknesses would have caught up with him.

It will rightly be objected here that we have assumed in this calculation that technical skills throughout an indefinite period remained only at their present or immediately impending level, whereas, surely, if civilization advances as we hope, medical and other skills will become capable of reducing ever further the damage done by the average genetic defect. This would cause a perpetual pushing-back of the time of attainment of equilibrium. However, as in the previous illustration, each step in saving more lives would be followed by a corresponding increase in the load eventually carried. So, for example, a saving of all but one in a hundred that would originally have been lost would lead upward toward a hundred fold increase in the load, to be attained only after some hundreds of thousands of years. At that point and for a long time before that, each individual would be endowed by nature with a unique assortment of many hundreds of cryptic, as well as conspicuous, inherent defects and would therefore constitute a special case. He would have to be given a superlatively well-chosen combination of treatments, training, and artificial substitutes just to get by. The job of ministering to infirmities would come to consume all the energy that society could muster for it, leaving no surplus for general cultural purposes. Yet, even at that stage, *most* of man's genes would still be fairly normal, so new mutations, always arising at the rate of some 20 per cent in every generation, would plague him even further.

Long before such an "advanced" stage of the genetic cul-de-sac was reached, however, this medical utopia would probably be subjected to such great strains as to throw men back toward more primitive ways of life. Many would find themselves incapable of such ways. To be sure, the difficulty then would in a sense be "self-rectifying." But so late and forced a rectification would be likely to cause the loss of much that had previously been gained.

Only on paper could techniques be advanced indefinitely to avoid ever reaching the equilibrium at which the incidence of genetic deaths equals that of new mutations. An indefinite continuance of the process would, on paper, require it to proceed until virtually every one of the ten thousand or more genes in every chromosome set had been badly damaged or incapacitated. Judging by present data on mutation rates, this would happen within a period of a few million years, provided that medical men during this period had been able to work with the kind of perfection they desire. What would this situation imply? It would imply that the then existing germ cells of what were once human beings would be a lot of hopeless, utterly diverse genetic monstrosities.

Surely the refashioning of these pitiful relics into human form would be a far more difficult task than the synthesis of human beings out of raw materials selected for the purpose. It would also be more difficult than the construction of robots. The marvelous techniques that such an age would possess would hardly be used for so preposterous and fatuous a purpose as the preservation of a genetic continuity that had lost its meaning.

Returning from these fantasies, we see that it is absurd to assume that environmentalist techniques alone, dealing purely with the phenotype, can in the long run keep ahead of mutations so as indefinitely to enhance or even preserve human well-being. The assumption bears a close analogy to that made by anti-Malthusians when they suppose that means can be found of continuing to increase the earth's supply of food, goods, and living room to accommodate a population that keeps on growing at its present rate. That rate, some 1.75 per cent per year, would amount to a thousand fold increase every four hundred years. The accumulation of mutation is an inordinately slower process than the expansion of population. Yet in either case the process will inevitably come to a halt too late, as a result of the misery and disorganization which it brings about, unless it is forestalled through long-range foresight that exercises a conscious control over reproduction. In the case of the population problem, that control need be exercised only over the total quantity of reproduction, whereas, to meet the mutation problem, the control must be exercised in a qualitative way—a much more difficult and still more important matter.

A favorite cliché with those who do not understand this situation is the statement that, by definition, natural selection must always be acting and must always be favoring the fitter. This statement overlooks the fact that the degree of genetically occasioned difference in reproductive rate—that is, the intensity of selection—can be far less in some situations than in others. But the major point disregarded here is that what is fitter in the immediate acts of life is not always fitter for a group or a species as a whole in the long run. In such a case the group is running a race toward debasement and sometimes toward extinction, in this respect following the great majority of species of the past. In the case of man, the trick factor in this connection is a very unusual one: culture. Although culture did serve to sharpen salutary types of human selection in the past, as we have seen, it has now reached a point at which its very efficiency, when not yet involving foresight in regard to genetics, has placed upon society the burden of supporting almost indiscriminately the ever increasing genetic failings of its members.

If, in accordance with the above cliché, we define fitness in the narrow (but erroneous) sense, by the criterion of leaving a larger number of immediate offspring, then, of course, later generations of man must, by definition, by increasingly fit. Yet this type of fitness is no longer the same as fitness in regard to the qualities conducive to the well-being and survival of mankind in general. In fact, it seems not unlikely that in regard to the human faculties of the highest group importance—such as those needed for integrated understanding, foresight, scrupulousness, humility, regard for others, and self-sacrifice—cultural conditions today may be conducive to an actually lower rate of reproduction on the part of their possessors than of those with the opposite attributes. Is it not too often true that today, when birth control is available, those persons most lacking in perspective or dominated by superstitions taboos or unduly egotistical or unmindful of others' needs or shiftless or bungling in techniques are the very ones with the largest retinue of children, whether legitimate or otherwise? These considerations suggest the possibility that a much faster-acting and more serious cause of genetic deterioration than the previously discussed accumulation of detrimental mutations occurring in the wake of relaxed selection is an actual reversal of selection in regard to those psychological traits that are of the highest social importance. Objective data are badly needed on this question.

The relaxation of selection that I have considered, as well as its postulated reversal, have been made possible by the fact that, under modern culture, the damage done by social as well as other defects of individuals is increasingly borne by society as a whole. This is *in itself* a great step forward. However, its potential consequences for the biological evolution of man have not been considered seriously enough, even though Darwin himself called attention to them.

We need not set forth in detail here the mistakes made by many of those who in the past have taken this problem seriously: their great overestimation of the speed of the genetic processes in question, their even greater underestimation of the efficacy of cultural influences in the shaping of men's minds, and the notorious support that some of them gave to the vicious doctrines of racism. Along with the highly essential repudiation of these mistakes, there has been a tendency to go to the opposite extreme of throwing out also the hard core that is really valid. It is so easy to shrug off effects that take thousands of years to become evident. And, as in the case of erosion, they are by that time accepted as a part of the order of nature.

Statesmen, economists, social scientists, and men of affairs are seldom interested in such remote matters. Yet, as evolutionists, we know that the greatest and most creative, as well as the most destructive, operations of the living world have been of this creeping, secular character. Unless we are willing to remain, when viewed in larger perspective, helpless creatures of circumstances, we must take these insidious operations into account, master their principles, and devise ways of dealing with them.

The Protection of Our Genetic Heritage

The crux of the problem is the interference with salutary types of selection in man that has arisen incidentally as a by-product of the widespread and increased effectiveness of mutual aid when it utilizes the tools supplied by science. What means can be used to protect our genetic heritage from this paradoxical situation? Occasional reactionary voices are to be heard calling upon us to reduce our mutual aid in the name of "rugged individualism," "private enterprise," or the like, and others are asking for a moratorium on science and even for a return to a fancied golden age.

However, it has been exactly the combination of intelligence with co-operative behavior that has made culture possible and raised men above beasts, and these propensities brook no stopping point. The enormous advances opening to men in consequence of the further extension of science (representing intelligence) and of a world-wide social organization (representing mutual aid) so utterly overshadow, in their potential effects within the next few hundred years, the damage that may be done in that period to men's genetic constitution that none but the unbalanced would consider now giving up, for genetic reasons, the march of civilization.

It must unfortunately be conceded, however, that even tomorrow suddenly, or within the century ahead of us gradually, a return to more primitive conditions might take place. It would not represent the voluntary choice of the majority of the world's inhabitants but the consequence of miscalculation or faulty organization. And if it did occur, men would endeavor to work their way up to a stage even more rational and co-operative than our present one. If they attained it, they would once more find themselves confronted with our genetic dilemma.

Thus far, men have usually been able to find an answer to the difficulties into which the application of a little science has gotten them, and that answer has been found in better science. Similarly, they have extricated themselves from difficulties occasioned by imperfect mutual aid through the application of improved mutual aid. Both factors are involved in the present case, and the solution requires a suitable combination of the two.

The main role of science in this matter is the discovery of the situation, but it will be important to get much more information concerning its details, both qualitative and quantitative, and concerning the kinds of strategies that would be effective in meeting it. However, the choice and implementation of these strategies is largely bound up with men's attitudes in regard to mutual aid.

Although the mores of our society approve the extension of society's aid to individuals for the purpose of saving their lives and thereby enabling them to reproduce, they do not yet, reciprocally, recognize a duty on the part of individuals to exercise their reproductive functions with due regard to the benefit or injury thereby done to society. So long as illegitimacy is avoided, the individual is not considered to be under any genetic obligation but deems it his right to have as many or as few children as he personally wishes. This being the case, his choice in the matter is largely determined by irrational factors and by shortsighted aims. Such practices worked out well enough genetically only so long as, in matters of survival, the families or the small groups were in large measure on their own.

What is most needed in this area of living is an extension of the feeling of social responsibility to the field of reproduction: an increasing recognition that the chief objective in bringing children into the world is not the glorification of the parents or ancestors by the mere act of having children but the well-being of the children themselves and, through them, of subsequent generations in general. When people come to realize that in some measure their gifts, as well as their failings and difficulties—physical, intellectual, and temperamental—have genetic bases and that social approval or disapproval will be accorded them if they take these matters into account in deciding how much of a family to beget, a big step forward will have been taken in the motivation of human reproduction.

It can become an accepted and valued practice to seek advice, though not dictation, in these matters, even as it is today in matters of individual health. Although no one enjoys admitting his faults, he can learn to take pride in exercising humility and ordering the most important of his biological functions—reproduction—in such ways as to win the approbation of himself and his fellows. This is, to be sure, a higher type of mutual aid, a superior moral code, than exists at present, but it can be just around the corner for people who from early youth have had the facts of genetics and evolution made vivid to them and who have been imbued with a strong sense of their participation in the attainment of human well-being.

Those for whom it seems wiser, in the interests of the coming generation, not to play as active a part as they might in producing people plagued by the very shortcomings that they have seen bring trouble to themselves can still find plenty of meritorious and satisfying work to give them a sense of fulfilment. Nearly everyone is above average at something and should be given opportunity to exercise his aptitudes in ways that do him credit and aid society. He may be a skilled teacher or custodian of the young, although, because of some unfortunate defect or combination of weaknesses, he is less suitable than the average for personal reproduction. If instructed from the beginning regarding the nature of the human effort and the interdependence of man, he can take this realization in his stride and devote himself to kinds of activity in which he can take pride.

That is not to say, of course, that there can be hard-and-fast rules and that any one or a few known genetic defects should enjoin a person from reproduction. Everyone has many minor and some more serious inherent imperfections. In conscious decisions, as in the process of primitive natural selection, it is the total balance of these that should count, and the answer need seldom be an all-or-none one.

Contrariwise, for those clearly better endowed, this kind of social motivation should lead them to place special importance on creating children, even though in such cases especially there will be great temptation to expend disproportionately much effort in other directions. It is true that economic and other aids could be of value here. However, the underlying mores will inevitably exert the more powerful influence. The main job, if the situation is to be rectified, consists in laying the foundations for these mores.

This does not mean that we can expect ever to specify just what would be the optimal number of children for any given individual to have, in view of his genetic constitution. Natural selection would never have succeeded in the past if such precision had been necessary. Environmental influences always complicate, sometimes inextricably, the determination of developed traits. It is enough, from a long-range point of view, if only the trend is in a salutary direction. And that is the result that must be sought in the attempt to protect man's most invaluable possession—his own genetic material.

Two developments of our present period are powerful positive influences toward the needed change in motivation. One is the sudden realization of the damaging effects of radiation on heredity. This, by reason of having been made a political football, has done more to arouse the public and its leaders to the fact that our genetic constitution requires protection than all the propaganda that eugenicists have ever put forth. Characteristically, the danger has been greatly exaggerated in some quarters, for ulterior purposes quite unconnected rationally with the matter at issue, and has been just as unjustifiably dismissed or played down in other quarters, where there were other axes to grind. Nevertheless, the over-all effect of the controversy has been highly educational and has helped to make people far more genetics-conscious than they ever were before. It so happens that this same radiation problem is one of the *proper* faces of the ax which is here being ground. Thus it is fitting to take advantage of the receptivity created by political circumstances to awaken the public to the more general need for a reformation of attitudes toward reproduction.

The other relevant development of our time is the menace of overpopulation. Even publicists are at last becoming alarmed at the smothering of cultural advance and the disaster to democratic institutions that it can bring about in a generation or two if unchecked. An absolute check will require not only that birth-control techniques be made available but also that large masses of people execute an about-face in their attitudes toward having children. They must recognize that to have or not to have children, and how many, should be determined primarily by the interests of the children themselves—that is, of the next and subsequent generations. If this change in outlook is effected—as it must be sooner or later—it is a relatively short step to the realization that the inborn equipment of the children also counts mightily in their well-being and opportunity for happiness.

The Genetic Offensive

Thus far I have emphasized conserving the genetic goods we have. As in most defensive operations, it is a dreary, frustrating business to have to keep racing merely to stay put. Nature did better for us. Why can we not do better for ourselves? For both psychological and material reasons, the best defensive in this, as in other matters, is the offensive.

A man finds little incentive to take steps to ward off a hidden danger that is a thousand or ten thousand years away. But when there is a definite possibility of tangible improvement in the conditions of life within a period such that he or his wards can directly experience it, then a man's efforts can be enlisted far more effectively.

Why should we or anyone else who has become aware of the marvelous advances that have been made in biological evolution consider it sufficient merely to keep things as good as they are if we can do much more than that? Certainly, the majority of mankind have come, within a few generations, to set their objectives much higher in the case of cultural evolution, now that they have learned how astonishingly amenable it can be to their own contrivings. They are also making considerable progress in the genetic reshaping of organisms of service to them. It seems almost inevitable that, if civilization avoids its present opposite pitfalls of mutual extermination and overpopulation, men will in the not too distant future want to utilize also, for their own benefit, the vast, though more unwieldy, possibilities of their own biological advancement. But, to make this advance, false gods will have to fall along the way.

It is sometimes objected that our cultural evolution has superseded our biological evolution. Nothing in either of these processes is inherently exclusive of the other, even though, as we have seen, culture is now developing in a way that does tend to run counter to biological progress. It is also asserted that biological progress is no longer necessary, since so much more rapid, radical, and diverse improvements can be effected by cultural means—not only with physical and chemical techniques and educational and sociological methods but ultimately too, no doubt, in ways of modifying embryological, including neurological, patterns.

The vast potentialities of cultural advance (subject to the mutational limitations previously discussed) are not to be denied. On the contrary, they need to be emphasized even more. But we should not think of phenotypic and genotypic operations as rivals. They can do best when they proceed in the same direction, even as they did in the formative days of our species. It is culture that chiefly distinguishes the scientist from the witch doctor, but it is genes that distinguish man from protozoön or virus and make his mastery of culture possible. In fact, the genetic and the cultural advances cannot even be considered additive; they are related more like the factors in a product formed by multiplying one by the other. Thus a small improvement in the genes may, in effect, work out as an enormous advance when there is already a high type of culture. Conversely, when acting with a high genetic endowment, a relatively small cultural advance can attain far more significance than otherwise.

The same change in attitude toward reproduction that is needed to insure the preservation of our genetic heritage is also the necessary basis for its improvement. If once it is accepted that the function of reproduction is to produce children who are as happy, healthy, and capable as possible, then it will be only natural for people to wish each new generation to represent a genetic advance, if possible, over the preceding one rather than just a holding of the line. And they will become impatient at confining themselves to old-fashioned methods if more promising ones for attaining this end are available. As the individualistic outlook regarding procreation fades, more efficacious means of working toward this goal will recommend themselves. In time, children with genetic difficulties may even come to be resentful toward parents who had not used measures calculated to give them a better heritage. Influenced in advance by this anticipation and also by the desire for community approval in general, even the less idealistic of the parental generation will tend increasingly to follow the genetic practices most likely to result in highly endowed children.

But before discussing these questions of biological means, it is important to consider ends—or, rather, objectives. We may start by a brief dismissal of the contention sometimes raised that one cannot recognize merits higher than one's own or lift one's self by one's bootstraps. If this were true, there would be no use in self-criticism, and persons of exceptionally high ability would never be identified except by each other. However, given the thesis that men can come to realize in some measure their own limitations and can conceive of beings superior to themselves, this question must be faced: By what criteria should they decide what changes would be desirable? For without some consensus on this ultimate question of values, men's efforts in biological—and, for that matter, also in cultural—directions can only be at cross-purposes.

Values

In the past, questions of values were passed upon for men, first, as a result of the natural selection of mutations that gave them their native predilections and the capability of having these modified by association; second, as a result of the natural selection of those developing cultures that led to the most dependable survival and multiplication. As we have seen, the combination of these processes resulted in a great increase in men's proclivities for communication and other types of co-operation and in the intelligence that made the co-operation more effective. At the same time, cultures were selected in such wise as to stress, although with very varied forms of expression, men's sense of the desirability of these functions and the attitudes lying behind them. We still have these feelings today and have a better realization than ever of their primacy for us subjectively and of their objective utility in survival. At the same time, of course, we also value many subsidiary motivations, such as those stemming from zest in the meeting of challenges, joy in accomplishment, the urge to solve problems, the desire for approval, sexual and familial love, the appreciation of different kinds of beauty, and so forth. All these overlapping propensities, as well as others, contribute to the success of the rationally guided co-operative activities that promote the survival and expansion of the group, a group that is coming ever more to comprise the species as a whole.

Despite the carpings and quibblings of some philosophers, the most generalized rational formulation of human aims that most persons concerned with the subject can agree upon is the promotion of the greatest over-all happiness. We need not define happiness more precisely here than as the sense of fulfilment derived from the attainment, or from approaching the attainment, of whatever is deeply desired. (Any one of a number of other terms might, of course, be substituted for "happiness" here, provided that it be defined in this way.) Granted that in given cases one man's meat can be another's poison and that some find fulfilment in actually giving pain to others, most men recognize that, from a longer-range standpoint, these interpersonal disharmonies are undesirable, in that they allow less over-all happiness. In the same class would come a one-sided attachment to subsidiary aims—such as the satisfaction of pride in the building of pyramids or in the unlimited accumulation of luxuries—that reduces the capacity to contribute to the over-all welfare and thereby hinders the survival, expansion, and long-term happiness of the group.

The will to self-development—hedonism, the urge to achieve—functionalism, the ideal of service—altruism, and a spiritual attitude toward existence—consecration, all these modes of approach to living, when followed up logically, become finally resolved into the pursuit of the same objective. The reason that self-development can be included among these orientations is that man is a naturally social animal in whose individual personality a major role is played by his regard for others. The final objective, likewise, may be thought of and designated in different ways, such as human happiness, richness of life, welfare, increasing survival, or advancement, since these are all diverse aspects of one great combination that in practice remains inseparable.

All this by no means implies that men ordinarily carry the remote, abstract-seeming goal of the common good in sight in their day-by-day activities. It is human to have very varied interests and desires—in fact, to crave variation, curiosity being one form of this craving—and to take the satisfaction of each kind of desire as an end in itself. This exceptional versatility in strivings, complemented by diverse capabilities, has been a highly important factor in man's success, for it has fostered development along many lines, and so varied opportunities could be taken advantage of and adapted to. More than any other creature, man's genius is generalized and his intelligence and potentialities protean. This dispersal of his efforts works out in the end to enhance his control over his own affairs and over his environment, in both his individual interest and that of his species.

Naturally, however, there are many points at which the pursuit of some personal desire as an end seems to lead (or actually does lead) to a different course of action from that which would be followed by directly navigating according to the distant North Star of over-all human happiness. In such a case, we are likely to say that there are conflicts between individual and social aims. But when we take into account the way a man is built, we see that he can render better service to his group if he is well-rounded and not plagued too much by deprivations and repressions. Moreover, a man's intelligence and his devotion to more distant aims are never sufficient, by themselves, to lead him into all the byways that he will find by trial to be fruitful. He must therefore compromise between aims and activities that seem more purely personal and those that are more obviously social and thus attain a harmonious balance among them. Yet, provided that his personality is intelligently integrated and strongly social, his over-all gratification will be especially deep if, in general, his life follows a course that appears to him to be in line with the highest welfare of mankind. Then, even if circumstances prevent the objective success of his endeavors, he may nevertheless, by this touchstone, attain a sense of personal achievement.

As this discussion implies, the balance reached by different people can be very different, and some may not acknowledge, at least to themselves, any social obligations at all. The form and level of the balance will depend not only on the very complex genetic basis provided by the given individual but also on his particular training and experiences and the culturally based attitudes that he has derived from his social group. Moreover, the groups themselves still differ widely in their degrees and kinds of socialization.

This question must then be raised: What genetic and cultural backgrounds are conducive to the highest success, as judged by the ulterior criterion of their promotion of over-all happiness? First, it is evident that the distribution of relative strengths of different drives that was most appropriate to the success of people when they were divided into many small, nearly autonomous groups is far from that most suitable for men organized into a vast society engaged in scientific and technological advances, mechanized production, transportation and communication, predominantly common interests, and rational, democratically guided decisions. Surely in this society most of us could do better if, by nature as well as by training, we had less tendency to quick anger, blinding fear, strong jealousy, and self-deceiving egotism. At the same time we need a strengthening and extension of the tendencies toward kindliness, affection, and fellow feeling in general, especially toward those personally far removed from us. These impulses should become sufficiently dynamic to issue in helpful action. As regards other affective traits, there is much room for broadening and deepening our capacity to appreciate both natural and man-made constructions, to interpret with fuller empathy the expressions of others, to create ever richer combinations of our own impressions, and to communicate them to others more adequately.

Another direction in which an advance is needed is in those traits of character that lead to independence of judgment and its necessary complement, intellectual honesty. We need to strengthen the drive to see things through to as near the bottom as possible and also the drive to co-ordinate the elements rationally. Just as important are the will and ability to take fair criticism with good grace and, further, to search and criticize ourselves until we recognize and discard, if need be publicly, judgments based on wishful or faulty thinking or on defective data. Of course, a great deal of all this may be taught, but there seem also to be great inborn differences in the facility and degree with which such emotion-fraught mental operations are learned and in the strength of feeling behind them.

Turning now to more purely intellectual matters, it is obvious that tomorrow's world makes desirable a much greater capacity for analysis, for quantitative procedures, for integrative operations, and for imaginative creation. With more and more of the daily grind taken over by automation, the human being will be increasingly freed for higher mental jobs; yet most of our population today would be by nature ill-adapted for such activities, even if they had the desire to pursue them.

How ignoble and inadequate for its potentialities is a society in which the material operations are conducted by the utilization of the equations of Willard Gibbs, Rutherford, Einstein, and Planck, while most of the human beings who are served thereby have hardly a glimmering of what is involved—while seeking to turn these forces chiefly to such purposes as broadcasting television commercials, football games, burlesque shows, and revival meetings, or seeing Europe in a week or accurately dropping H-bombs nine thousand miles away. Missed are the opportunities for those profound stirrings which would be theirs if they could and would follow the inner workings of these forces that the combined efforts of a relatively few among them have put at their disposal. Missed also is the thrilling awareness of what vastly enlarged possibilities a more rational use of these and other powers could open up to everyone. If men are not to be mere cogs in their work and pawns in their play, they must have deeper and broader vision, as well as a more virile, broadly based comradeliness. Then their machines and science can give them increasing freedom for further achievement and further savoring of the bounties of our expanding universe instead of deeper enslavement in routines.

How are men to attain the higher intelligence and enhanced fellow feeling and sensitivity that will better fit them to the modern world?

Certainly there can and must be reforms in the ways and mores of society, and especially in the bringing-up and teaching of children, that will work major improvements. Some enthusiasts for biological techniques even hope that some day, by means of suitable elixirs, the growing brain may be influenced favorably during its embryonic development. But we cannot rest in the precarious hope of such a miracle. And it would be beyond reason to expect that cultural methods alone, powerful though they are, could bring to the average man that ability to understand, appreciate, and exploit the forces of nature and artifice which is today reserved for specialists in their respective fields. Yet a considerable measure of that ability, in every field at once, must become the property of the common man if he is to enter into the great cultural inheritance that will make him a really voluntary agent, led by the mind and not by the nose.

Correspondingly, a man's nature must also have at its very core a genuine warmth of feeling for his fellows if, despite the personal difficulties, compromises, and renunciations that inevitably beset everyone, he is to derive an adequate sense of fulfilment from his and their joint day-by-day efforts and achievements and also from the larger contemplation of grand-scale human progress. This, too, is a situation which calls for not only cultural methods but also genetic ones.

That genetic methods could be effective is illustrated by the vast individual differences in native intellectual capacity, temperament, and emotional pattern that exist among human beings, even as among other higher animals. Studies of twins and people brought up in institutions and foster homes have shown clearly the high, though far from absolute, importance of their heredity in the determination of psychological, as well as so-called physical, traits. Undoubtedly, even seemingly minute features of the personality can be strongly influenced by the genes. More important, there are abundant instances of extremely high mental ability, of a generalized kind, reappearing conspicuously in some members of families while missing other members. Certainly, there is already genetic material on hand, recognizable through its expressions, which, if conferred on the population at large, could enable men in general to find freedom and release by engaging in great co-operative, as well as individual, assaults against the seeming inexorabilities of the outer world and their own stubborn natures and by giving the feelings thus engendered creative and artistic expressions.

For achieving this end, which, after all, would be only another beginning, it is imperative that men think through their values to the point at which they recognize the primary importance for themselves of these two essentials: deeper mental insight and the feelings that give them joy in common action and in individual creation that can be shared. At present they are likely to place equal or greater emphasis on nonessentials. If genetic methods are to be used, men must come to realize that at this stage of their development they still have many divisive tendencies, provincial attachments to styles, features, and peculiarities of their own particular group. They must learn that such predilections, about which they could now wrangle endlessly, must be discounted and set as much in the background of their thoughts as possible, in order that emphasis may be placed on the key faculties mentioned—which, after all, count the most for everyone. No people, no caste or class, has a monopoly on these essentials. But there is everywhere a dearth of them, in relation to present needs. Let men become more conscious of this dearth and, conversely, of the richer life that later generations may have in proportion to the degree to which that dearth is remedied. In the meantime, no one need fear that there will be a danger of men's really salutary diversities becoming wiped out. And later, in a wiser, kindlier age, men may more safely and calmly consider how this spice of life—variety—can be turned to better account.

Motivations in Reproduction

It cannot be denied that the technological and social innovations of our age, combined with the influence of the scientific world view, are weakening the hold of ancient taboos and superstitions, loosening the rigidities of the family system, and greatly liberalizing the attitude of large numbers of people in regard to matters of sex. Within the twentieth century, in fact, the change has been so pronounced as to justify the application of the term "sexual revolution" to this situation, even though the ostensible, officially proclaimed standards are still much the same as in Victorian times. As yet this revolution has been of a predominantly individualistic kind. It has concerned itself chiefly with making people freer of the natural reproductive consequences of sex and therefore less rigorously bound sexually, as well as less burdened with excessive childbearing and child-rearing. But it has not concerned itself with the converse matter—allowing the decisions leading to parenthood to become more rational from a genetic viewpoint and to be guided in greater measure by the type of native endowment that would be valuable for the children themselves. A further break with tradition is necessary before such a viewpoint can lead to perceptible changes in this respect.

The way to such a reform in viewpoint is paved by several groups of circumstances. One of these is the sexual revolution itself. A second is the trend toward a more social outlook in general, brought about by the increasing degree of co-operation in our society consequent upon the impact of scientific and technological advances on our economic and political system. A third, more specific, influence is brought to bear as the pressure of too rapidly increasing population causes increasing numbers of people to realize that the interests of the potential children themselves must be taken into account in decisions whether to reproduce and that the implementation of these decisions requires their using artificial controls for preventing reproduction where advisable. At the same time, sterility is becoming increasingly remediable. Finally, and still more pertinent, although not yet operative, will be the realization that artificial means are already at hand, and others nearly at hand, whereby the likelihood of the child's receiving a superior genetic endowment can be greatly improved without individual sex practices being thereby interfered with. It is this fourth factor that must increasingly turn the scales.

Nevertheless, it remains true that some long-intrenched attitudes, especially the feelings of proprietary rights and prerogatives about one's own germinal material, supported by misplaced egotism, will have to yield to some extent. This feeling does not represent a natural instinct, since there are primitive tribes yet alive who do not have even the concept of biological fatherhood and others that, although having it, readily and without their parental relationships being affected thereby, adopt, confer, or exchange infants. That is, their egotism does not extend to their stirps. Actually, they are more logical in this respect than are most of the persons in more advanced societies who pride themselves on their ancestry or inborn constitution. For that is the one thing they themselves have been least—in fact, not at all—responsible for. Moreover, paradoxically, this kind of pride is today likely to be more intense in persons who are less, rather than more, fortunately endowed.

To more than balance the necessary weakenings of this time-worn vanity in regard to one's stirps, other feelings will tend to develop that are of equal or greater potency. Among them will be justifiable pride in acomplishment of a far more exacting and laudable kind than that of procreation: namely, having made children of especially high endowment possible and having brought them up. Deep attachments to these children will develop and a justified sense of identification. These new reproductive mores will come into being only very gradually. At first, only those with freer and more daring spirits will venture on these alien-seeming paths (to be indicated in the following two sections). There will be no clear break, and in the same family children of choice and children of tradition will grow up side by side in mutually helpful familial association. But "nothing succeeds like success," and the successes in these instances will often be outstanding.

PRESENTLY AVAILABLE GENETIC TECHNIQUES

Long before this point is reached, many warning voices—even among geneticists, or perhaps among them more especially—will be raised, protesting that we cannot predict the results of any matings in such heterogeneous material as an existing human population. We have, for such purposes, the merest smattering of knowledge regarding the genes concerned; and what the more important ones—those concerned with the traits we are here the most interested in—will do in new combinations is anyone's guess. We could not proceed according to the principles governing simple Mendelian differences, and we should be practicing not genetics but a dangerous hoax.

This argument sets up straw men and then knocks them down with straw swords. The multifactorial basis of most phenotypic differences in human and other natural cross-breeding populations is undisputed —especially in regard to differences in traits of importance and therefore long subject to natural selection—although, of course, some gene differences produce major effects, along with many more that produce minor ones. We can seldom hope, in practice, to know just what genes are concerned, nor, even if we knew them, could we say just what their effects in postulated combinations would be. Nor can we control or predict what combination a given zygote will have. We will readily admit, moreover, that the effects are seldom exactly additive. We are well aware that sometimes, in genetic effects, $2 + 2$ makes -3 or even -6.

Now if that were the whole story, natural selection would not have been more effective in sexually reproducing organisms than in asexual ones, and the method of sexual reproduction would not have been retained in animals or plants. It is unlikely that life could have advanced even as far as the stage of triploblastic animals. (The points involved are discussed in my various 1958 papers.) But if the world had somehow been supplied with wild animals, plants, and primitive men anyway, then these men, without knowledge of genetics, would never have succeeded in producing the races of domestic animals and cultivated plants highly adapted to their purposes that they transmitted to us. Even modern breeders, if supplied with these improved races, would never have been able to develop the still more improved strains that we have today, despite their knowledge of Mendelian principles.

What made all these advances possible was, for one thing, the fact that in the great majority of cases a given gene difference exerts a similar effect (e.g., the increase or the decrease of a given trait) in one genetic combination as in another one. In other words, although the effects are not accurately additive, they do work out, on the whole, in the expected directions. Second, in the comparatively rare instances in which they fail to do so, longer-term selection usually eliminates the "four-flushers" (the seemingly helpful variants that turn out to be detrimental when in otherwise desirable combination) long before these frauds have become irrevocably established. For these reasons, selection carried out on a purely empirical basis does work.

It works in spite of the frequent impossibility of discriminating between genetically and environmentally based effects. It is true that where selection has "mistakenly" favored a merely phenotypic deviant—one that was occasioned only by environmental circumstances—no genetic progress is made thereby, but no consistent harm is done either; for these "mistakes" occur in diverse directions so as ultimately to cancel one another. This situation leaves the determination of the over-all result to the remaining cases—those in which some genetic basis did exist for their having been selected. By the accumulation of many such steps, the genetic composition is caused to move significantly in the direction of selection.

Even primitive man was intelligent enough to do much better than nature usually did, inasmuch as he effected changes in his domestic races, in the directions sought, at a pace much faster than that at which species ordinarily change when under natural conditions of breeding. This was because his manner of selecting was less haphazard and more single-tracked than nature's. The selection practiced by modern man, especially in the present century, has been much more rapid and effective still, even though it likewise has in the main had to proceed without analysis of the individual genes concerned. Where the inherited differences are multifactorial, as they usually are in the case of traits of importance in any organism, such as milk yield in cattle, gene identification is seldom feasible. Neither has it been necessary in order to achieve decided results from selection. Nevertheless, knowledge of the genetic principles at work has helped greatly in the guidance of selection processes, especially in their later stages.

General genetic knowledge has discredited futile practices based on such mistaken beliefs as the inheritance of acquired characters, telegony, etc. It has proved invaluable by supporting such useful practices as progeny testing; taking the environment into account and, where possible, controlling it more rigorously; being guided by measurements made among different classes of relatives and measurements of the correlation between different traits; making judicious use of both outbreeding and inbreeding; and so on. However, even these methods seldom play much part in the earlier generations of selection, which is when the most rapid progress is usually made, despite the selection being based mainly on superficial phenotypic evidence. Since the hereditary mechanism is known to be the same in man as in sexually reproducing organisms in general, there is no reason to doubt, on purely genetic grounds, the potential efficacy of an empirical type of selection for man also—consistent with general genetic considerations but not, at least in the early stages, based on knowledge of the particular genes concerned.

It should be mentioned again that in all selectional work in which the potentialities of the individual genes in further combinations are not known, some chosen for multiplication will later turn out, when in homozygous condition or in given groupings, to have undesirable or even lethal effects. This happens also in natural matings. To be sure, careful progeny testing could reduce such events to an even lower than natural frequency. But, in any case, over-all progress will take place in the direction of selection and can, especially in the earlier stages, be very rapid. Then, in later stages, the delicate task of screening for the most dependable and precisely suited genes can be better carried out, after the general level has been raised and a plentiful supply of promising genetic material has become available. In the meantime, the favorable phenotypic effects of this rise in the level can be enormous.

What now are the means that would make such positive selection possible in man? The most effective method presently feasible is, of course, artificial insemination. Many thousands of people have already been begotten in the United States by this procedure—a considerable proportion of them, although not most, by sperm of donors other than the husband when the husband was sterile. Here is an excellent opportunity for the entering wedge of positive selection, since the couples concerned are nearly always, under such circumstances, open to the suggestion that they turn their exigency to their credit by having as well-endowed children as possible.

Unfortunately, most of the physicians in such cases, deterred by the fear of public and legal censure and having little appreciation of genetic matters, seem to be chiefly concerned with hiding their operations and avoiding a conspicuous result or failure that they might be blamed for, and they furtively attempt to produce a child as nearly as possible like that which might have been born if the father had been fertile. With this aim, they choose a donor resembling the husband in physique and even in religion, free from readily detectable defects, of course, and they are careful to keep the identity of this donor as secret as possible. Often a medical student or intern, if he is discreet and close at hand, serves, for a consideration, as a multiple donor, without regard for the fact that U.S. Army I.Q. tests have indicated this group to have the lowest mental ratings of all professions tested. But cases have even been reported of out-of-work men, originally picked up in bars, who make an easy living by regularly selling their semen on what amounts to a mass scale. Thus opportunities for introducing a substantial genetic leaven into the population at large are flouted.

Insemination by an outside donor has now gone on for so long and become so prevalent that it is high time for it to come out into the open, even in individual instances. A judge in Chicago branded the procedure "adultery," but a number of precedents in which the children were begotten by consent of both husband and wife have been held to be legal. It is highly important for the genetic paternities of the children thus produced to be properly recorded. Not only would invaluable data be provided in human genetics and selection, but, more specifically, better judgments could thereafter be made concerning the genetic potentialities of the given donors and their descendants. It should be recognized that the couple concerned in such a case, as well as the physician, has performed a service to mankind meriting not disgrace but honor. With such an outlook, even before it was generally held, both physician and couple would be armed with better incentives to take genetic considerations into account. They would be encouraged to make the best use possible of such a chance to engender the most precious thing we know of: a worthy human being.

Perhaps university teachers in scientific subjects are, through their personal experiences, better aware than any other large group of persons of the enormous differences in over-all intelligence between the average and the best endowed and of the fact that training, though essential, cannot be an adequate substitute for high native endowment. For real competence in understanding and dealing with the world as known by science, only a mind that is truly exceptional in terms of the now existing distribution will suffice. However, general intelligence is complex and multifactorial, despite the existence of rare genes that individually have a decidedly enhancing influence on intelligence (shown by cases of long-persisting, sharply segregrating effects in given pedigrees). Because of this complexity, the progeny of individuals of exceptionally high intellectual endowment tend to exhibit considerable regression and variability, even among the better-endowed segregants. This by no means signifies that selection in this area is futile—it can, in fact, have a high over-all degree of success. Rather, it underlines the importance of choosing as donors individuals of the most outstanding native mental ability—that is, those at the extreme end of the positive tail of the distribution—so far as possible, and when they do not have serious defects in other directions. Moreover, those individuals should be preferred as donors whose relatives give considerable evidence that the superior qualities are highly inheritable.

Fortunately, such a high degree of selection as here indicated is made possible by the method of artificial insemination, by reason of the large number of spermatozoa that each individual usually produces. Recently the technique of freezing spermatozoa to very low temperatures was introduced into practice, and, thanks to the accumulation of sperm thus made possible, the frequency of successful conceptions has been raised considerably above that following either natural insemination or artificial insemination of the more usual type. The deep-frozen spermatozoa can be stored virtually indefinitely without deterioration. Thus a considerable supply can in time be gathered from a chosen donor and preserved for any desired length of time. For purposes of selection, there would in such cases be enormous advantages in postponing the use of most of this supply until, say, twenty years after the donor's decease (see Muller, 1935; Hoagland, 1943). In retrospect, after the personal attachments and animosities aroused during the donor's life had faded, much less biased judgments could be made concerning his actual merits as well as shortcomings. During this "probationary period," a limited but significant amount of progeny testing could be carried out to support a sounder estimate of the donor's genetic potentialities.

Such a procedure would also afford considerable psychological advantages, for it would eliminate that bane of present-day gynecologists who practice artificial insemination—the fear of intrigue arising between the donor and the woman concerned if either one learns the identity of the other. The motivation for possible jealous reactions of the husband would at the same time be greatly diminished. Moreover, both members of the couple, as well as all other persons, would be much more likely to recognize the donor's exceptional worth, a worth that would usually put him out of a class with living competitors. Finally, the chief present objections to having the identity of the donor known would be removed. Instead, all interested persons would wish to have the relationships concerned entirely above-board.

As was recently pointed out by Dr. Richard Meier, of the Mental Health Research Institute of the University of Michigan, another method of genetic upgrading is the outright adoption of children by those who otherwise cannot or do not wish to bear children or wish to have less than the average quota of their own genetic progeny. Since such practice does not involve as radical a departure from present-day customs and attitudes as does artificial insemination, it might recommend itself more readily in wide circles. This procedure presupposes, of course, that couples of high native endowment would be willing to bear more children than they could bring up and to give them out for adoption. Those consenting to do so would truly be socially minded.

Perhaps in the long run, though, this method would interfere more with people's personal lives and occasion more risk of undesirable personal entanglements than the half-adoption represented by artificial insemination. Certainly the outright adoption method here suggested would not allow selection to be as rigorous, even though the eggs as well as the sperm are to some extent specially selected. Then, too, this method would lack the psychological advantage, operative in the earlier transitional period of feelings on the subject, of at least half the genetic material having been derived from those who are to bring up the child. But there is no doubt that the method would be better adapted than artificial insemination for some situations and could thus serve a useful function.

Technical Advances in the Offing

We are surely just around the corner from other advances in artificial techniques concerned with reproduction that might extend the possibilities of positive selection much further. For example, there have as yet been only a few abortive attempts to cultivate either male or female germ cells outside the body. An energetic program of research on the subject would probably be successful within a few years in enabling spermatogonia, at least, to be multiplied indefinitely in vitro and to be induced, when desired, to undergo the processes of maturation into spermatozoa. If this could be done, then spermatogonia instead of spermatozoa might be preserved in the deep-frozen state, a technique that has already proved successful with some types of somatic cells. Later, at any desired date, the spermatogonia could be multiplied and caused to mature so as to furnish an unlimited supply of mature spermatozoa from an originally small amount of material derived from any given donor. Only our present superstitious attitudes prevent such research from being actively pursued today.

As for the female germ cells, means are already known whereby the multiple release of mature eggs can readily be effected within the female with the aid of pituitary hormones. Only a little research would be required to develop methods of flushing out these eggs from the female reproductive tract, to be fertilized in vitro with chosen sperm and then implanted in selected female hosts at the appropriate stage of their reproductive cycle. This procedure is parallel to artificial insemination. It permits the multiple distribution of eggs of a highly selected female into diverse recipient females, yet allows the child to be derived, on its paternal side, from the recipient's husband.

It is not unlikely that techniques involving mature eggs could be combined with deep freezing to allow indefinitely prolonged storage. Thus similar advantages, selectional as well as psychological, might be gained for the female germ cells as are already available, even though not yet in use, for spermatozoa.

Such techniques applied to the eggs could, of course, be combined with artificial insemination by sperm of a donor other than the recipient's husband when the couple desired it. This would be another method equivalent to outright adoption. However, it would afford the opportunity for far more powerful selection than adoption of the relatively primitive sort previously discussed and would avoid the possible psychological and social difficulties that might attend such adoption.

Still another possible development, perhaps somewhat further off technically, is parthenogenesis. It is not commonly realized that Gregory Pincus some twenty years ago succeeded in producing several vigorous, fertile female rabbits by artificial parthenogenesis. Genetic "markers" were present that showed these rabbits to be of purely maternal origin, although (as proved by breeding tests of them) heterozygous for at least some of the genes for which the mother also had been heterozygous. The inference could be drawn from this situation that in these cases diploidization had occurred through the union of the egg nucleus with one that would normally have passed off in a polar body. The great majority of the eggs that had been stimulated to begin parthenogenetic development, however, had remained haploid and, obviously in consequence of this, failed to give viable embryos. Unfortunately, the induction of diploidization had not been brought under control when research along this line was discontinued.

A more promising type of parthenogenesis might be achieved by an extension of the technique used by Briggs and King (1952) with frogs' eggs. They succeeded in obtaining normal development after implanting nuclei from cells of frog blastulae and gastrulae into recently fertilized eggs whose own nuclei had been extirpated by micromanipulation. Before this method, which affords many interesting possibilities for embryological research, could become of eugenic importance, it would, of course, have to be extended to mammalian and eventually human fertilized eggs. Methods would have to be developed of successfully implanting into them, with subsequent development, diploid nuclei derived not from embryos but from the immature germ cells—spermatogonia or oögonia—of adults, or possibly even from certain generalized types of somatic cells of adults. On theoretical grounds, this extension does not seem at all far-fetched.

This type of parthenogenesis would make possible the production of multiple progeny that resemble their genetic parent and one another about as closely, phenotypically, as identical twins reared apart resemble one another and that are about as alike genetically as identical twins. Here, then, there would be an extremely high order of predictability regarding the nature of the children and virtually no regression except that occasioned by environmental factors, such as the manner of bringing up the children. Of course, the high general intelligence of the progenitor would in most cases be capable of being applied in very diverse directions in his separate manifestations, as we might term them. Yet in the rearing of such children and in the later decisions of their lives, despite the diversity of the situations, invaluable guides would be available in the already existing knowledge concerning the progenitor's proclivities, character traits, and modes of reaction to different circumstances. Often he would himself have been able to give pertinent advice along such lines. All this would usually make it possible to avoid, in these individuals' upbringing and later career, many of the mistakes inadvertently made the first time. Thus, far better opportunity could be provided to these outstanding persons for making the most of their unusual potentialities.

When one considers how much the world owes to single individuals of the order of capability of an Einstein, Pasteur, Descartes, Leonardo, or Lincoln, it becomes evident how vastly society would be enriched if they were to be manifolded. Moreover, those who repeatedly proved their worth would surely be called upon to reappear age after age until the population in general had caught up with them. In this way, then, mankind would be able to reap the benefit of that alternation of asexual reproduction (for reliably multiplying types of tested worth) with techniques of sexual reproduction (for trying ever new combinations), an alternation that has been so advantageous in some other classes of organisms. Later generations will look with amazement at the pitifully small amount of research now being carried on to open up such possibilities, even though for years specialists have realized that they lie just around the corner.

Just as our economic and political system is inevitably, although too slowly, being modified to fit our present technological capabilities of large-scale automatic production, despite the fervor with which men try to cling to their ancient preconceptions of how business and government should operate, so too on the biological side of human affairs the time-honored notions of how reproduction should be managed will gradually give way before the technological progress that is opening and will further open up new and more promising possibilities. Practices that today are confined to couples afflicted with sterility will be increasingly taken up by people who desire to improve their reproductive lot by bestowing on themselves children with a maximal chance of being highly endowed, and thereby to make an exemplary contribution to humanity. Making some sacrifice in the matters of traditional feelings of vainglory about the idiosyncrasies of their personal stirps and braving the censure of the old-timers about them, they will form a growing vanguard that increasingly feels more than repaid by the day-by-day manifestations of their solid achievements as well as by the profound realization of the value of the service they are rendering.

But recently there has been a tendency in some circles to by-pass the arguments for positive selection by countering that, instead, means will be found of making direct alterations or substitutions of a desired kind in the genetic material itself while leaving it in the main unchanged. Such proposals range from the idea of substituting individual chromosomes or parts of chromosomes, derived from selected donors, in the chromosome set of a given person's germ cell to the idea of inducing by a mutagenic agent or ultrafine manipulative process a given chemical change in a given gene. It would be very rash to deny that some day such extraordinary feats may be possible, but they are definitely not around the corner in the sense that the developments discussed above are.

Dismissing for the moment the unparalleled advances in technique that would here be necessary, we must take into account that such procedures would also require the most minute knowledge of the role played by individual genes in the inordinately complex economy of the human organism, including its highest, most multifactorial functions, such as general intelligence. It would also be necessary to know the locations of these genes in the genetic map. And if mutations were actually to be induced or nucleotide substitutions made, we should have to know the internal constitution of these genes and the functions in them of each of their tens or hundreds of thousands of nucleotides. Though the nucleotides are of only four kinds, it is their precise arrangement in line that counts. It is not likely that all this will be worked out before men are on a much higher genetic, as well as cultural, level than they are today.

Present evidence indicates that there are something like four billion nucleotides exactly ordered like the letters in enormous words in each haploid chromosome set. Any of these, by having a different nucleotide, of the remaining three possible types, substituted for it, might cause a change, sometimes significant and sometimes not, in the individual's phenotype. It is preposterous to suppose that, in the foreseeable future, knowledge would be precise enough to enable us to say what substitution to make in order to effect a given, desired phenotypic alteration—not that this would *never* be possible. But to suppose that, after it had become possible, men would still be bound by the reproductive traditions of today, preferring this ultra-sophisticated method of improvement to the readily available one of selecting donor material free from the given defect or already possessing the desired innovation—that would be a calumny on the rationality of the human race. It would be like supposing that in some technically advanced society elaborate superhighways were constructed to carry vehicles on enormous detours to avoid defiling hallowed domains reserved in perpetuity for their millions of sacred cows.

If human superstitions are really so unchangeable and subject to such inviolable taboo, then it is very unlikely that humanity will succeed in progressing to the stage where it understands that otherwise most marvelous organization under the sun—its own constitution. But many customs and attitudes *have* changed, and today we see them changing rapidly. It would be very strange if in this age of exploding knowledge and technique our reproductive practices remained immune to reformation.

More Distant Prospects

Evolution in the past has been for the most part a matter of millions of years. In this larger view, what we have been discussing is but a matter of today—the step we are just about to take. So great are the present psychological impediments to this step, arising out of our traditions, that we have not had time to consider the enormous vistas beyond.

The rapid upgrading of our general intelligence must be accompanied and co-ordinated as closely as possible with a corresponding effort to infuse into the genetic basis of our moral natures the springs of stronger, more genuine fellow feeling. At the same time, especially interested groups will see to it that diverse abilities and proclivities of specific types will here and there be multiplied, both those of a more purely intellectual nature and those making possible more far-reaching and poignant appreciation of the varied kinds of experiences that life may offer. As all these genetic resources of mankind grow richer, they will increasingly be combined to give more of the population many of their benefits at once. Observation shows that these faculties are not antagonistic but rather mutually enhancing. Finally, increasing attention can be paid to what is called the physical side: bettering the genetic foundations of health, vigor, and longevity; reducing the need for sleep; bringing the induction of sedation and stimulation under better voluntary control; and increasing physical tolerances and aptitudes in general.

In many physical respects there are optimal degrees of development for given types of organisms, beyond which other functions tend to be too much interfered with. Yet these optima are seldom absolute, for there are often ways of breaking through the seeming limits by means of novel developments that open new directions for solving the old problems. These directions of inquiry are so advanced that we may leave them to the more competent minds of the future to tackle.

But, at least so far as intelligence is concerned, there are no indications that we are now approaching any physiologically set limit or optimum. It is quite evident that we could benefit indefinitely by a continued increase in our mental powers: to enable us to analyse more profoundly; to recognize more readily common features when they lie deeply buried; to grasp more and more elements of a situation at once and co-ordinately; to see more steps ahead; to think more multidimensionally; and to imagine more creatively. Here, too, it is to be hoped that new breakthroughs will eventually be found; otherwise, limits will in time appear. If only we take the first, most obvious steps that we have here been considering, we shall be preparing the way for making our successors capable of planning ahead much further and more soundly than we ourselves of this fumbling generation can.

It is so easy to sit smugly back in the conceit that we have now reached nearly the acme of biological evolution and that, except for eventually bestowing on everyone the genetic advantages already enjoyed by the most favored, we can hereafter confine our advances to cultural evolution, including the manipulation of things outside our own genetic constitutions. It is true that cultural evolution, in this broad sense, is far more diversified, rapid, and explosive, both figuratively and literally, than biological evolution can be. There is every reason to extrapolate, along with "science-fiction" enthusiasts (despite their frequent unbalance!), that if men do not destroy one another, they will cultivate the deserts, jungles, poles, and oceans, extend their domain successively to ever more distant worlds, and perhaps even, as first suggested by Bernal, build colonies in empty space. Along with the increasing understanding and mastery over physicochemical forces that such expansion implies, there will be corresponding advances in the biological and social realm. That is, there will be spectacular progress in means of reshaping and controlling bodily structures and functions by operations repeated in each generation anew and also in means of interrelating people psychologically to achieve higher, more harmonious, and more constructive interactions of their feelings, thoughts, and doings. Yet all this does not mean that genetic advances beyond the stage represented by the happiest possible combinations of the best endowed of present-day humanity would be either supererogatory, unimportant, or relatively limited.

If their genetic constitution is so unimportant for beings with such advanced means of extragenic control as envisaged for our successors of, say, five hundred years from now, why could they not just about as well use apes or even lowlier creatures instead of men and duly reconstruct and train them? Or, if the genetic difference between apes and men is really so important in determining their amenability to profit by culture and to contribute to it, why would not beings as far beyond present-day men genetically as we are beyond the apes be inordinately better suited still for exploiting the benefits of culture? The biological distance from apes to men is a relatively slight one, yet how potent! And do we hastily-made-over apes really believe that, having attained this makeshift form, further steps of this kind are to be despised? Our imaginations are woefully limited if we cannot see that, genetically just as well as culturally, we have by our recent turning of an evolutionary corner set our feet on a road that stretches far out before us into the hazy distance.

Of course, the genetic changes that would be desirable for us in the future are not, in the main, developments like fur, wings, photosynthetic ability, or anything else that would be less effective or less adaptable than our own artifices. They would not be replacements for cultural devices but the very opposite: means of better gearing together the biological and the cultural, of making still more out of our culturally enhanced propensities, and of more effectively advancing our culture. As we have seen, this evolutionary trend actually began with the advent of man, although he was unaware of it as a long-term phenomenon. But now we may carry it forward consciously and with ever longer foresight.

Although the most important genetic advances for any creature who creates a culture are obviously those that suitably extend and enhance his psychological faculties, it is a mistake to conclude that corporeal changes of diverse kinds would be of little account—such as, for example, further developments of the senses, on the one hand, or of effector organs, on the other hand. Unfortunately, spelling out radical possibilities along these lines is likely to provoke more ridicule than understanding at the present stage of purblindness in this field. For evolutionary biological developments would, in general, appear utterly fantastic to a group in whom they had not yet occurred, whereas, of course, after the event they are taken for granted as being the only reasonable arrangement.[1]

It would be presumptuous to try to specify here just what form the long-range developments are likely to take and in what sequence they will occur. Intellectual and moral developments of the types mentioned as most important now will probably continue to occupy the center of the stage for a long time to come, if not indefinitely. The faculties here concerned could probably be extended and enhanced in many ways which our present ignorance of their structure does not permit us even to guess at.

Only after we humans have advanced considerably toward the higher level to which the rough-and-ready empirical methods now available can raise us, will we be in a position to make firmer, more

[1] For this reason I will give only one illustration of how a biological, genetically based series of developments could be important to a species with an already highly advanced culture. This example has to do with the means of personal communication. At present these are largely confined, in man, to gestures and to some form of speech (including its derivative, writing). Speech, marvelously effective though it has been in the long run, is pitifully slow and plodding compared with the inner flow of ideas. It has special tendencies to be misleading, and it requires forcing everything into an inappropriate one-dimensional order. Wishful thinkers, mystics, and pseudo-scientists have long dreamed of telepathy to escape from the bondage of spoken forms. However, telepathy is a species of magic for which, so far as can be seen, no biological possibility exists. It might therefore appear unlikely that any biological kind of communication better than speech could arise in evolution.

Yet such a development, involving a combination of already known biological phenomena, is in fact conceivable. Many different types of organisms have developed light-emitting tissues. In some, such as the squid, the spatial and temporal pattern of the luminescence may be very complex and subject to considerable voluntary control. In some fish the light is projected through a special lens with reflector. However, a more eyelike structure, if its retinal surface were provided with luminous cells, might serve even better. The main development, then, would be the suitable connection of the visually imaginative portion of the brain, point by point, with the luminescent layer in such wise that the visual patterns imagined in the cerebrum become transferred into luminescent patterns that can be seen by other individuals. In effect, this would be the reverse of the transference that occurs in seeing, when the image produced by light focused on the retina is transmitted, via impulses in the fibers of the optic nerve, so as to give rise to a corresponding pattern of stimulated cells in the occipital cortex. Now the actively radiating pattern in the luminescent layer could either be projected onto a screen in a darkened inclosure outside, where it might in turn be subject to amplification or recording, or, alternatively, it could be seen directly by a close-up inspection of the transmitting "eye" by the receiving eye.

In addition to direct visual representations of things, subject to motion as in a cinema, there would doubtless be diverse symbolizations of conceptual thoughts. All in all, such communication could be raised to a level of facility, directness, speed, multidimensionality, comprehensiveness, and precision so much higher than is afforded by speech as to inaugurate far more advanced processes of communing, of learning, of thinking itself, of appreciation, and of resulting action. But for bringing these potentialities to fruition, further improvements of a neurological and psychological nature would be in order, even as was the case with speech.

definite plans envisaging longer-range possibilities. Only then, when we have developed superior intelligence and greater co-operativeness, can we expect to reach a workable degree of agreement on these plans. Only then can we begin to use more exact methods and to co-ordinate them better. It is too early for blueprints.

If we are to preserve that self-determination which is an essential feature of human intelligence, success, and happiness, our individual actions in the realm of genetics must be steps based on our own personal judgments and inclinations. They should be as voluntary as our other major decisions of life. Although these decisions are all conditioned by the mores about us, these mores can be specifically shaped and channelized by our own distinctive personalities. The immediate job, then, is to make a start at getting this genetic "Operation Bootstrap" incorporated into our mores, by precept and, where feasible, by example. But we must remember that the highest values to be sought in it are, in essence, those so long proclaimed but seldom actualized: wisdom and brotherhood, that is, the pursuit of "the true and the good." When it is realized that the genetic method offers simply an additional but indispensable approach toward this ancient ideal, then our voluntary genetic efforts, scattered and disjointed though they must now be, will tend in a common direction.

There are sure to be powerful attempts to pull in diverse directions, in genetic just as in other matters, but we need not be afraid of this. The diversities will tend to enrich the genetic background, increasing the resources available for recombination. These partial attempts can then be judged by their fruits, and these fruits, where sound, will be added to our bounty.

It seems highly unlikely that, in a world-wide society at an advanced level of culture and technology, founded on the recognition of universal brotherhood, such diversities would proceed so far and for so long as again to split humanity on this shrunken planet into semi-isolated groups and that these groups would thenceforth undergo increasing divergence from one another. It is because man is potentially master of all trades that he has succeeded. And if his culture is to continue to evolve indefinitely, he must retain this essential plasticity and with it the feeling that all men are, at bottom, of his own kind.

Through billions of years of blind mutations, pressing against the shifting walls of their environment, microbes finally emerged as men. We are no longer blind; at least, we are *beginning* to be conscious of what has happened and of what may happen. From now on, evolution is what we make it, provided that we choose the true and the good. Otherwise, we shall sink back into oblivion. If we hold fast to our ideal, then evolution will become, for the first time, a conscious process. Increasingly conscious, it can proceed at a pace far outdistancing that achieved by trial and error—and in ever greater assurance, animation, and enthusiasm. That will be the highest form of freedom that man, or life, can have.

Bibliography

ANONYMOUS. 1939. "Artificial Insemination and Illegitimacy," *Jour. Amer. Med. Assoc.*, CXII, 1832.

———. 1959. "My Business Is Making Babies," *Man's Adventure,* II, 38.

BACON, ALLAN. 1959. *Man's Next Billion Years.* New York: Exposition Press.

BERNAL, J. D. 1929. *The World, the Flesh and the Devil: Three Enemies of the Rational Soul.* New York: E. P. Dutton & Co.

BREWER, HERBERT. 1935. "Eutelegenesis," *Eugenics Rev.,* XXVII, 121.

———. 1939. "Eutelegenesis," *Lancet,* I, 265.

BRIGGS, R., and KING, T. J. 1952. "Transplantation of Living Nuclei from Bastula Cells into Enucleated Frogs' Eggs," *Proc. Nat. Acad. Sci.,* XXXVIII, 455.

BUNGE, R. G., and SHERMAN, J. K. 1953. "Fertilizing Capacity of Frozen Human Spermatozoa," *Nature,* CLXXII, 767.

DARWIN, CHARLES. 1871. *The Descent of Man.* London: J. Murray.

DRUMMOND, HENRY. 1904. *The Ascent of Man.* London: Hodder & Stoughton.

GALTON, FRANCIS. 1883. *Inquiries into Human Faculty and its Development.* London: Eugenics Society, 1951.

HARDIN, GARRETT. 1959. *Nature and Man's Fate.* New York: Rinehart & Co.

HOAGLAND, H. 1943. "The Chemistry of Time," *Scient. Monthly,* LVI, 56.

HOAGLAND, H. and PINCUS, G. 1942. "Revival of Mammalian Sperm after Immersion in Liquid Nitrogen," *J. Gen. Physiol.* XXV, 337.

HUXLEY, J. S. 1941. *Man Stands Alone.* New York: Harper & Bros.

———. 1943. *Evolutionary Ethics.* London: Oxford University Press.

KELLICOTT, WILLIAM ERSKINE. 1911. *The Social Direction of Human Evolution.* New York: D. Appleton & Co.

KROPOTKIN, PETER. 1902. *Mutual Aid a Factor of Evolution.* New York: McClure, Phillips & Co. Latest ed. with Foreword by ASHLEY MONTAGU. Boston: Extending Horizons Press, 1955.

MACKINNON, JOHN G. 1960. "What Chance for a Well Born Race?," *The Humanist,* XX, No. 2 (in press).

MEIER, RICHARD L. 1957. "World Population Problems in Relation to Global Peace and Welfare." Address to American Humanist Association, Regional Conference, Ann Arbor, Michigan, December 7.

———. 1958. "Is Birth Control Enough?" *Humanist,* XVIII, 69 (based in part on address of December 7, 1957, but without the discussion of adoption).

MEIER, RICHARD L. 1959. *Modern Science and the Human Fertility Problem.* New York: John Wiley & Sons.

MULLER, H. J. 1935. *Out of the Night: A Biologist's View of the Future.* New York: Vanguard Press. (Fr. trans. J. ROSTRAND, Paris: Gallemard, 1938.)

——— and twenty co-signers. 1949. "The 'Geneticists Manifesto,'" *J. Heredity,* XXX, 371.

———. 1950. "Our Load of Mutations," *Am. J. Human Genet.,* II, 111.

———. 1953. Review of Charles Galton Darwin's *The Next Million Years, New York Herald-Tribune, Book Review,* January 11, p. 3.

———. 1955. "Life," *Science,* CXXI, 1.

———. 1957. "Man's Place in Living Nature," *Scient. Monthly,* LXXXIV, 245.

———. 1956. "Genetic Principles in Human Populations," *Am. J. Psychiat.,* CXIII, 481.

———. 1957. "Possible Advances of the Next Hundred Years: A Biologist's View," in *The Next Hundred Years: A Scientific Symposium,* p. 33. New York: Jos. E. Seagram & Sons.

———. 1958. *Man's Future Birthright.* Durham, N.H.: University of New Hampshire.

———. 1958. "Human Values in Relation to Evolution," *Science,* CXXVII, 625.

———. 1958. "Evolution by Mutation," *Bull. Am. Math. Soc.,* LXIV, 137.

———. 1958. "How Much Is Evolution Accelerated by Sexual Reproduction?" *Anat. Rec.,* CXXXII, 480.

———. 1959. Address in *The Future of Man,* p. 33. New York: Jos. E. Seagram & Sons.

———. 1959. "Prospects of Genetic Change," *Amer. Scientist,* XLVII, 151.

PINCUS, GREGORY. 1939. "The Development of Fertilized and Artificially Activated Rabbit Eggs," *Jour. Exper. Zool.,* LXXXII, 85.

———. 1939. "The Breeding of Some Rabbits Produced by Recipients of Artificially Activated Ova," *Proc. Nat. Acad. Sci.,* XXV, 557.

———. 1951. "Fertilization in Mammals," *Scient. American,* CLXXXIV, 44.

PINCUS, GREGORY, and SHAPIRO, H. 1940. "Further Studies on the Activation of Rabbit Eggs," *Proc. Amer. Phil. Soc.,* LXXXIII, 631.

ROSTAND, J. 1959. *Can Man Be Modified?* New York: Basic Books Inc.

SWIM, H. E., HAFF, R. F., and PARKER, R. F. 1958. "Some Practical Aspects of Storing Mammalian Cells in the Dry Ice Chest," *Cancer Research,* XVIII, 711.

SIR CHARLES GALTON DARWIN

CAN MAN CONTROL HIS NUMBERS?

When I was honoured by the invitation to make a contribution to this symposium I was embarrassed by the fact that all the other contributors were professional experts in the various subjects associated with evolutionary theory, whereas my own claims could at best be classed as those of an amateur. I have therefore chosen a subject where perhaps I can get on closer terms with the rest, because forecasting on incomplete data is related to statistical theory, a subject of some of my earlier studies.

Most of the contributions to the symposium are concerned with the way our knowledge has expanded during the past century, and it seemed it would not be uninteresting to attempt an estimate of the probable state of the world at the times when there might be celebrations of the second and later centenaries of 1859. Interesting contributions to the subject of man's future have been given by Huxley and Muller, and I certainly cannot aspire to making criticisms of their work. My own aim has been to deal with a far shorter range of time than they do, though I shall permit myself a few comments on the remoter future, too.

The Present Numbers

It can be taken as established by the demographers that our present world population of more than two and a half billion will almost surely have become at least five billion by the end of the twentieth century. No famines or pestilences on any reasonably probable scale can affect this, and war of the old type would also be quite unimportant. Even an atomic war would hardly be likely to make a great difference by its direct effects, but it must be recognised that there would very probably be a breakdown of world economics, with con-

SIR CHARLES GALTON DARWIN is equally renowned as physicist and as author-lecturer on the prospects of scientifically controlling man's future destiny in such matters as overpopulation and effects of increased longevity. He was formerly Director of the National Physical Laboratory and, with R. H. Fowler, developed the Darwin-Fowler system of statistical mechanics. He is a Fellow of the Royal Society and a member of the American Philosophical Society.

sequent killing of many more—perhaps even half the world—by famine. However, I do not propose to pay consideration to atomic wars because of the present great uncertainties about them.

According to expert agricultural opinion, it should be possible to feed these five billion. It may call for the enforcement of better farming methods in many places, and also for great outlay on irrigation schemes. Also it may require what may be called the charitable transfer of food from parts of the world where there has been overproduction to other parts suffering from shortage. According to the experts it should be possible in this way to raise food production to double the present quantities, and so to feed the doubled population. This, of course, is not as happy a result as might appear at first sight, because we have to remember that even now half the world is undernourished.

The world, then, seems to be capable of dealing with the problems of the next fifty years without taking any very radically new kind of action, but what will happen then? Why should not the population tend to double again in the following fifty years up to ten billion, and this would certainly strain the resources of the finite area of our Earth. We have got onto the Malthusian spiral of geometrical increase, and we must ask whether anything can be done to prevent our relapse into the hard conditions of most ancient periods of history when the escape from the spiral was through recurrent famine, pestilence, and massacre.

It will be seen that quite a new feature has now entered into our outlook on human life. In the old days population used to fluctuate about roughly constant numbers, being held there by natural selection. In judging whether some past epoch had been a good or a bad one, it was quite reasonable to make the estimate by merely counting heads; a good epoch would be one when numbers were increasing, a bad when they were diminishing. But now we are free from the ruthless action of natural selection, and we are faced with the prospect that increase will be a bad thing, because it may lead to a disastrous lowering of world conditions. We are being forced to make a revolutionary change in our standards of value, and in view of the conservatism of the human mind, there may well be difficulties in persuading a majority of human beings of the necessity of this revolution.

The New Heredity

In his contribution Huxley has propounded the view that the way of the world has been radically changed through the emergence of man's mind. He claims that there can be no further really important biological evolution of the old kind among animals. In particular there is no possibility of any animal emerging as superior to ourselves, for the reason that we should see the threat and exterminate the animal before things had gone too far. Indeed it does seem likely that the most extreme evolution that will occur among animals in the future will not be among wild animals but among domesticated ones, where man's control, from generation to generation, in changing their forms, can operate much more quickly and continuously than ever would natural selection.

Huxley then goes on to claim, I think rightly, that for the future a new kind of evolution will emerge which he calls psycho-social. Man will evolve less through his genetic nature, than because he has the capacity of sharing his knowledge with his fellows so that the processes of human life are controlled in a manner radically different from anything that has gone before. The human race has indeed discovered how to make certain types of acquired character heritable through the processes of education and mutual instruction, and this is a tremendous revolution.

In his interesting contribution Muller follows up the same subject, and examines its genetic consequences in some detail. He emphasizes the formidable difficulties with which we shall be faced on account of the recurring development of deleterious mutations. Indeed, if his subject is thought of as a proposition in general and not merely in human biology, it is hardly an exaggeration to say that the Mendelian laws of heredity absolutely require a very severe form of natural selection for their successful operation. Only so will the perpetually recurring deleterious mutations be eliminated, so that opportunity will be given for the much rarer beneficent ones to come into play. He takes it for granted that we need to have a world from which ruthless natural selection is eliminated, and he shows how, by close attention to genetic principles—many of them already very nearly within our reach—we might hope to keep within bounds the evil effects of mutation.

We may all agree with these views of Huxley and Muller, subject to the condition that man really does succeed in freeing himself permanently from natural selection in the old sense of the term. Man can now aspire to the complete mastery of nature, but subject to the one condition that he can master himself. I shall later discuss in more detail the prospects for this mastery, but here I will only point out the extreme urgency of the matter, for unless the problems are all solved within half a dozen generations, population pressure is likely to be so great, that there will be a return to the old conditions of the struggle for life. The evolution may still be mainly of the psycho-social type, but it will have none of the pleasing rather utopian qualities which we might have hoped for.

Nature and Nurture

It is appropriate to examine the new type of heredity that has emerged more closely. I use the cliché of this title in the sense that Nature is meant to cover the purely genetic qualities of our race, while Nurture applies to the qualities we derive from education and social contact. The term "culture" is sometimes used for this purpose, but it tends to have an emotional significance which I want to avoid.

To judge from the study of fossilized brain-cases there is no clear indication that mankind has grown in intelligence since the evolution of *Homo sapiens*. His Nature has made little further contribution to his status in the world, and yet this has been fantastically altered as judged by the standard of his numbers. Leaving aside such things as the invention of tools and of fire (which preceded the emergence of *Homo sapiens*) his first great increase derived from the invention of agriculture ten thousand years ago, and this gradually increased his numbers by a factor of perhaps five or ten. Five thousand years ago he invented civilization, which again gave an increase on a similar scale. But these two multiplications have been entirely put in the shade by the increases of the past two centuries due to the Scientific Revolution, for during this short space he has multiplied his numbers a further five times, and these increases are still continuing at an even greater rate. Considered merely by the standard of numbers, Nurture has proved itself immensely more important than Nature.

In spite of these quite overwhelming results from Nurture, I must confess that I believe that in the long run Nature is more important. Thus Nurture has contributed these three great inventions, but when we consider the lesser details of its effects they show an instability that is disappointing. Each of us undoubtedly owes most of our conduct and of our creeds to education, but in many important matters we tend to hold quite different opinions from those of even the preceding generation, and this hardly seems to accord with any obvious law of heredity. In the present changing conditions of the world, with the rapidly increasing fields of knowledge, the departure from the views of our fathers may not be surprising, but the weakness of Nurture heredity is not limited to this phase of our experience, as may be illustrated by an example taken from past history.

There can be no doubt that one of the most important things inculcated by Nurture has been religion, and therefore one might hope it would be one of the most durable. Now though it may be argued that the Christian doctrines have endured for nineteen centuries, there can be no doubt that the enthusiasms associated with those doctrines have changed every few centuries. These enthusiasms were the things for which men were ready to die, and there seems no uniform thread running through them at all. Thus the important things for the Reformation were quite different from those for the Crusades, and these again were quite different from the curious doctrinal heresy-hunting campaigns of five centuries before.

This example seems to suggest that the new kind of heredity working on Nurture has none of the permanence of the Mendelian type working on Nature. In spite of the immense importance that most people attach to religion, it seems that its enthusiasms only endure for less than say five centuries. The heredity of Nurture thus seems rather to resemble the cruder old idea that each generation will tend to revert half way back towards the normal, so that in say ten or fifteen generations its effects will have become negligible.

In studies of history generalization is notoriously dangerous, because history never really repeats itself, but it would surely be interesting for historians to attempt to examine this intensely important subject in the hope that there might emerge something like principles of heredity in the evolution of opinion and conduct. But if I am at all correct in the example I have taken, the conclusion is that there is little of permanence in Nurture heredity, always excepting the three great examples of agriculture, civilization and science which I have cited and the possibility that one day some genius may make a new invention of similar importance.

Anyone wanting to press a new good cause on his fellows is always in danger of thinking that, if only he could persuade the world, everything would become perfect, but this must not blind us to the fact that *Homo sapiens,* like any other animal species, is likely to maintain the general characteristics of his hereditary Nature nearly unchanged for something like a million years. Thus a certain fraction of mankind—and not a very small one—tends to turn to crime, and it is to be doubted if the proportion varies very much. It may be true that in times of high prosperity there is less of what may be called the hungry man's crime, from the simple fact that no one is hungry. But there is much crime that cannot be excused by this stimulus, and it is to be doubted if this other type has become any rarer. Are not bank-robberies and fraud and crimes of violence just as common as they ever were? Is it not likely that there will be criminals who continue to disgrace the brave new world we are all hoping for, and that they will not respond to the benevolent treatment planned for their conversion?

I certainly do not aspire to make any definite judgment in this matter of the general rivalry between Nature and Nurture, but I have been attempting to set forth the case that, contrary to the hopes of many people, man's Nature will continue to dominate the world.

Birth Control

When any species of higher animal succeeds in maintaining its numbers in the next generation it does so mainly by the possession of three instincts: the instinct of self-preservation, the sexual instinct, and the parental instinct. As to the first of these I need not speak. In regard to the other two there is considerable variation in that some animals produce quite a large brood, of which few survive, while others may produce only three or four young in the course of their whole lives. Of course, any animal must on the average produce at least more than two offspring if its numbers are to be maintained.

Man is endowed with the same instincts, and I propose to continue calling them "instincts" even though the word may have acquired some more technical meaning in modern psychology. Indeed, in some respects he has these instincts more strongly than have most animals. Thus most mammals and birds become sexually inclined during only part of the year, whereas man and the monkeys have no relaxation from the instinct all year round. As to the parental instinct, it has, of course, a very different quality from the sexual in that it has to maintain its vigour so long as the young still need protection. With most animals this signifies a few months, but for man it means something like twenty years. Both sexual and parental instincts have been maintained by natural selection. Thus anyone with a weak sexual instinct would be apt to beget few children, and again any parents who are not driven to care for their children by the affection which is the conscious working motive of the parental instinct, will lose a greater fraction of them. Since we have to believe that instincts are heritable, it is evident that these qualities will be possessed by a population to the degree that may be required in order to ensure the maintenance of its numbers.

Until a short time ago these two instincts sufficed to maintain human populations, but the ingenuity of man has contrived to find and to exploit a gap in his equipment of instincts by the recent developments of birth-control. Thus the sexual instinct can be fully satisfied without paying the price that used to be inevitable. Again the parental instinct in most people seems to acquire its full force only after the birth of the child, and it appears that it can be more or less satisfied by lavishing all the parental affection on even a single child.

If I may be permitted so to put it, by the invention of contraception, the species *Homo sapiens* has discovered that he can become the new variety *"Homo contracipiens,"* and many take advantage of this to produce a much reduced fraction of the next generation. We have found out how to cheat nature. However, it would seem likely that in the very long run nature cannot be cheated, and it is easy to see the revenge it might take. Some people do have a wish for children before they are conceived, though for most of them it has not the strong compulsion of the two instincts. There will be a tendency for such people to have rather more children than the rest, and these children will tend to inherit a similar wish and so again to have larger families than do others. In succeeding generations there will be some who inherit the wish to an enhanced extent, and these will contribute a still greater proportion of the population. Thus the direct wish for children is likely to become stronger in more and more of the race and in the end it could attain the quality of an instinct as strong as the other two. It may well be that it would take hundreds of generations for the progenitive instinct to develop in this way, but if it should do so, nature would have taken its revenge, and the variety *Homo contracipiens* would become extinct and would be replaced by the variety *Homo progenetivus.*

All this, of course, will happen only if the practice of birth-control becomes so prevalent that, through it, population numbers should actually tend to decrease.

The Short Term

In attempting a long term forecast, much consideration would have to be given to the possible evolutionary changes in man, but for the short term, say one of five or six generations, this difficulty does not arise, because there is no time for heredity to make any modifications in human nature. Two centuries hence man can be taken to be practically identical in his nature with present-day man.

I have already alluded to the revolutionary change that must affect our mentality with the realization that increase in numbers is now likely to be an evil and not a good as it used to be in the past. In those days the judgment depended on the antithesis between life and death, but with the development of birth-control the antithesis has fortunately been changed to one merely between life and non-life and this should be much more acceptable. There seems really no alternative to the development of birth-control as the only humane way of avoiding the threatened evils.

Birth-control is already a widely accepted practice, but most of the various methods are expensive, laborious, and unattractive. For there to be any prospect of its coming into world-wide use, something much better is an absolute requisite. This provokes strongly the question whether nearly enough study is being given to the matter, as contrasted with all the immense and costly research that is being done on other medico-biological problems, for example, on cancer. However, the work that is being done shows promise, and though success in the research has not yet been achieved, it looks to be not far off. We may hope that in the course of a few years there will be something, perhaps a "pill," which would be easy to use, easy to obtain, emotionally acceptable, and without undesirable collateral effects. If the attempt to achieve this should fail, we cannot hope that birth-control will make any really important contribution to the population problem, and we must fear that the increases will continue up to the point where natural selection will again play its ruthless part.

However, I shall assume that this is not so, and that soon we shall possess a really acceptable contraceptive pill. Even then, however, the problem will not have been solved, for large scale factories must be built to make the pill, and, more formidable still, there would be need of a vast educational campaign to instruct the whole world, which means dealing with everybody between the ages of 15 and at least 45 years, a total of perhaps a billion people all told. It would seem optimistic to expect that anything like this could be accomplished in under fifty years, and by that time the five billion of mankind will be already feeling the pressure of their numbers.

The Administration of Control

There will remain the formidable problem of administration, and the central difficulty in this is that the artificial control of numbers would have a natural instability. Thus suppose that half the nations of the world succeeded in finding a way of limiting their numbers, while the other half refused to do so. In a few decades the limiters would be in a serious minority, and without going into the details of the matter, it is hard to believe that in the long run they could stand up against the vigour of the much more numerous non-limiters, trained as they would be in the hard battle for mere life. It is an open question whether the limiters would be conquered from above or from below, but a conquest from below by the boundless provision of cheap labour would be just as effective from the present point of view as the more usual type of conquest.

It would seem inevitable from these considerations that in the struggle for life a refusal to limit numbers gives a positive advantage.

This raises the important point that even now the Roman Catholics forbid some of the proposed practices, and there are also many peoples who regard it as a proof of virility to produce a large number of children. We must hope that both these difficulties may be overcome, but here is still the danger that new creeds of the same kind might arise in connection with such an intimate and emotional matter as family-planning. Indeed, it is hardly too much to say that a firm belief that contraception is a sin would have a strong positive value in the struggle for life between different communities.

On the other hand, it must be recognized that, if anything can be done, now is the time for it, largely because of what I have called the gap in our instincts, through which we can satisfy our sexual wishes and our parental affection while making only an incomplete replacement of our numbers. In consequence of this gap, our emotions would not be much aroused by any limitations imposed on the numbers of our children. Thus there are already examples where it has proved easy to control numbers in one direction or the other through legislation. A few years ago France became anxious about its decreasing numbers, and by the provision of children's allowances in the taxation scheme the process was at once reversed. In the case of Japan the danger was the opposite, and their terrifying increases have been stopped at least to a large extent by the legalisation of abortion, under which something like a million operations a year have been performed without apparently causing discontent.

An interesting feature about control by legislation is that it would be easy to give it a eugenic direction. At the present time equalitarianism is so rampant in political thought that this would commend itself to few legislators, but there can be little doubt that if any country should carry out a eugenic policy for even a few generations, that country could dominate all its neighbours by the sheer increase in the ability of its people. Moreover, it would not be difficult to do this. There is no need to give thought to the particular qualities that are desired, because the aim would not be to produce highly exceptional people, but merely to raise the average of intelligence, relying on the operation of chance to produce the exceptions from among this raised average. Thus people earning large salaries are likely to be rather abler than others, and much could be accomplished by merely arranging the system of taxation so that these people should be induced to have more children than the rest. Such a policy would be quite contrary to all political thinking in democratic countries at the present time, but there can be little doubt that the first country to embark on it, and to maintain it for a few generations, would reap a rich reward against its rivals.

There is an opposite aspect to this matter, and it draws attention to a condition to which we have already been exposed for a good many years. It has been the educated, intelligent and prudent people who have hitherto practised birth-control most, and these must therefore have been making a smaller contribution to the next generation than the contribution of the less prudent and the less intelligent. Any system of purely voluntary birth-control is all too likely to be adopted most frequently by such people, and so we are continually exposing ourseves to the danger of lowering the average of the intelligence of our nations. Even if no approval is given to a positive eugenic policy, it should be possible for legislation to counteract this negative tendency.

The Forecast

In the light of these views I will attempt a forecast of the state of the world in the next century or two. I need hardly say that I realise that this is an over-ambitious task to undertake, and I would emphasize that all forecasting only deals with probabilities. I am giving what I regard as the more probable things that will happen, with no attempt at assigning any degree of certainty to them.

The central problem of the world, at any rate after the next fifty years, will be over-population. It will be mitigated to a considerable extent by increasing use of birth-control, but there will be no time for this to develop to a degree that will remove the problem. As an example, it is hardly possible that it should reduce the five billion of fifty years hence to four billion.

Food production will be greatly increased to match these numbers, but it will remain true that half the world—and this, of course, means a greatly increased number—will still be undernourished. The principle will continue to hold that however much food is produced there will always be too many mouths asking for it.

Political habits of thought are very conservative, and in the course of two centuries they cannot be expected to change very much. At the present time the mutual jealousies between countries dominate political thought, and this jealousy will increase rather than diminish under the hardening conditions of life.

This has the consequence that a single world-government, so ardently hoped for by idealists, will not be achieved. However, it is worth glancing at one of the formidable difficulties it would have to face if it could be created. One of its main tasks, perhaps the most important of all, would be the control of population numbers in the various regions of the world. But government requires not merely benevolent good will; it must also be able to enforce its rule by sanctions. What would the government do if it discovered that in some region the population was intentionally being increased beyond the numbers apportioned to it? It would seem that the ultimate sanction would have to be to kill off the excess. Is it likely that such an extreme step would ever be undertaken? But if it were not, the consequence would be that the world-government would have failed in its main purpose.

In the light of this, each country will tend to adopt its own policy about the control of numbers. It will in fact be an easier task than it would be to do this now, because one of the effects of the harder conditions of life will be to diminish individual personal liberty in favour of the state, and already it has been seen that much can be done about controlling numbers by legislation.

In the far future the instability inherent in the control of numbers may have a dominating effect, but during the short period contemplated it will not have time to exert this effect. Thus some countries, probably those already most prosperous, will succeed in limiting their numbers, and so will be able to retain much of the present good life. Others will fail to do so, or perhaps either on principle or through the ambitions of power politics they will refuse to attempt it.

The world will thus be divided by the jealousy of the unprosperous directed against the prosperous. Under these conditions it is hard to believe that wars can be avoided, but it is to be hoped that they will be small wars of the old type, and not the major atomic wars which are so much in our minds at present. For a time at all events the superior equipment and culture of the countries with limited population should suffice to defend them against their more numerous opponents.

In the overpopulated countries many of the characteristics of our civilization will survive, but they will be chiefly the superficialities because life will be too hard to permit the peoples to go deeper. On the other hand, in the countries that have succeeded in limiting their numbers progress will continue. New discoveries will be made which may tend to ease the life not only of these countries but of the whole world. Scientific knowledge will continue to advance. The torch of learning will still burn, and the great names of the past will still be honoured.

I am very fully conscious that the views I have expressed run entirely counter to many of the optimistic hopes of the present age. I myself see little prospect of escape from the return to hard conditions of life, and much of my motive in setting my views down is the hope that they may be contradicted by others who have a deeper knowledge than I can claim of the laws of nature.